# Low Earth Orbit Satellite Design

**SPACE TECHNOLOGY LIBRARY**
Published jointly by Microcosm Press and Springer

---

**The Space Technology Library Editorial Board**

Managing Editor: **James R. Wertz** (Microcosm, Inc., El Segundo, CA)

Editorial Board: **Roland Doré** (Professor and Director International Space University, Strasbourg)
**Tom Logsdon** (Senior member of Technical Staff, Space Division, Rockwell International)
**F. Landis Markley** (NASA, Goddard Space Flight Center)
**Robert G. Melton** (Professor of Aerospace Engineering, Pennsylvania State University)
**Keiken Ninomiya** (Professor, Institute of Space & Astronautical Science)
**Jehangir J. Pocha** (Letchworth, Herts.)
**Rex W. Ridenoure** (CEO and Co-founder at Ecliptic Enterprises Corporation)
**Gael Squibb** (Jet Propulsion Laboratory, California Institute of Technology)
**Martin Sweeting** (Professor of Satellite Engineering, University of Surrey)
**David A. Vallado** (Senior Research Astrodynamicist, CSSI/AGI)
**Richard Van Allen** (Vice President and Director, Space Systems Division, Microcosm, Inc.)

More information about this series at http://www.springer.com/series/6575

George Sebestyen • Steve Fujikawa
Nicholas Galassi • Alex Chuchra

# Low Earth Orbit Satellite Design

George Sebestyen
McLean, VA, USA

Nicholas Galassi
Bethesda, MD, USA

Steve Fujikawa
Crofton, MD, USA

Alex Chuchra
Arnold, MD, USA

Space Technology Library
ISBN 978-3-319-68314-0     ISBN 978-3-319-68315-7  (eBook)
https://doi.org/10.1007/978-3-319-68315-7

Library of Congress Control Number: 2017955031

© Springer International Publishing AG 2018
This work is subject to copyright. All rights are reserved by the Publisher, whether the whole or part of the material is concerned, specifically the rights of translation, reprinting, reuse of illustrations, recitation, broadcasting, reproduction on microfilms or in any other physical way, and transmission or information storage and retrieval, electronic adaptation, computer software, or by similar or dissimilar methodology now known or hereafter developed.
The use of general descriptive names, registered names, trademarks, service marks, etc. in this publication does not imply, even in the absence of a specific statement, that such names are exempt from the relevant protective laws and regulations and therefore free for general use.
The publisher, the authors and the editors are safe to assume that the advice and information in this book are believed to be true and accurate at the date of publication. Neither the publisher nor the authors or the editors give a warranty, express or implied, with respect to the material contained herein or for any errors or omissions that may have been made. The publisher remains neutral with regard to jurisdictional claims in published maps and institutional affiliations.

Printed on acid-free paper

This Springer imprint is published by Springer Nature
The registered company is Springer International Publishing AG
The registered company address is: Gewerbestrasse 11, 6330 Cham, Switzerland

# Preface

*Space Mission Analysis and Design*, published in 1991, was the first book that comprehensively treated space and spacecraft system design. It is still the most comprehensive book on that subject. While later editions do cover some spacecraft hardware design issues, the book's main emphasis is on the analysis and design of space missions, rather than the design of spacecraft hardware itself.

Since 1991, interest in LEO spacecraft has increased tremendously, as has the number of young workers in the field. With the advent of CubeSats, the number of satellites being built and launched today has skyrocketed.

This book, *Low Earth Orbiting Satellite Design*, is intended to complement *Space Mission Analysis and Design*. It focuses on the design of the spacecraft hardware and software, and it intends to provide the new crop of space enthusiasts with the tools they need to design hardware and software for space. At the end of the book, a spacecraft design problem is given. The reader is encouraged to complete the design posed by the problem with the spreadsheets and equations provided in the book, thereby affirming all that has been learned over the course of this text.

McLean, VA, USA  
December 2016

George Sebestyen

# Contents

| | | | |
|---|---|---|---|
| **1** | **The Space Environment** | | 1 |
| | 1.1 The Environment | | 1 |
| | | 1.1.1 The Earth Magnetic Field | 1 |
| | | 1.1.2 Solar Energy | 3 |
| | | 1.1.3 Residual Atmosphere | 3 |
| | | 1.1.4 Gravity and Gravity Gradient | 5 |
| | 1.2 The Earth and Spacecraft Coordinate System | | 5 |
| | 1.3 Other Space Environmental Matters | | 7 |
| **2** | **Satellite Missions** | | 9 |
| | 2.1 Satellite Orbits | | 9 |
| | 2.2 Satellites Today | | 10 |
| | 2.3 Satellite Imaging | | 13 |
| | | 2.3.1 Imaging Payload Fundamentals | 15 |
| | | 2.3.2 The Telescope | 16 |
| | | 2.3.3 Image Quality | 18 |
| | | 2.3.4 Adequacy of the Light Input | 19 |
| | | 2.3.5 Image Integration (Exposure) Time | 21 |
| | | 2.3.6 Pointing to a Target on the Ground | 23 |
| | | 2.3.7 Swath Width | 26 |
| | | 2.3.8 Spacecraft Agility and Targeting | 28 |
| | | 2.3.9 Imaging Spacecraft Attitude Sensing, Control Requirements | 28 |
| | | 2.3.10 Data Quantity and Downlink Data Rate | 29 |
| | | 2.3.11 An Imaging Scenario | 30 |
| | 2.4 Satellite Constellations | | 31 |
| | | 2.4.1 Present Constellations | 31 |
| | | 2.4.2 Coverage and Gaps | 33 |
| | | 2.4.3 Other Satellite Constellation Considerations | 37 |

## 3 Orbits and Spacecraft-Related Geometry ... 39
3.1 Acceleration of Gravity, Velocity, Period. ... 39
3.2 Position of Spacecraft as a Function of Time ... 40
3.3 Spacecraft Elevation, Slant Range, CPA, Ground Range ... 42
3.4 Pointing to a Target on the Ground From the Spacecraft. ... 47
3.5 Ballistic Coefficient and On-Orbit Life ... 49
3.6 Computing the Projection of the Sun on Planes on the Spacecraft ... 51

## 4 Electric Power Subsystem Design ... 55
4.1 Required Orbit Average Power (OAP). ... 56
4.2 Battery Capacity and Battery System Design ... 57
    4.2.1 Battery Capacity. ... 57
    4.2.2 Battery Choice ... 59
4.3 Solar Arrays Configuration ... 60
4.4 Beta Angle Vs. Time ... 65
4.5 Solar Cells and Cell Laydown ... 65
4.6 EPS Block Diagram. ... 66

## 5 Spacecraft Communications ... 69
5.1 Frequency Allocation. ... 69
5.2 Modulation Types ... 71
5.3 Bit Error Rate (BER) and Forward Error Correction (FEC) ... 72
5.4 Link Equations. ... 73
5.5 Spacecraft Antennas. ... 76
    5.5.1 The N-Turn Helix Antenna ... 76
    5.5.2 Half Wave Quadrifilar Helix Antenna ... 77
    5.5.3 The Turnstile Antenna ... 78
    5.5.4 The Patch Antenna. ... 78
    5.5.5 Horn Antennas ... 79
    5.5.6 Dish Antennas ... 80
    5.5.7 Intersatellite Links and Steerable Antennas. ... 81
    5.5.8 Phased Arrays. ... 82
    5.5.9 Deployable Antennas ... 82
5.6 Increasing Throughput by Varying Bit Rate or Switching Antennas. ... 82
5.7 Geometrical Constraints on Space-to-Ground Communication ... 84
5.8 RF Subsystem Block Diagram. ... 85

## 6 Spacecraft Digital Hardware ... 87
6.1 Computer Architecture. ... 87
6.2 Computer Characteristics and Selection ... 89
6.3 Spacecraft Computers Available Today ... 89

## 7 Attitude Determination and Control System (ADACS) ... 91
7.1 ADACS Performance Requirements Flowdown ... 91
7.2 Description of the Most Common ADACS Systems ... 93
    7.2.1 Gravity Gradient Stabilization. ... 93

|  |  | 7.2.2 | Pitch Bias Momentum Stabilization | 95 |
|---|---|---|---|---|

- 7.2.2 Pitch Bias Momentum Stabilization ... 95
- 7.2.3 3-Axis Zero Momentum Stabilization ... 97
- 7.2.4 Magnetic Spin Stabilization ... 98
- 7.3 The ADACS Components ... 99
  - 7.3.1 Reaction Wheels and Sizing the Wheels ... 99
  - 7.3.2 Torque Coils or Rods: Momentum Unloading ... 100
  - 7.3.3 Star Trackers ... 102
  - 7.3.4 GPS Receivers ... 105
  - 7.3.5 Other ADACS Components ... 106
  - 7.3.6 The ADACS Computer and Algorithms ... 106
  - 7.3.7 ADACS Modes ... 107
- 7.4 Attitude Control System Design Methodologies ... 108
- 7.5 Integration and Test ... 112
- 7.6 On Orbit Checkout ... 114

## 8 Spacecraft Software ... 115
- 8.1 Functions and Software Architecture ... 116
- 8.2 Performing Each Function or Module ... 118
  - 8.2.1 Initialization of the CDH Processor, Hardware, and Operating System ... 118
  - 8.2.2 Executing Scheduled Events ... 118
  - 8.2.3 Stored Command Execution ... 119
  - 8.2.4 Housekeeping ... 120
  - 8.2.5 Management of the On-Board Electric Power System ... 120
  - 8.2.6 Management of the On-Board Thermal Control System ... 121
  - 8.2.7 Telemetry Data Collection ... 121
  - 8.2.8 Communications Software ... 122
  - 8.2.9 Attitude Control System Software ... 123
  - 8.2.10 Uploadable Software ... 123
  - 8.2.11 Propulsion Control System Software ... 124
- 8.3 Software Development ... 124

## 9 Spacecraft Structure ... 127
- 9.1 Introduction ... 127
- 9.2 Requirements Flow-Down and the Structure Design Process ... 128
- 9.3 Structure Options, Their Advantages and Disadvantages ... 130
- 9.4 Structure Materials and Properties ... 136
- 9.5 Fasteners ... 137
- 9.6 Factors of Safety ... 138
- 9.7 Structural Analyses ... 139
  - 9.7.1 Structural Analysis Overview ... 139
  - 9.7.2 Structural Analysis Steps in Detail ... 140
- 9.8 Weight Estimate ... 154

## 10 Deployment Mechanisms ... 161
- 10.1 Deployment Devices ... 162
  - 10.1.1 Hinges ... 162

|  |  | 10.1.2 | Deployable Booms. | 162 |
|---|---|---|---|---|
|  |  | 10.1.3 | Large Deployable Antennas. | 164 |
|  | 10.2 | Restraint Devices. | | 165 |
|  |  | 10.2.1 | The Explosive Bolt Cutter. | 165 |
|  |  | 10.2.2 | Electric Burn Wires. | 166 |
|  |  | 10.2.3 | Solenoid Pin Pullers. | 167 |
|  |  | 10.2.4 | Paraffin Pin Pushers. | 168 |
|  |  | 10.2.5 | Motorized Cams or Doors. | 168 |
|  |  | 10.2.6 | Separation System. | 168 |
|  |  | 10.2.7 | Dampers. | 169 |
|  |  | 10.2.8 | Fluid Dampers. | 169 |
|  |  | 10.2.9 | Magnetic Dampers. | 170 |
|  |  | 10.2.10 | Constant Speed Governor Dampers. | 170 |
|  | 10.3 | Choosing the Right Mechanism. | | 170 |
|  | 10.4 | Testing Deployables. | | 171 |
| **11** | **Propulsion** | | | 173 |
|  | 11.1 | The Basics. | | 173 |
|  | 11.2 | Propulsion Systems. | | 176 |
|  |  | 11.2.1 | Cold Gas Propulsion System. | 176 |
|  |  | 11.2.2 | Hydrazine Propulsion System. | 178 |
|  |  | 11.2.3 | Other Propulsion Systems. | 179 |
|  | 11.3 | Propulsion System Hardware. | | 179 |
|  | 11.4 | Propulsion Maneuvers. | | 181 |
|  |  | 11.4.1 | Maneuvers for Spacecraft in a Constellation, Maintaining and Getting to Station. | 181 |
|  | 11.5 | Other Propulsion Requirements. | | 185 |
| **12** | **Thermal Design**. | | | 187 |
|  | 12.1 | The Thermal Environment. | | 188 |
|  | 12.2 | Heat Absorption. | | 191 |
|  | 12.3 | Heat Rejection. | | 192 |
|  | 12.4 | Heat Generated by the Spacecraft Electronics. | | 192 |
|  | 12.5 | Tools Available for Altering Spacecraft Thermal Performance. | | 193 |
|  |  | 12.5.1 | The Impact of Surface Finishes. | 193 |
|  |  | 12.5.2 | Thermal Conduction | 194 |
|  |  | 12.5.3 | Conducting Heat across Screwed Plates or Bolt Boundaries. | 195 |
|  |  | 12.5.4 | Heat Pipes. | 195 |
|  |  | 12.5.5 | Louvers. | 196 |
|  |  | 12.5.6 | Heaters. | 196 |
|  | 12.6 | Constructing a Thermal Model of the Spacecraft. | | 197 |
|  | 12.7 | A Point Design Example. | | 197 |
|  | 12.8 | Thermal and Thermal Vacuum Testing. | | 199 |
|  | 12.9 | Model Correlation to Conform to Thermal Test Data. | | 200 |
|  | 12.10 | Final Flight Temperature Predictions. | | 200 |

| | | | |
|---|---|---|---|
| **13** | **Radiation Hardening, Reliability and Redundancy** | | 203 |
| | 13.1 | Radiation Hardening | 203 |
| | | 13.1.1 Total Dose | 203 |
| | 13.2 | Reliability | 206 |
| | 13.3 | Redundancy | 208 |
| **14** | **Integration and Test** | | 209 |
| | 14.1 | Component Level Testing | 209 |
| | | 14.1.1 The "Flat-Sat" | 211 |
| | 14.2 | Spacecraft Level Tests | 211 |
| | 14.3 | Environmental Testing | 212 |
| | | 14.3.1 Vibration Tests | 212 |
| | | 14.3.2 Thermal Test | 218 |
| | | 14.3.3 Bakeout | 219 |
| | | 14.3.4 Thermal Vacuum Test | 220 |
| **15** | **Launch Vehicles and Payload Interfaces** | | 223 |
| | 15.1 | Present Launch Vehicles | 223 |
| | 15.2 | Launch Vehicle Secondary Payload Interfaces | 225 |
| | 15.3 | Secondary Payload Environment | 229 |
| | | 15.3.1 Vibration Levels | 229 |
| | | 15.3.2 Mass Properties | 231 |
| | | 15.3.3 Insertion, Separation and Recontact | 231 |
| | | 15.3.4 RF Environment | 231 |
| | | 15.3.5 Acoustic Environment | 232 |
| | | 15.3.6 Shock Environment | 232 |
| | | 15.3.7 Additional Spacecraft Environmental and Other Factors | 233 |
| | 15.4 | Analyses, Documentation and Other Factors | 233 |
| **16** | **Ground Stations and Ground Support Equipment** | | 235 |
| | 16.1 | Ground Stations | 235 |
| | 16.2 | Ground Support Equipment | 238 |
| | 16.3 | Ground Station Manual and Operator Training | 238 |
| | 16.4 | Other Ground Station Matters | 239 |
| **17** | **Spacecraft Operations** | | 241 |
| | 17.1 | Ground Station Functions for Spacecraft/Payload Operation | 241 |
| | | 17.1.1 The Map and Access Time Interval Display | 242 |
| | | 17.1.2 Telemetry Monitoring | 243 |
| | | 17.1.3 Spacecraft Command Generation | 245 |
| | | 17.1.4 Anomaly Discovery and Resolution | 245 |
| | | 17.1.5 Archiving TTM and Data | 246 |
| | 17.2 | Data and Data Rate Limitations | 246 |
| | 17.3 | Other Ground Station Operations | 246 |
| | | 17.3.1 Post Launch and Checkout | 246 |
| | | 17.3.2 Test Plans and Reports | 247 |

|  |  |  |  |
|---|---|---|---|
| | 17.3.3 | Manning the Ground Station | 247 |
| | 17.3.4 | Cost of Spacecraft Operations | 247 |
| | 17.3.5 | Operator Training and the Spacecraft Simulator | 248 |
| | 17.3.6 | Mission Life Termination | 248 |
| | 17.3.7 | Ground Station Development Schedule | 248 |

## 18 Low Cost Design and Development . . . . . . . . . . . . . . . . . . . . . . . . . 249
- 18.1 Approach to Low Cost . . . . . . . . . . . . . . . . . . . . . . . . . . . . . . . . . . 249
- 18.2 The Contract Should Focuses on Functional Rather than Technical Specifications . . . . . . . . . . . . . . . . . . . . . . . . . . . . . . . 250
- 18.3 Experienced, Small Project Team . . . . . . . . . . . . . . . . . . . . . . . . . 250
- 18.4 Vertical Integration . . . . . . . . . . . . . . . . . . . . . . . . . . . . . . . . . . . . 251
- 18.5 Short Schedules and Concurrency of Development and Manufacturing . . . . . . . . . . . . . . . . . . . . . . . . . . . . . . . . . . . . 251
- 18.6 Make Major Technical and Cost Trade-Offs Rapidly and Decisively . . . . . . . . . . . . . . . . . . . . . . . . . . . . . . . . . . . . . . . . 252
- 18.7 Production Coordinator to Expedite Manufacturing . . . . . . . . . . 252
- 18.8 Do Not Try to Save Money in Testing . . . . . . . . . . . . . . . . . . . . . 253
- 18.9 Holding Program Budget Responsibility Tightly . . . . . . . . . . . . 253
- 18.10 Conclusion . . . . . . . . . . . . . . . . . . . . . . . . . . . . . . . . . . . . . . . . . . . 254

## 19 Systems Engineering and Program Management . . . . . . . . . . . . . . . 255
- 19.1 Introduction . . . . . . . . . . . . . . . . . . . . . . . . . . . . . . . . . . . . . . . . . . 255
- 19.2 Top Level Requirements . . . . . . . . . . . . . . . . . . . . . . . . . . . . . . . 255
- 19.3 Requirements Flowdown . . . . . . . . . . . . . . . . . . . . . . . . . . . . . . . 256
- 19.4 Multiple Approaches . . . . . . . . . . . . . . . . . . . . . . . . . . . . . . . . . . 257
- 19.5 Trade Studies . . . . . . . . . . . . . . . . . . . . . . . . . . . . . . . . . . . . . . . . 257
- 19.6 Selection of a Point Design . . . . . . . . . . . . . . . . . . . . . . . . . . . . . 258
- 19.7 Concept of Operations . . . . . . . . . . . . . . . . . . . . . . . . . . . . . . . . . 258
- 19.8 Preliminary Design Review (PDR) . . . . . . . . . . . . . . . . . . . . . . . 258
- 19.9 Interface Control Documents (ICDs) . . . . . . . . . . . . . . . . . . . . . . 259
- 19.10 Detail Design . . . . . . . . . . . . . . . . . . . . . . . . . . . . . . . . . . . . . . . . . 259
- 19.11 Critical Design Review (CDR) . . . . . . . . . . . . . . . . . . . . . . . . . . 259
- 19.12 System and Mission Simulations . . . . . . . . . . . . . . . . . . . . . . . . . 260
- 19.13 Test Bed and "Flatsat" . . . . . . . . . . . . . . . . . . . . . . . . . . . . . . . . . 260
- 19.14 Statement of Work . . . . . . . . . . . . . . . . . . . . . . . . . . . . . . . . . . . . 260
- 19.15 The Work Breakdown Structure . . . . . . . . . . . . . . . . . . . . . . . . . 260
- 19.16 Cost . . . . . . . . . . . . . . . . . . . . . . . . . . . . . . . . . . . . . . . . . . . . . . . . . 273
- 19.17 Scheduling . . . . . . . . . . . . . . . . . . . . . . . . . . . . . . . . . . . . . . . . . . . 274
- 19.18 Critical Path . . . . . . . . . . . . . . . . . . . . . . . . . . . . . . . . . . . . . . . . . . 274
- 19.19 Schedule Slack . . . . . . . . . . . . . . . . . . . . . . . . . . . . . . . . . . . . . . . 275
- 19.20 Earned Cost . . . . . . . . . . . . . . . . . . . . . . . . . . . . . . . . . . . . . . . . . . 275
- 19.21 Cost to Complete Calculation . . . . . . . . . . . . . . . . . . . . . . . . . . . 275
- 19.22 Requirements Creep and Engineering Change Proposal . . . . . . . 276
- 19.23 Reallocating Budgets, Cost Management . . . . . . . . . . . . . . . . . . 276

| | 19.24 | Documentation | 276 |
|---|---|---|---|
| | 19.25 | Test Plans and Test Reports | 277 |
| **20** | **A Spacecraft Design Example** | | 279 |
| | 20.1 | The Spacecraft Mission Requirements | 279 |
| | 20.2 | Derived Technical Requirements | 279 |
| | 20.3 | Preliminary Design | 282 |
| | 20.4 | Design Steps | 283 |
| **21** | **Downloadable Spreadsheets** | | 285 |

Appendix 1: Tensile Strengths of SS Small Screws .................... 287

Appendix 2: NASA Structural Design Documents Accessible
at http://standards.nasa.gov ..................................... 289

Appendix 3: Temperature Coefficients of Materials ................... 291

Appendix 4: Hohmann Transfer Orbit ................................ 293

Appendix 5: Elevation and Azimuth from Spacecraft
to Ground Target for Various CPA Distances ..................... 295

Appendix 6: Beta as a Function of Time (Date) ...................... 297

Appendix 7: Eclipse Duration ...................................... 299

Glossary .......................................................... 301

References ........................................................ 309

Index ............................................................. 313

# About the Authors

**Alex Chuchra** received a Master of Science degree in Mechanical Engineering. He worked in spacecraft thermal analysis and design during his entire career, first at RCA/GE/Martin Marietta and later at Swales and Orbital-ATK. After becoming an independent contractor, he went on to work on many spacecraft at Ford Aerospace, Fairchild, Lockheed Martin, and the NASA Goddard Space Flight Center. He worked on the thermal design of the GOES-R series, Quik TOMS, Indostar, CGS Wind and Polar, GPS-2R, TOPEX, GOES-1 series, Satcom-Ku, and several other spacecraft. He is a contributing author of the *Aerospace Corporation Spacecraft Thermal Control Handbook*. He imparts to the reader an understanding of the design drivers, the bounding thermal cases, and the way these influence spacecraft design.

**Steve Fujikawa** received a Master of Science degree in Mechanical Engineering from Stanford University and a Bachelor of Science degree in Mechanical and Aerospace Engineering from Cornell University. His expertise is in spacecraft Attitude Determination and Control Systems, and he has contributed to their development at many large aerospace primes and most agencies of the Federal Government, including NASA and the DOD.

Mr. Fujikawa is the Founder and President of Maryland Aerospace, Inc., which manufactures CubeSats, SmallSats, subsystems, and components, including integrated ADACS systems for cost-sensitive applications.

**Nicholas Galassi** received his Bachelor of Science degree in Mechanical Engineering from Worcester Polytechnic Institute. He spent the first decade of his career performing structural analysis of military aircraft such as the F/A-18 and A-6F attack fighters and AV8B Harrier jump jet. He has spent the last three decades performing analysis of space structures. His experience includes managing engineering groups and providing analysis support to over 18 space programs.

Mr. Galassi founded his own analysis consulting company, which has supported NASA, DOD, and private spacecraft programs. His support on this book provides guidance to the engineer in analyzing spacecraft and instrument and component structures for a typical spacecraft mission.

**George Sebestyen** received his Doctor of Science degree in Electronics from the Massachusetts Institute of Technology. He has done research in radar, digital computer technology, speech bandwidth compression, signal processing, and the automation of video image analysis. He managed the research activities of a division of E-Systems and later of Litton Industries. After a stint in the Office of the Secretary of Defense, where he was an Assistant Director for Tactical Warfare for the Director of Defense Research and Engineering, he returned to industry.

He was first Vice President of Systems Engineering at Sanders Associates (now BAE), then General Manager of the radar and ordnance division of the company, and later Group Vice President of the Federal Systems Group. This Group consisted of six divisions with a very broad range of products in most aspects of electronics. The Group developed and manufactured (1) Air Force and Navy ECM equipment for aircraft; (2) Sonobuoys and ASW signal processors, displays, and oceanographic buoys; (3) Army air search radars; (4) IR countermeasures for aircraft and helicopters; (5) communication intercept and DF systems for aircraft, ships, and submarines; and (6) command and control systems.

Sebestyen went on to become Vice President of the Engineering Division of the Boeing Aerospace Company. This 8200-people division and its test facilities provided engineering for strategic and tactical missiles and airborne early warning aircraft. The division was also responsible for a number of large spacecraft programs.

Sebestyen then started his own company, *Defense Systems, Inc.*, which built surveillance systems and spacecraft. The company, now part of Orbital-ATK, built 34 different spacecraft in the range of 50–3300 lbs.

This book is based on his experiences and is intended to help spacecraft hardware designers and builders.

# Book Overview

This book is intended for the practical, hands-on spacecraft designer community. It does not provide a comprehensive treatment of the Earth environment, Astrodynamics or Spacecraft Mission Design. Only what is necessary for spacecraft hardware design is covered. The book assumes that the spacecraft mission has already been defined and specified. In the last chapter, a sample problem is given for the reader to solve.

***Chapter 1 Space Environment*** describes the magnetic field, acceleration of gravity and solar and Albedo radiation. These are quantified so that the satellite designer can use these in developing the electric power and attitude control and propulsion systems, and also calculate atmospheric drag forces and tipoff torques that the ADACS and propulsion systems must overcome.

***Chapter 2 Satellite Missions*** describes various types of orbits, satellites and the growth of the number of small LEO satellites. This chapter also contains a detailed description of an imaging satellite and imaging satellite design. Design equations and graphs are provided. The chapter ends with a discussion of satellite constellations, ways of deploying the individual satellites to their stations and station keeping.

***Chapter 3 Orbits and Spacecraft-Related Geometry*** begins by defining the spacecraft orbital elements, obtaining the satellite ground track, ground station access times and elevation angles during each pass. It covers the most often used geometrical relationships, orbit period, pass durations for various minimum elevation angles, how a ground station off the spacecraft ground track would see the spacecraft and how pass durations and peak elevation angles would be affected by the ground station Closest Point of Approach (CPA) to the ground trace.

***Chapter 4 Electric Power Subsystem Design*** details the procedure for obtaining the required Orbit Average Power (OAP) and Battery Capacity. It discusses how to obtain the instantaneous and orbit average power generated by a set of solar panels, then ends with a block diagram of an electric power subsystem.

***Chapter 5 Spacecraft Communications*** discusses frequency allocations for spacecraft, modulation types and Forward Error Correction. It presents the Eb/No required for different Bit Error Rates and Modulation Types. It develops the RF link equations as a function of elevation angle for a typical spacecraft antenna and a large dish ground station. Various antennas and their gain patterns are described. The chapter concludes with ways of increasing the spacecraft throughput by varying data rates as the link margin varies, and by switching between high and low gain antennas.

***Chapter 6 Spacecraft Digital Hardware*** briefly describes C&DH, ADACS and Image or Experiment Processing digital computer selection.

***Chapter 7 Attitude Determination and Control*** describes the various ADACS systems and how mission requirements are used to flow down to ADACS requirements. It then discusses the various kinds of attitude control systems, gravity gradient, pitch bias momentum, 3-axis zero momentum and spin stabilized systems. It describes the components of ADACS, software development, integration and test and on-orbit checkout.

***Chapter 8 Spacecraft Software*** describes software architectures, the functions the spacecraft software performs and how it does so. The chapter concludes with a discussion of software development methods.

***Chapter 9 Spacecraft Structure*** describes the various structure configurations, their advantages and disadvantages and the detailed structural analysis procedure.

***Chapter 10 Deployment Mechanisms*** describes common mechanisms for solar panel and antenna deployment and for launch vehicle separation.

***Chapter 11 Propulsion*** describes cold gas and hydrazine propulsion systems, computing the delta V required for rapid deployment of a spacecraft to its station, and station keeping within a constellation.

***Chapter 12 Thermal Design*** describes the thermal balance equation, methods of altering the spacecraft thermal performance by application of surface finishes and by creating conductive and radiative paths. Methods of constructing a thermal model of a spacecraft, making temperature predictions, correlating model predictions to Thermal Vacuum Test data are described next. The chapter concludes with a point design example.

***Chapter 13 Radiation Hardening, Redundancy and Reliability*** are discussed. Shielding, use of redundancy and computing reliability are also discussed.

***Chapter 14 Integration and Test*** describes the process of integrating the spacecraft and discusses component and system-level test procedures.

***Chapter 15 Launch Vehicles and Spacecraft Interfaces*** describes the characteristics of several launch vehicles and the vibration environment that the spacecraft will see. The spacecraft must be designed to accommodate these. The mechanical interfaces to spacecraft as primary and secondary payloads are also described.

Book Overview                                                                 xix

***Chapter 16 Ground Stations and Ground Support Equipment*** describes the functions of Ground Stations and those aspects of the ground station operations that the spacecraft designer must know.

***Chapter 17 Spacecraft Operations*** covers the functions and some of the displays used during the operation of the spacecraft.

***Chapter 18 Low Cost Design and Development*** describes procedures and program management techniques that lead to reduced spacecraft cost.

***Chapter 19 Systems Engineering and Program Management*** describes the systems engineering process from requirements flowdown to the subsystem specifications. It also gives examples of tradeoff analyses. The development documentation process is described, and the various reports, like PDR and CDR are described. A Statement of Work and a very detailed example of the Work Breakdown Structure is given.

***Chapter 20 A Spacecraft Design Example*** develops the technical requirements from a customer statement of what the spacecraft is to do, and describes the spacecraft Preliminary Design process for the design of the spacecraft hardware. It is encouraged that the reader perform his or her own point design to these requirements.

***Chapter 21 Downloadable Spreadsheets*** describes 29 calculations frequently needed in spacecraft design. These spreadsheets can be downloaded and adapted to the specific requirements of the designer.

# Chapter 1
# The Space Environment

Space environment and related matters as they apply to spacecraft hardware design will be discussed in this chapter.

## 1.1 The Environment

### 1.1.1 The Earth Magnetic Field

The Earth magnetic field is approximately a dipole of 30,000 NanoTesla (nT) strength at the surface of the Earth at 0° magnetic latitude. It varies inversely with the cube of the radius from the center of the Earth, and it also varies with the magnetic latitude. A simplified dipole model of the scalar magnetic field (MF) in nanoTesla is in Eq. (1.1); and the picture of the magnetic field around the Earth and its variations with altitude, H, magnetic latitude, φ, are shown in Fig. 1.1a, b. R is the radius of the Earth. An accurate model of the field is given in Appendix H of Reference 71, based on the International Geomagnetic Field Model, from which the field can be computed as a function of altitude, Latitude, Longitude and time.

$$\mathbf{MF(nT)} = \mathbf{30{,}000}^{*} \left(\mathbf{R}/(\mathbf{R}+\mathbf{H})\right)^{3*} \left[\mathbf{1} + \mathbf{3}^{*}\left(\sin(\varphi)\right)^{2}\right]^{0.5} \quad (1.1)$$

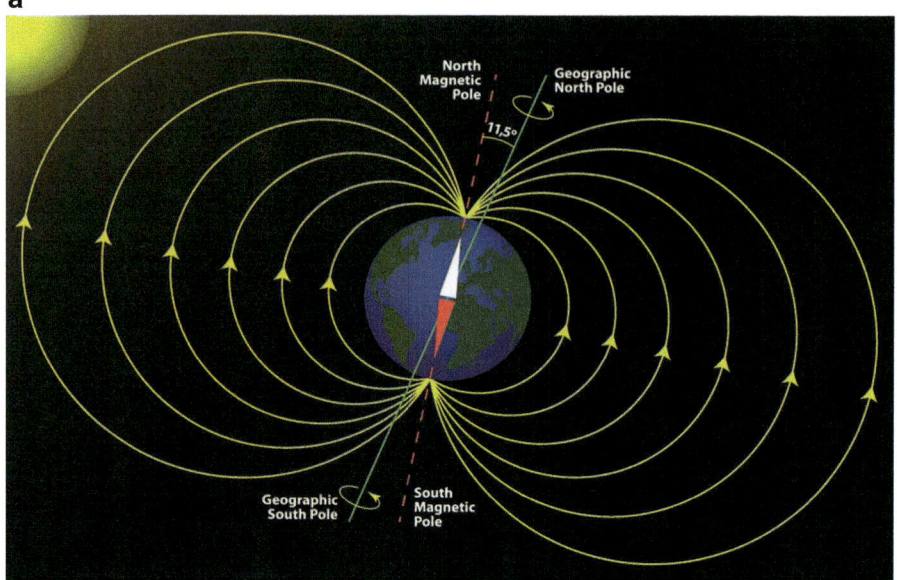

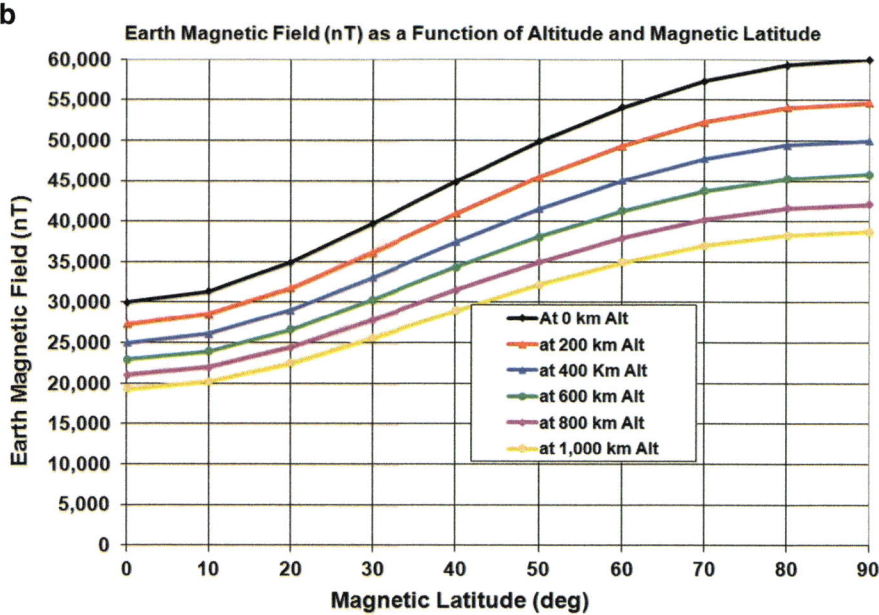

**Fig. 1.1** (**a**) The earth magnetic field. (**b**) The earth magnetic field vs. altitude and magnetic latitude

1.1 The Environment

The direction of the magnetic field at midlatitudes is nearly parallel to the surface of the Earth, as seen from the picture of Fig. 1.1a. It "sucks in" towards the Poles. The property that the field is nearly parallel to the Earth at midlatitudes is used by some CubeSat spacecraft to align one dimension of the spacecraft with the local horizon. The spacecraft includes a horizontal magnet. At midlatitudes, the magnet aligns itself with the magnetic lines and provides a degree of stabilization to the spacecraft. When the spacecraft approaches the Polar region, it tends to first "suck in" then flip.

The magnetic field of Earth varies with time. Models representing the field are updated every 7 years. To illustrate the variation of the magnetic field, consider the motion of the Poles. In 2015, the magnetic North Pole was located at 86.27° N and 159.18° W. The South Pole was at 62.26° S and 136.59° E. The North Pole was drifting North at about 40 miles per year. It was also getting weaker.

### 1.1.2 Solar Energy

The heat sources seen by a spacecraft are direct radiation by the sun, Earth-reflected radiation of the Sun (Albedo) and outgoing long wave Earth radiation. The sun incident power varies between 1322 and 1414 W/m$^2$, with a median of 1367 W/m$^2$.

The Earth-reflected power (Albedo) and Long Wave outgoing radiation from Earth vary with altitude and position around the Earth. Albedo radiation is mostly in the visible range and has a mean reflectance of 0.3, meaning that 0.3 times the incident sun energy is reflected from Earth omnidirectionally. The reflection coefficient is less 0.25 near the Equator and about 0.7 toward the Poles.

The challenge faced by most spacecraft designs is how to get rid of the large amount of incident heat.

### 1.1.3 Residual Atmosphere

Atmospheric density is a function of altitude and solar sunspot activity. Figure 1.2 shows the atmospheric density as a function of altitude for High, Low and Medium sun spot numbers.

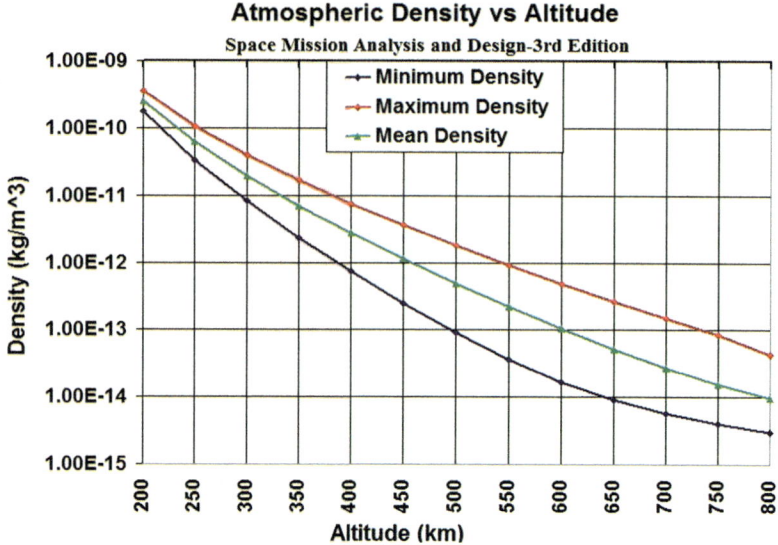

**Fig. 1.2** Atmospheric density varies with altitude & sun spot activity

Atmospheric density at any given altitude varies significantly, depending on sunspot activities. Sunspot activity follows a seven-year cycle and is shown in Fig. 1.3. The red line is a projection.

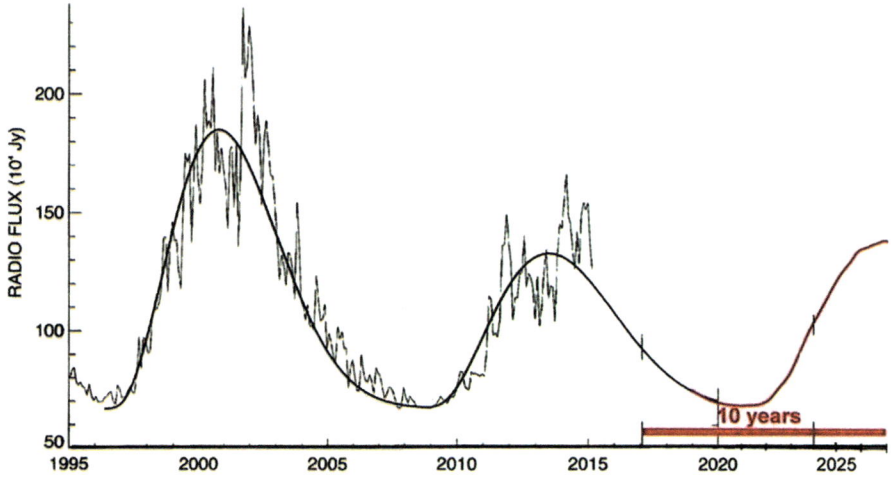

**Fig. 1.3** Sunspot activities (NASA)

At low orbit altitudes (below ≈ 650 km), atmospheric density results in significant drag on a spacecraft, causing it to lose altitude. Atmospheric drag can also induce torques if the spacecraft Center of Gravity (CG) and its Center of Pressure

(CP) are not near each other. Since sunspot activity has a major impact on atmospheric density at low altitudes, it is important to take it into consideration over the mission life of the spacecraft.

### 1.1.4 Gravity and Gravity Gradient

The acceleration of gravity at Earth's surface is 9.806 m/s², and it varies inversely with the square of the altitude. This functional relationship is shown in Eq. (1.2), where G(H) in m/s² is the acceleration of gravity at altitude H. The gradient of the acceleration of gravity is an important quantity, as it determines how well spacecraft with gravity gradient booms will function. These spacecraft utilize the gravity gradient to create a restoring force for a pendulum formed by the spacecraft mass, the gravity gradient boom and the tip mass. This subject will be covered in more detail in the ADACS chapter. More detailed descriptions of the Space Environment are in references 3, 36 and 72.

$$G(H) = 9.806^* \left[ 6378.14 / (6378.14 + H) \right]^2, \text{ H is Altitude in km} \quad (1.2)$$

## 1.2 The Earth and Spacecraft Coordinate System

The position of a spacecraft in orbit at any instant is defined by the six Keplerian orbital parameters, given below. Instead of Eccentricity and Semimajor Axis, sometimes Apogee and Perigee are used to describe the size and shape of the orbit. Figures 1.4 and 1.5 illustrate the orbital elements.

| | | | |
|---|---|---|---|
| 1 Semi-Major Axis | a | Half of the major diameter of the orbit $a = R + (H_A - H_P)/2$, where $H_A$ is the Apogee, $H_P$ is the Perigee. R is the Earth radius |
| 2 Eccentricity | e | $e = (H_A - H_P)/a$ |
| 3 Inclination | i | The angle between the orbit plane and the Equatorial plane |
| 4 Argument of Perigee | ω | The angular distance between Ascending Node and Perigee |
| 5 Time of Perigee Passage | T | Time |
| 6 Longitude of Ascending Node | Ω | The difference (degrees) between the Vernal Equinox and the Longitude where the spacecraft crosses the Equator going North |

**Fig. 1.4** Orbital elements

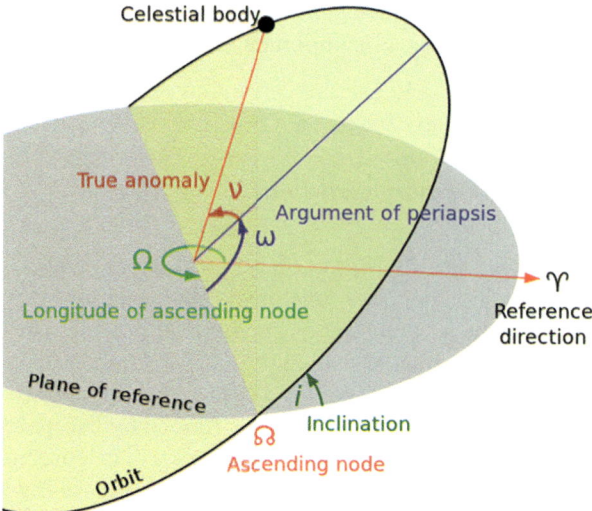

**Fig. 1.5** The orbital elements

Sometimes, orbital elements are expressed in Two-Line Elements. An example is below:

```
THOR ABLESTAR R/B1
1 00047U 60007C  96198.95303667 -.00000008 +00000-0 +24803-4 005026 2
2 00047 066.6626 011.9766 0252122 190.4009169.1818 14.34618735877842
```

Line 2 contains the relevant orbital elements. They are given below:

| Column Number | Number | | | |
|---|---|---|---|---|
| First | Last | Charact | Description | |
| 1 | 1 | 1 | Line Number Indentification | |
| 3 | 7 | 5 | Catalog Number | |
| 9 | 16 | 8 | Inclination (degree) | 1 |
| 18 | 25 | 8 | Right Ascension of Ascending Node | 2 |
| 27 | 33 | 7 | Eccentricity w Assumed Leading Decimal | 3 |
| 35 | 42 | 8 | Argument of the Perigee | 4 |
| 44 | 51 | 8 | Mean Anomaly (degree) | 5 |
| 53 | 63 | 11 | Mean Motion (Revolutions Per Day) | 6 |
| 64 | 68 | 5 | Revolution Number at Epoch | |
| 69 | 69 | 1 | Checksum Modulo 10 | |

**Fig. 1.6** Two-line orbital elements

## 1.3 Other Space Environmental Matters

The Earth axis of rotation is tilted relative to its orbit plane around the Sun by 23.5°. This causes the Sun to move up and down relative to the Equatorial plane by up to ±23.5° during the year.

The Earth-Sun distance varies from 147,166,462 km (around January 3) to 152,171,522 km (around July 4), with an average of 149,597,870.7 km.

The images of the Sun and the Moon both subtend an average of ≈0.52°. Optical instruments on spacecraft need to take this into account to assure they do not point to the Sun or Moon unintentionally.

The angular positions of stars are maintained in a Star Catalog by the Goddard Space Flight Center (GFSC). This catalog is matched against the star field seen by a Star Tracker on a spacecraft to establish the spacecraft position and attitude.

# Chapter 2
# Satellite Missions

## 2.1 Satellite Orbits

The most frequent orbits are shown in Fig. 2.1. Polar orbits are those where the plane of the orbit passes through the poles. They have inclinations of 90° and are usually circular. Because the Earth rotates under the orbit, these satellites can survey the entire Earth.

Beta angle is the angle between the Sun line and the orbit plane. In a Beta = 0° Polar orbit, the Sun will see the orbit edge on. For Beta = 90° (top left figure), the Sun will see the orbit at normal incidence. Figure 2.1 also shows a Polar orbit at Beta = 45° and a 60° inclination orbit, used when mid- latitude coverage is required. In the top three figures, a nadir pointing four-sided spacecraft is also shown.

Sun Synchronous orbits are near Polar orbits inclined so that the spacecraft sees the same points on Earth at the same time each day. For example, a 97.8° inclined orbit at 540 km will describe exactly 15 orbits per day, and each day the spacecraft will cover every point on Earth at the same time. Such orbits are very useful for imaging, as the orbit will ensure that the spacecraft will be over a specified area of the Earth during (say) mid-morning hours.

Elliptical orbits that dip low at Perigee are often used by spacecraft that must be at low altitudes over designated areas of the Earth, but could not survive long at the low altitude. For this reason their orbit is elliptical, so that they would spend a lot of time at higher altitude, thereby reducing average drag and maximizing surveillance time. A special case of the elliptical orbit is the Molnya orbit, which is highly elliptical. It has an Apogee of 40,000 km, and its orbital period is 12 h. A spacecraft in this orbit can survey the Northern hemisphere for half a day at a time.

The GPS satellite orbit is at an altitude of 20,180 km at an inclination of 55°. The 32 spacecraft provide continuous worldwide coverage for navigation systems.

While this book does not address Geostationary Satellites, circular orbits at 35,786 km have 24-h orbit periods, so the satellite appears to hang over a single point on Earth. Such orbits are used for commercial radio and television broadcasts.

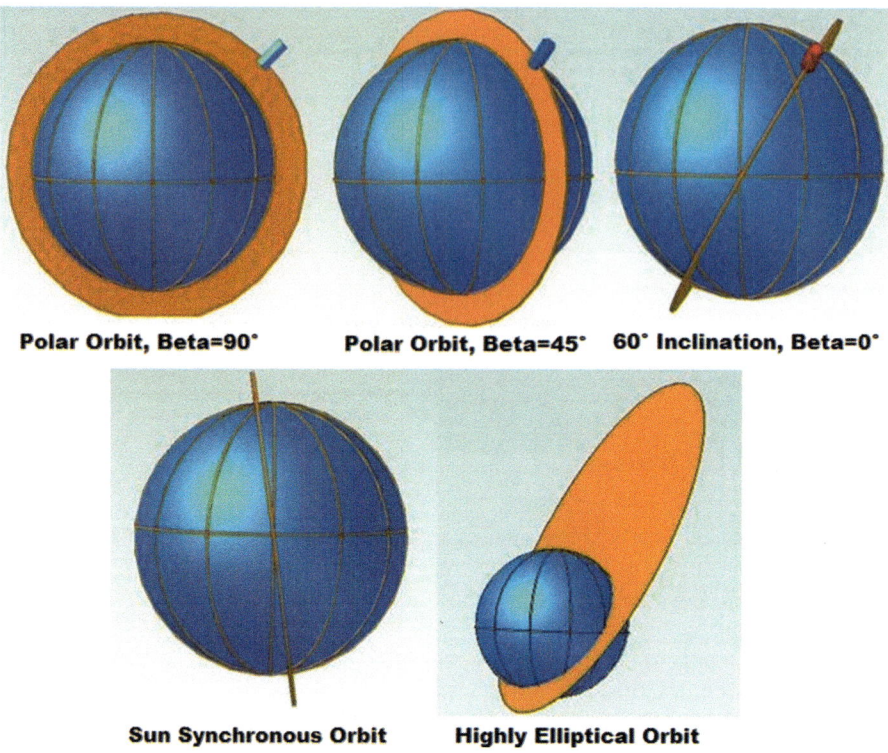

**Fig. 2.1** Polar, inclined, sun-synchronous and elliptical orbits seen from the sun

## 2.2 Satellites Today

According to Wikipedia, there were about 1100 active satellites in orbit in 2015, and about 2600 that no longer work. More than 50 countries have spacecraft programs of one sort or another. In 2015, a total of 86 spacecraft were launched, 32 of which were GEO and 44 that were LEO. Additionally, 33 Cubesats were launched in 2014. This number is not included in the previously mentioned satellites launched or on orbit. So far, some 300 CubeSats have been launched, and there are another 150 or so in the pipeline.

The GEO satellite missions are mostly for communications or television. There are some MEO satellites (GPS) that provide navigational capabilities to its users. Almost all of the rest are LEO satellites. Their missions range from weather, science, communications, Earth observation and imaging. The number of small LEO satellites is increasing dramatically. It is expected that in the period from 2014 to 2023 an average of 115 small LEO satellites will be launched per year.

Some representative LEO satellites are illustrated in the following figures:

2.2 Satellites Today

**Fig. 2.2** 180 lbs, 89.5°, 756 × 887 km orbit, gravity gradient stabilized communications satellite

**Fig. 2.3** Two 170-lbs polar orbit digital store-forward communications satellites

The satellite in Fig. 2.2 has an explosive bolt separation system. Figure 2.3 shows horizontal mating of the satellites with the launch vehicle. Two satellites are launched, one on top of the other.

The spacecraft below illustrate the great variety of LEO spacecraft, orbits, methods of stabilization, types of propulsion systems and methods of deploying a variety of solar arrays and antennas (Figs. 2.4–2.7).

**Fig. 2.4** 190 Ibs Polar orbit, 800 km altitude radar calibration satellite with 17 deployable antennas

**Fig. 2.5** 400 Ibs, 822 km Polar orbit, 3-axis pitch bias momentum stabilized satellite. Gimbaled dish antenna & deployable solar panels

**Fig. 2.6** 900 lbs, 548 km, 40° Orbit satellite with hydrazine propulsion, 3-axis stabilized, 3 reaction wheels, deployable solar panels with solar array drive

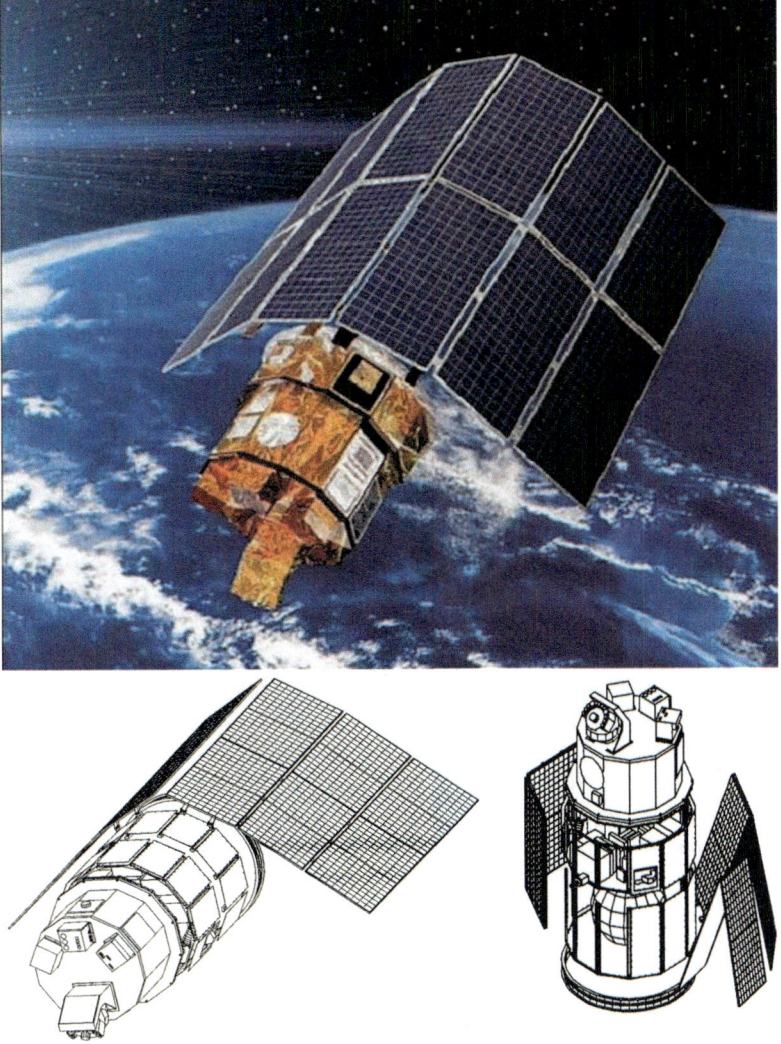

**Fig. 2.7** 579 lbs, 430 × 1375 km, 68.9° Orbit 3-axis stabilized satellite with 3 reaction wheels, hydrazine propulsion, deployable solar panels

## 2.3 Satellite Imaging

Imaging spacecraft are used for remote sensing and for taking images or strip photographs of points on Earth. Remote sensing satellites are typically in Polar or near-Polar orbits to cover the entire Earth. Often, near-Polar orbits are also sun synchronous so that any given point on Earth is covered at the same time each day. Figure 2.8 illustrates such an orbit and several imaging spacecraft.

**Fig. 2.8** Imaging spacecraft orbit and several imaging spacecraft

Typical Imaging Orbit

IKONOS in 681 km orbit
0.82 m panchromatic
3.2 m multispectral
Swath up to 60° Off-Nadir
20 km/sec cross track rate
Revisit every 3 days

WorldView-2 in 770 km orbit
0.46 m panchromatic
1.8 m multispectral
Swath up to 40° Off-Nadir
<3.7 days revisit

There are basically two different types of imaging spacecraft, shown in Fig. 2.9. A few imaging spacecraft "fly" horizontally, with the telescope parallel to the horizon. These use a 45° elliptical mirror in front of the telescope to image the Earth. To change the image position, only a light mirror needs to be moved in pitch and roll. The spacecraft itself does not change attitude and can be pitch bias stabilized, saving the cost and weight of 2 reaction wheels. One advantage of this type of spacecraft is that pointing agility is easily achieved. Another is that it presents a small cross section, so aerodynamic drag is reduced, and aerodynamic torques are eliminated because the spacecraft CG and Center of Pressure are essentially colocated. One of the disadvantages, however, is that the spacecraft is substantially longer to make room for the gimbaled mirror in front of the telescope.

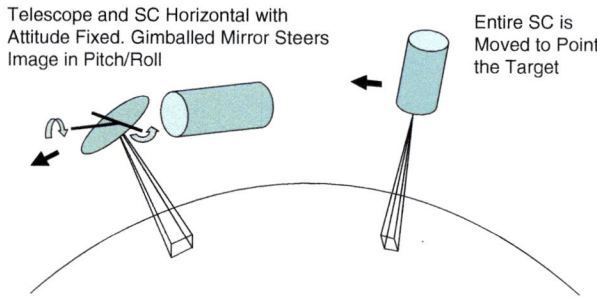

**Fig. 2.9** Two types of imaging spacecraft

2.3  Satellite Imaging

Most imaging spacecraft point to the target on the ground. To change aim point, the spacecraft attitude must be changed. This requires a 3-axis stabilization system with large enough reaction wheels to achieve the aim point agility required by the mission.

### 2.3.1  *Imaging Payload Fundamentals*

Most imaging spacecraft contain a telescope that looks at the Earth. The area on the Earth imaged is centered at the telescope aim point. Ground resolution achievable is defined by the telescope Diffraction Limit, given in the equation at the top of Fig. 2.10. Diffraction limit (DL), telescope aperture (D), orbit altitude (H) and wavelength ($\lambda$) are all in meters. For a spacecraft at 540-km altitude with a 35-cm diameter telescope, the DL is given for three wavelengths in the table in Fig. 2.10. The DL varies between 0.753 m and 1.223 m, depending on the wavelength.

Geometrical resolution on the ground is usually defined as the Ground Sample Distance (GSD) and is given by the equation at the bottom of Fig. 2.10. The spacecraft GSD (for nadir-pointing) depends on the spacecraft altitude (H), telescope focal length (FL) and camera pixel size (P). At an off-nadir angle, $\phi$, the GSD increases as $1/\cos(\phi)$.

A numerical example is given in Fig. 2.10 for $\phi = 0$, H = 540 km, camera pixel size of $5.5*10^{-6}$ m, telescope FL = 3 m and telescope aperture of 35 cm. The GSD = 0.99 meter.

A telescope with a 10-cm aperture may be flown on a CubeSat, while a telescope with a 35-cm aperture may be the largest that experimenters might fly. Note that the geometrical resolution, determined by the focal length of the telescope (for a given pixel size camera) is well matched to the diffraction limit in these examples. In other words, trying to increase the focal length to get better resolution is not worth the effort, as the diameter establishes the diffraction limit that cannot be bettered. Note that F# = FL/D $\approx$ 8.57. The camera pixel size, in this example, is 5.5 µm, such as the Truesense (Kodak) KAI-16050 16 MP CCD array.

**Diffraction Limited (DL) Resolution = DL=1.22*λ*H/D**

| Wavelength (λ) in meters | 4.0 x 10⁻⁷ | 5.0 x 10⁻⁷ | 6.5 x 10⁻⁷ |
|---|---|---|---|
| Telescope Aperture (m) | 0.35 | 0.35 | 0.35 |
| Diffraction Limit (m from 540 km) | 0.753 | 0.941 | 1.223 |

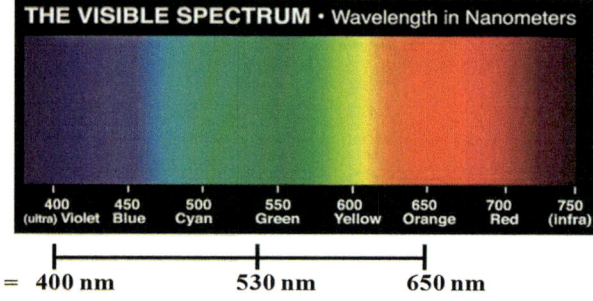

GSD = (Pixel Size)*Altitude/(Focal Length)
Focal Length = (Pixel Size)*Altitude/GSD
Example: 0.99=5.5*10^-6*540,000/3

**Fig. 2.10** Diffraction limit as a function of wavelength

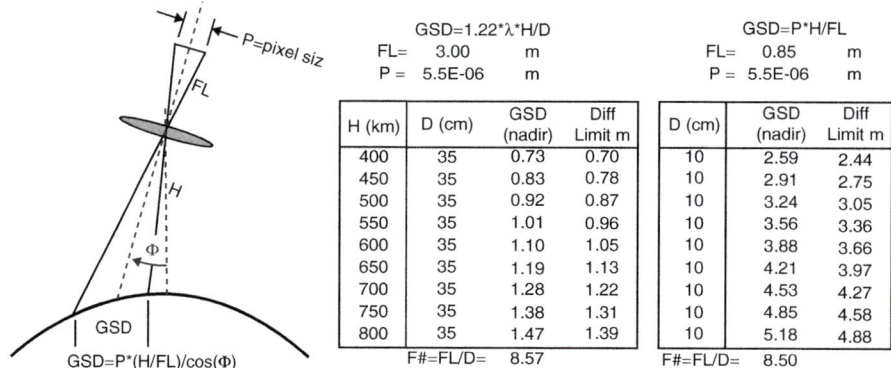

**Fig. 2.11** Geometrical resolution and diffraction limit

## 2.3.2 The Telescope

A picture of a typical Cassagrain telescope is shown in Fig. 2.12. The Primary Mirror (PM) reflects the incident light, which is reflected by the Secondary Mirror (SM) to the camera Focal Plane. First, however, the light passes through corrector lenses, shown in the central tube. Since the Cassagrain Telescope folds the optical path in two, the actual length of the telescope is about half of the focal length.

## 2.3 Satellite Imaging

The telescope is made of nearly zero temperature coefficient materials to ensure that the telescope remains in focus over the temperature variations it will encounter in space. The determination of how much of a dimensional change or elongation is permitted for the telescope to remain in focus is a very complicated matter and will not be addressed here. However, according to a simple rule of thumb, the change in the FL due to temperature variations should not cause the diameter of a spot to increase much over half the size of the camera CCD pixel. For a 35-cm diameter, nominally 3.0 m FL telescope and $5.5*10^{-6}$ pixel size, this is a FL change of about 0.047 mm. Because the optical path is folded, the telescope structure must not change dimensions by more than 0.023 mm. The thermal design of the telescope must achieve such dimensional stability over the temperature range the telescope will encounter in space.

This is a very challenging task. The temperature coefficients of different materials often used for constructing telescopes are Aluminum, Graphite Epoxy and Invar 36. Their Temperature Coefficients are, respectively, 22.2 pp/m/C°, 2.1 pp/m/C° and 1.2 pp/m/C°. The telescope length is approximately 1.5 m.

The fractional elongation of the telescope structure permitted is 7.66 ppm. If made of Aluminum, the maximum temperature change permitted is only 0.345 C°. If made of Graphite Epoxy, the temperature could change by 3.64 C°; and if made of Invar 36, the temperature could change by 6.38 C° before the telescope would defocus. The thermal designs of the spacecraft and telescope are a major engineering challenge.

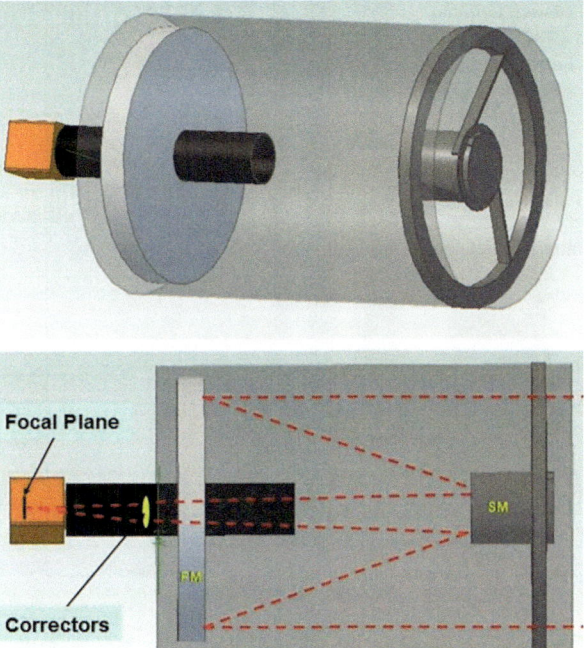

**Fig. 2.12** Cassagrain telescope with camera and ray paths

## 2.3.3 Image Quality

From an optical design point of view, the performance of a telescope is typically characterized by the Modulation Transfer Function (MTF). MTF is the spatial frequency response of the optical system. As in electronics, where a filter is characterized by its transfer function in the frequency domain, the MTF describes the ability of the optical system to pass or transfer an image.

Figure 2.13 shows increasing frequency sinusoidal and rectangular amplitude bar patterns as inputs to an optical system. The spatial frequency is measured in line pairs/mm. The output for each pattern and the system frequency response are shown below. Note that for a frequency response amplitude >10% of the low frequency response, the image is reasonably good and usable. MTF = 10%, for this example, occurs at 70 lp/mm. The dimension of a line pair is 1/70th of a mm, or 14 µm. This is the resolution of this optical system. If a CCD camera with 5.5-µm pixel size at the focal plane is used, the sharpest detail in the image would be greater than 2 pixels in size. A good design should provide an output that is better matched to pixel size.

For a typical telescope, MTF > 10% is often used to specify the response required for an input image of the size of the GSD. Of course, MTF = 10% should be achieved everywhere in the Field Of View of the image, not just at the center.

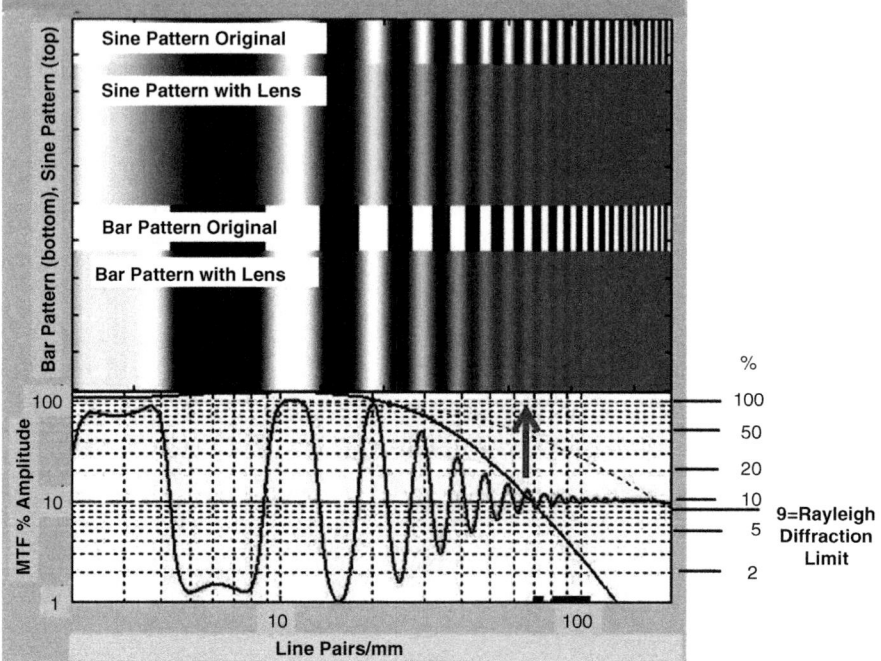

**Fig. 2.13** Modulation Transfer function (MTF) a spatial frequency response of an optical system (Normal Koren)

## 2.3 Satellite Imaging

Another measure of the quality of a system, from the point of view of a human observer and his ability to interpret an image, is the National Image Interpretability Rating Scale, NIIRS. Since that takes into account many factors other than the telescope design, it will not be discussed here.

### 2.3.4 Adequacy of the Light Input

We covered the subject of resolution. Now, we must determine how long an exposure time is required to get a good picture. The image Signal-to-Noise Ratio (S/N) is determined by comparing the intensity of the image to the intensity of the background noise, Albedo and CCD electron noise. This is accomplished by performing a radiometric analysis of the system (telescope, sun angle, time of day, season, latitude, etc.). An example of a radiometric analysis for a 35-cm aperture, 2.8-m focal length telescope looking at the Earth from an altitude of 600 km at specified time and latitude, and using a camera exposure time of 150 μsec (one pixel spacecraft travel time) is shown in Fig. 2.14.

The CCD sensor noise is 16 electrons per pixel, and the sensor saturates at 30,000 electrons. For various Sun elevation angles, the figure shows the number of image electrons per pixel. If a 10 bit A/D converter is used to digitize the image, at 90° Sun elevation angle the image intensity is only 4.5 bits of the 10 bit converter range. Pixel noise is 16 electrons/pixel; and the S/N per pixel varies from 21.84 dB (at 90° elevation) to 12.03 dB (at 24.86° elevation). The latter is not good enough for a good picture.

The bottom left of the figure shows the time of day that the camera is usable and produces enough light for a good picture. Although the fraction of the sensor dynamic range occupied by the image is small, the S/N is adequate for a good picture.

As seen from the green region of the figure, acceptable images can be produced from about 7 AM to about 4 PM. If we increase the image exposure time to 750 μsec, see Fig. 2.15, the usable time of day increases by 1 h at each end of the day.

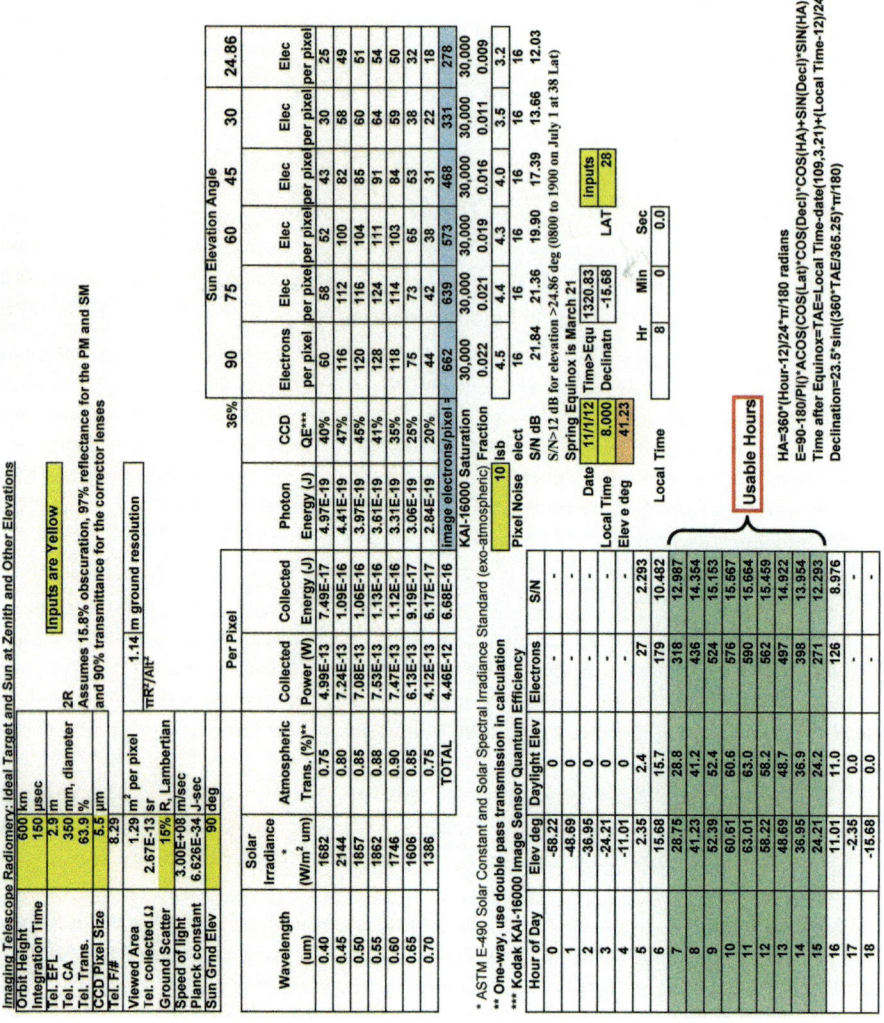

**Fig. 2.14** Radiometric analysis to determine the image S/N

## 2.3 Satellite Imaging

**750 μsec Exposure Time Gains One Hour Useful Time at Each End of the Day**

| Hour of Day | Elev deg | Daylight Elev | Electrons | S/N |
|---|---|---|---|---|
| 0 | -58.22 | 0 | - | - |
| 1 | -48.69 | 0 | - | - |
| 2 | -36.95 | 0 | - | - |
| 3 | -24.21 | 0 | - | - |
| 4 | -11.01 | 0 | - | - |
| 5 | 2.35 | 2.4 | 136 | 9.283 |
| 6 | 15.68 | 15.7 | 894 | 17.472 |
| 7 | 28.75 | 28.8 | 1,591 | 19.976 |
| 8 | 41.23 | 41.2 | 2,180 | 21.344 |
| 9 | 52.39 | 52.4 | 2,621 | 22.143 |
| 10 | 60.61 | 60.6 | 2,882 | 22.556 |
| 11 | 63.01 | 63.0 | 2,948 | 22.654 |
| 12 | 58.22 | 58.2 | 2,812 | 22.449 |
| 13 | 48.69 | 48.7 | 2,485 | 21.912 |
| 14 | 36.95 | 36.9 | 1,988 | 20.944 |
| 15 | 24.21 | 24.2 | 1,356 | 19.283 |
| 16 | 11.01 | 11.0 | 632 | 15.965 |
| 17 | -2.35 | 0.0 | - | - |
| 18 | -15.68 | 0.0 | - | - |
| 19 | -28.75 | 0.0 | - | - |
| 20 | -41.23 | 0 | - | - |
| 21 | -52.39 | 0 | - | - |
| 22 | -60.61 | 0 | - | - |
| 23 | -63.01 | 0 | - | - |

**Fig. 2.15** Increasing exposure time to 750 μsec adds 1 h of usable time to each end of the day

### 2.3.5 Image Integration (Exposure) Time

As seen from the foregoing results, it may be advantageous to increase the image exposure time from the 150 μsec 1 pixel fly-bye time. There are two ways of increasing exposure time:

(a) Impart to the spacecraft a negative pitch rate to increase the allowable dwell time per pixel
(b) Make the spacecraft point to the target, and keep pitch, roll or yaw slewing so that the telescope aim point remains fixed

Figure 2.16 shows the pitch rate needed to permit the effective exposure time to increase, as shown. A pitch rate of $-0.60°/sec$ would permit an exposure time of about 600 μsec, probably enough for most purposes. Of course, it takes time for the spacecraft to go from $0°/sec$ pitch rate to $-0.6°/sec$ pitch rate, and then return to 0 after the image was taken. The elapsed time for the entire maneuver can be substantial and reduces the number of images that can be taken per unit time.

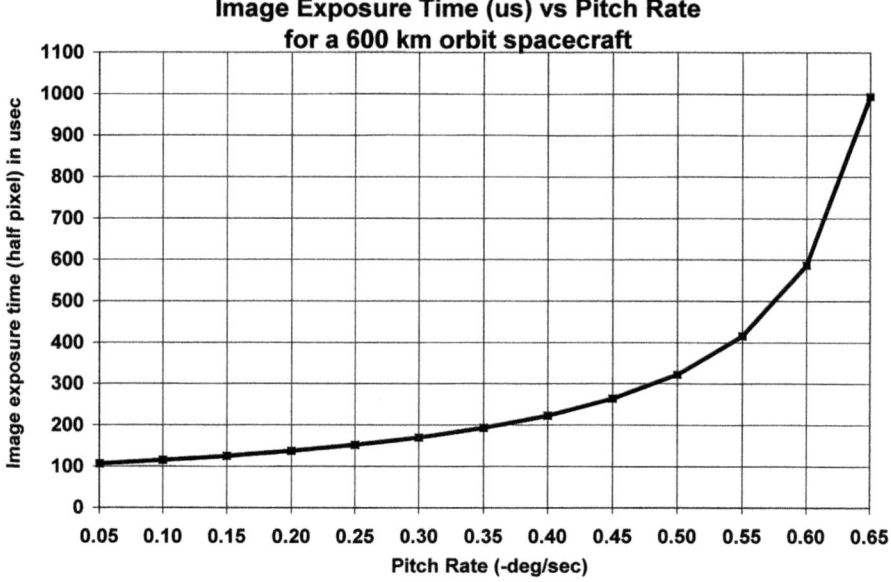

**Fig. 2.16** Establishing a negative spacecraft pitch rate increases allowable image exposure time

The other alternative for increasing the available dwell time over a pixel is for the spacecraft to point to the target and stay pointing to it until the image is taken. There are two ways to point to the target:

1. Pitch and yaw to point to the target, then keep changing pitch and yaw. For this case, the orientation of the image FOV keeps changing. It is more difficult to "stitch" successive pictures into a strip photo.
2. Roll the spacecraft to aim at a line parallel to the SC ground trace by an amount equal to the Spacecraft-Target CPA range, and continuously change the pitch to keep pointing to the target. In this case, the image orientation remains constant and parallel to the spacecraft ground trace. It is much easier to "stitch" consecutive images into a strip photo. In most cases, this second alternative is preferred.

## 2.3.6 Pointing to a Target on the Ground

The procedures for both methods of pointing the spacecraft to the target are described below and illustrated in Fig. 2.17:

| Computing SC Yaw and Pitch | Computing SC Roll and Pitch |
|---|---|
| 1. Compute SC-Target angle relative to Lon line | 1. Compute the great circle range between SC and Target |
| 2. Subtract SC heading | 2. From this, compute subtended angle |
| 3. Compute great circle range from SC to Target | 3. Determine the angle subtended by the CPA range |
| 4. From this, compute Earth subtended angle between SC- Earth Center and Target-to- Earth Center lines | 4. In the plane defined by Target-Earth Center and CPA-Earth Center lines determine Roll at CPA |
| 5. In the plane defined by the SC-Earth Center and the Target- Earth Center lines, we now have 2 lines and the included angle. From these compute Pitch. We have both Yaw & Pitch | 5. In the plane defined by the Roll angle, Target and SC, determine the Pitch angle. |

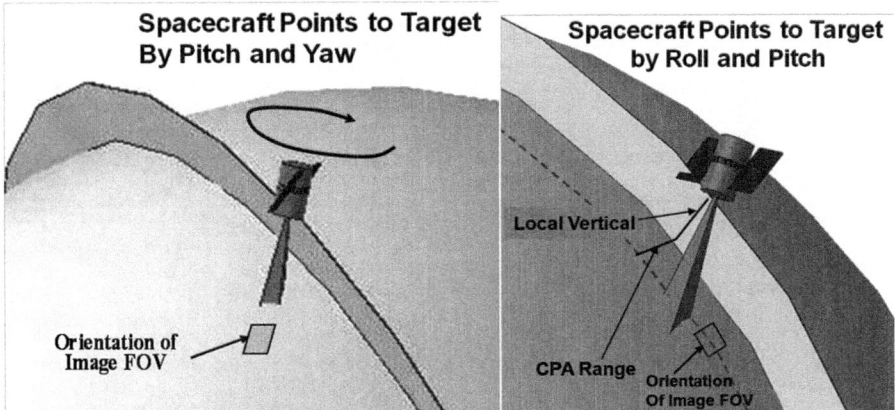

**Fig. 2.17** Spacecraft points to target either by pitch and yaw or by roll and pitch

The computations in the tables above are illustrated in Fig. 2.18.

Spacecraft Yaw and Pitch required to point to the target is computed as follows: The subsatellite point and target latitudes and longitudes are known and given for this example in the top part of Fig. 2.18. The orbit altitude and instantaneous satellite heading are also given. The SC to Target Azimuth is computed from the differences of Target-SC latitude and longitude. The spacecraft Yaw angle is then the difference between the SC-Target Azimuth and the instantaneous spacecraft heading.

The required Pitch is computed by first computing the ground range subtended angle, α, and then, by using the law of cosines, the slant range. From the slant range and α, the off-nadir angle or Pitch is computed by use of the law of sines.

For the case where pointing to the target is to be accomplished by a roll and pitch maneuver, first the angle, ζ, subtended by the cross track distance at CPA is computed. Then, the slant range at CPA is computed using the law of cosines. Finally, the off nadir angle or pitch is computed using the law of sines.

These results are plotted in Figs. 2.19 and 2.20 as functions of time.

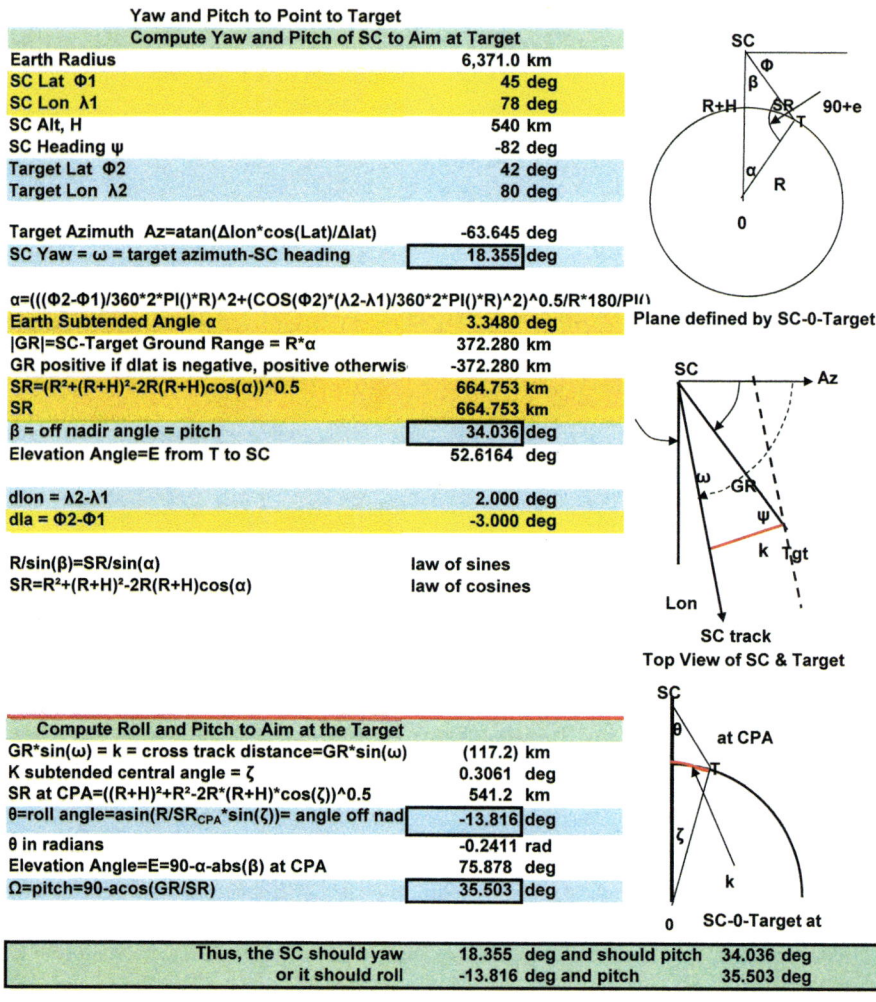

Fig. 2.18 Equations and an example, the spacecraft points to the target by yaw and pitch or by roll and pitch

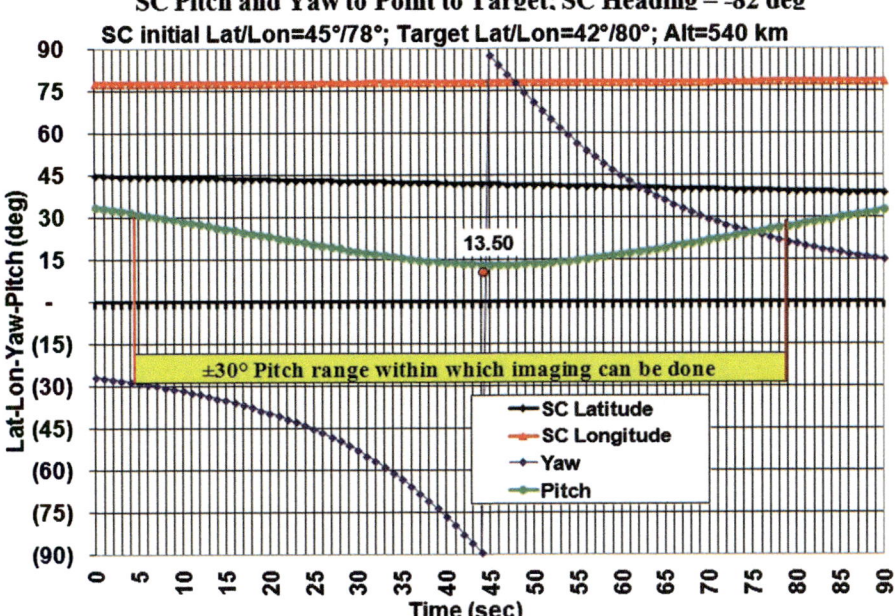

**Fig. 2.19** Spacecraft pitch and yaw as a function of time to keep the spacecraft pointing to the target

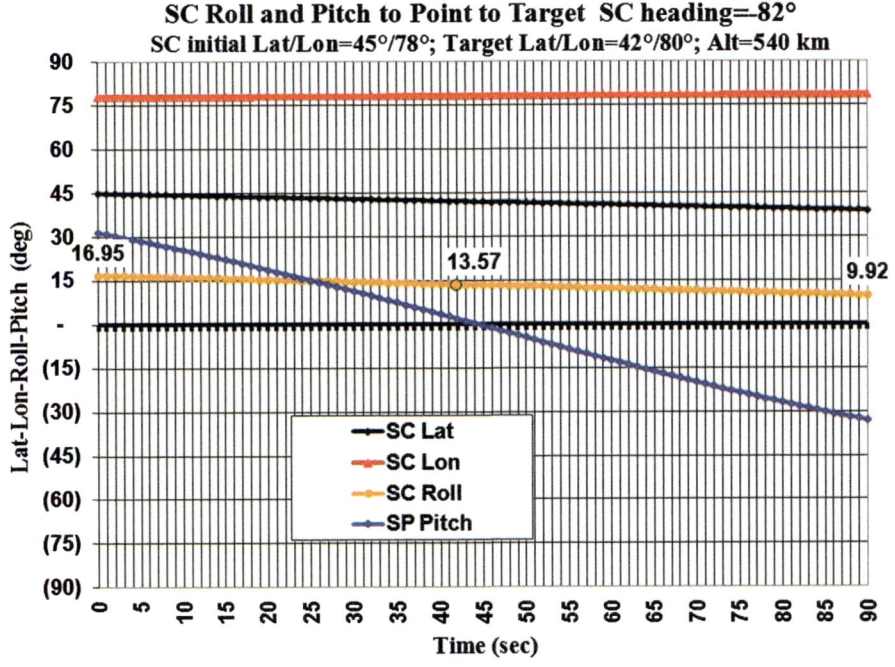

**Fig. 2.20** Spacecraft roll and pitch as a function of time to keep the spacecraft pointing to the target

## 2.3.7 Swath Width

Earth coverage provided by an imaging spacecraft depends on the maximum Off-Nadir look angle. For a 600-km orbit, Fig. 2.21 illustrates the maximum ground range visible from the spacecraft. At a 40° Off-Nadir look angle, the ground elevation from the target is about 45°. This is close to the lowest useful elevation angle for imaging. At this angle, the ground range to the target is about 525 km. Thus, the swath width of potential coverage by this spacecraft is about 1050 km. However, the distance between successive orbits is much greater, so there will be areas that the spacecraft cannot cover. Figure 2.22 shows the ground distance at a constant latitude between successive orbits as a function of latitude. It is evident that the spacecraft covers only a small fraction of the total area of the Earth. By choosing the orbit altitude, the rate at which successive orbits cover different areas of the Earth can be controlled.

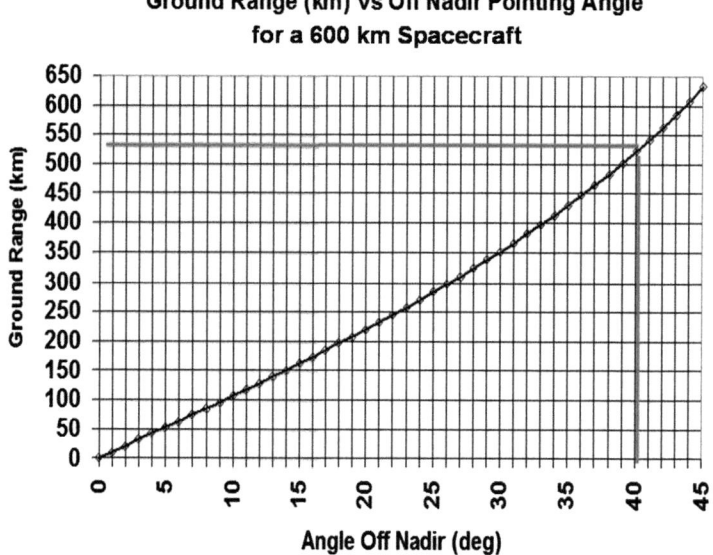

**Fig. 2.21** Swath width as a function of Off-Nadir look angle for a 600-km altitude spacecraft

## 2.3 Satellite Imaging

| Latitude | Spacecraft Altitude (A km) | | | | | | | | |
|---|---|---|---|---|---|---|---|---|---|
| | 400 | 500 | 600 | 700 | 800 | 900 | 1,000 | 1,100 | 1,200 |
| 0 | 2,576 | 2,633 | 2,691 | 2,749 | 2,807 | 2,866 | 2,926 | 2,985 | 3,045 |
| 10 | 2,537 | 2,593 | 2,650 | 2,707 | 2,765 | 2,823 | 2,881 | 2,940 | 2.999 |
| 20 | 2,421 | 2,475 | 2,529 | 2,583 | 2,638 | 2,693 | 2,749 | 2,805 | 2,852 |
| 30 | 2,231 | 2,281 | 2,330 | 2,381 | 2,431 | 2,482 | 2,534 | 2,585 | 2,637 |
| 40 | 1,973 | 2,017 | 2,061 | 2,106 | 2,151 | 2,196 | 2,241 | 2,287 | 2,333 |
| 50 | 1,656 | 1,693 | 1,730 | 1,767 | 1,805 | 1,842 | 1,881 | 1,919 | 1,958 |
| 60 | 1,288 | 1,317 | 1,345 | 1,375 | 1,404 | 1,433 | 1,463 | 1,493 | 1,523 |
| 70 | 881 | 901 | 920 | 940 | 960 | 980 | 1,001 | 1,021 | 1,042 |
| 80 | 447 | 457 | 467 | 477 | 488 | 498 | 508 | 518 | 529 |
| 90 | 0 | 0 | 0 | 0 | 0 | 0 | 0 | 0 | 0 |

Range (km) per Orbit at Constant Latitude

Period (min) = $P = 2*\pi*((A+R)^3 \mu)^{0.5}$

Earth Circumference at Altitude = $C = 2*\mu*(R+A)$

Earth Turns per Orbit (deg) = $OD = 360*P/1440$

Orbit-to-Orbit Distance at Lat (deg) = $R*\cos(Lat)*OD$

**Fig. 2.22** Range at constant latitude between successive orbits vs. latitude

For the case where the orbit period is an exact sub-multiple of a day, after a certain number of orbits, the initial orbit repeats exactly. Since the orbit period is a function of altitude, one can chose altitudes to make this happen. At other altitudes, the orbit will precess either to the East or to the West, depending on altitude. Figure 2.23 shows that at an altitude of ≈567 km the orbit will repeat exactly in 7 days. At an altitude of 590 km, the orbit precesses about 100 km to the West per day. Since the range between orbits at 38° latitude is about 2400 km, it will take about 240 days for the orbit to repeat exactly. But, since the swath width is about 1100 km, a given target can be imaged in about 13 days (2400-1100)/100 = 13. This is still a long time between image opportunities. Flying at 600 km increases the per day drift of the orbit to about 225 km/day, resulting in an opportunity to image the target in (2400-1100)/225 ≈ 5.7 days. The point of this example is that the choice of altitude has a significant impact on the utility of an imaging spacecraft, making it very important to carefully study the impact of altitude choice.

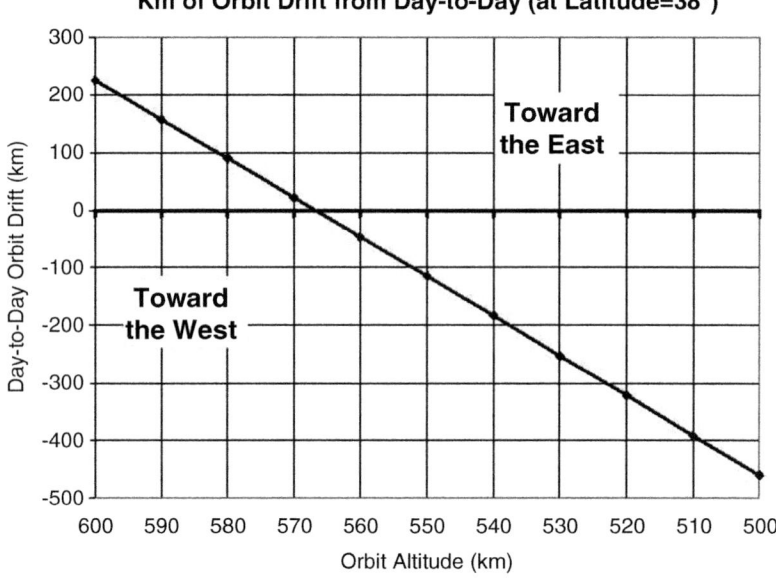

**Fig. 2.23** Orbit drift as a function of spacecraft altitude

## 2.3.8 Spacecraft Agility and Targeting

In a typical imaging scenario, the goal is to image as many targets as possible, but spacecraft agility limits the target set that can be imaged on a given pass. Targets must be within the swath as well as within similar Off-Nadir angles. For example, if a spacecraft can slew at an average rate of 4°/second (this is a very agile spacecraft), then (ignoring slew acceleration, deceleration and settling time) it takes about 40° in roll to go from a target 200 km on one side of the ground trace to another target 200 km on the other side of the ground trace. It takes, nominally, 10 s to accomplish this maneuver during which the spacecraft traveled about 70 km. Adding acceleration, deceleration and settling time to the time to slew 40°, the downrange distance the spacecraft travels between these two targets is much greater than 70 km. To determine spacecraft agility requirements, typical scenarios should be simulated. To render the spacecraft agile, the size of the reaction wheels or control moment gyros must be large. Sizing these will be covered in the chapter on Attitude Control.

## 2.3.9 Imaging Spacecraft Attitude Sensing, Control Requirements

Imaging spacecraft pose the greatest challenge to the attitude sensing and control subsystem. Positioning the center of an image with 100 m accuracy, from an altitude of 600 km, for example, requires the ability to sense spacecraft attitude with an

2.3 Satellite Imaging

accuracy of 0.00477° (16.92 arc seconds) and to control pointing with an accuracy of 0.00955°. Attitude control accuracy is typically a factor of 2 worse than attitude sensing accuracy. Star trackers can provide this accuracy. Modest cost small star trackers achieve accuracies on the order of 25 arc seconds; the best, large star trackers can achieve about 3 arc-seconds (0.00083°). With such a star tracker, the aim point from 600 km is accurate to 8.7 m.

The mission dictates the pointing accuracy required. For example, if the mission requires only that the target be contained in the picture taken, there is no need for the pointing accuracy to be greater than 10% of the image FOV. For a 10 MP image with a GSD of 1.0 meter, the image is 2.582 km by 3.873 km; and 10% of this is about 250 meters. An attitude sensor with 0.0119° (42.9 arc-sec) can control the spacecraft to the required pointing accuracy. Even a small, relatively inexpensive Star tracker can do this. If, on the other hand, the position of points on Earth needs to be determined with (say) 20-m accuracy, a star tracker with 0.0019° or 6.8 arc-sec accuracy is needed. The difference in cost, size and weight of these two Star Trackers is enormous.

Because a typical star tracker FOV is about 15° × 15°, in an imaging spacecraft application usually two star trackers are used.

In addition to pointing accuracy, as previously mentioned, the scenario also drives the required slew rate.

## 2.3.10 Data Quantity and Downlink Data Rate

Taking strip photographs places a large demand on both data storage and downlink data rate requirements. In the previous example, a 10 MP image at 1.0 m GSD was 2.582 km by 3.873 km. If the longer dimension is along the spacecraft ground track, and assuming a 10% overlap between successive pictures is required, about 2 pictures per second must be taken. Thus, the spacecraft accumulates image data at the rate of 20 MP/sec, or, if the image is digitized to 24 bits per pixel, the rate of image data accumulation is 480 Mbits per second (before FEC and Encryption).

If a strip photo is to cover 20 km along the spacecraft ground track, then the strip would contain 6 images, resulting in 1.44 Gbits of data. At 4 Mbps data rate, it would take 6 min - the duration of a typical pass - to get imagery from a single strip to the Ground Station.

By JPG compressing the images, the downlink time could be cut to 1 min (10 s per image). In a typical pass, a total of about 36 images could be downlinked.

If the mission requirement exceeds this capacity, there are several alternatives for increasing the spacecraft image downlink throughput. The downlink data rate could be increased. Additionally, near-Polar ground stations could be employed so that image data could be downlinked during each pass and transmitted to its destination via terrestrial means. Or a data relay, such as a Geostationary satellite, could be used to permit more continuous data dumps to ground.

## 2.3.11 An Imaging Scenario

Now let us go through a sample imaging scenario. Figure 2.24 shows England and Ireland. A 600-km nearly polar orbit spacecraft ground trace is shown in red, while the ±300 km swath limits, corresponding to a maximum of ±30° Off-Nadir angle roll, are shown by the yellow lines.

The 5 intended targets are shown in yellow, and straight lines connecting the targets are shown in green.

Target coordinates (latitudes and longitudes) are shown in Fig. 2.25. The time shown in this figure is the time in seconds the spacecraft will reach CPA to each target. The entire scenario is 140 s. The image longitudes are the distances in km from the spacecraft ground trace.

The last column shows the average roll rate in degrees/sec that the spacecraft has to be able to execute to slew from one target to the next.

The peak slew rate is between targets 3 and 4, where the average slew rate has to be −0.209°/sec. This is a slow slew rate that the spacecraft can execute.

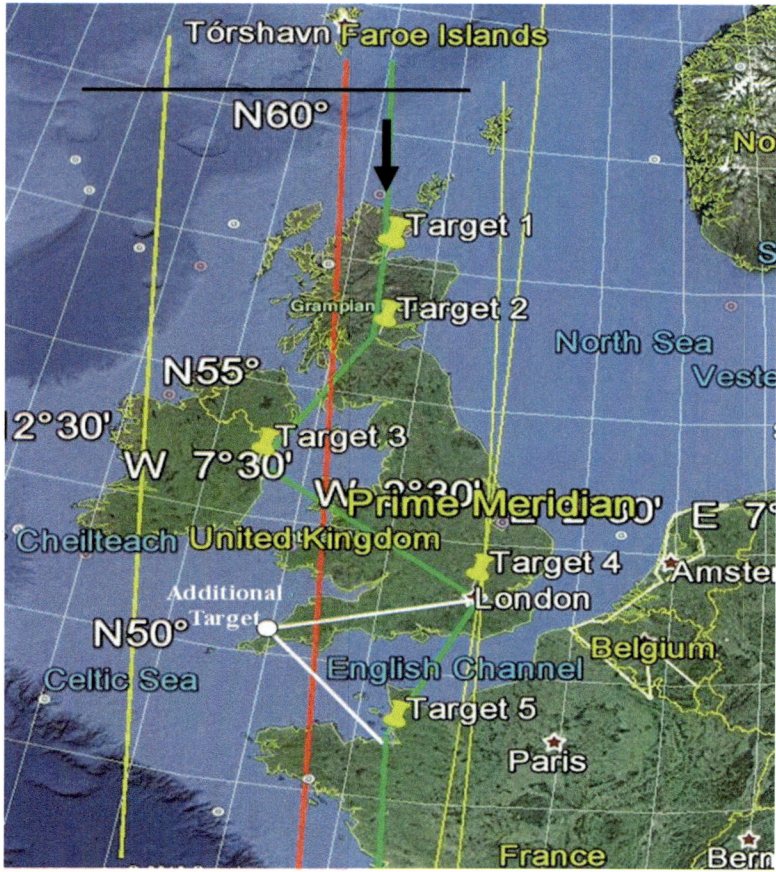

**Fig. 2.24** Imaging scenario shows 5 targets to be imaged, the spacecraft ground trace and the slew limits

2.4  Satellite Constellations

However, if an additional target, shown in white, is added, the maximum average slew rate would become 0.6°/sec, and the spacecraft may not be able to slew that fast. In this case, imaging the target on this pass may have to be omitted.

| Image | Image Lat (deg) | Image Lon (deg) | Sat Lon (deg) | Image Lat (km) | Image Lon (km) | Time (sec) | Delta Lon deg/sec |
|---|---|---|---|---|---|---|---|
| 1 | 57.616453 | 4.015961 | 1.500 | 0.000 | 150.003 | 0.00 | 0.000 |
| 2 | 56.165594 | 3.773844 | 1.451 | -161.508 | 143.983 | 23.07 | -0.010 |
| 3 | 53.479483 | 6.285197 | 1.311 | -299.015 | 329.544 | 65.79 | 0.059 |
| 4 | 51.656939 | 0.218353 | 1.109 | -202.884 | -61.496 | 94.77 | -0.209 |
| 5 | 48.760064 | 1.681853 | 1.200 | -322.477 | 35.360 | 140.84 | 0.032 |

**Fig. 2.25** Target Lat/Lon, Spacecraft CPA Times, Average roll rates

## 2.4  Satellite Constellations

### 2.4.1  *Present Constellations*

The purpose of using a constellation of satellites is to increase geographic coverage on the Earth surface or to reduce revisit time. There are several satellite constellations now, and many are planned for the future. Most present constellations are for navigation by use of satellites at MEO orbit, or for communications with satellites at LEO orbits. The major orbital characteristics of several constellations are listed in Fig. 2.26. Pictures of the orbits of GPS and Globalstar are shown in Fig. 2.27a, b.

| Name | Purpose | Country | SC | Planes | SC/Plane | Alt (km) |
|---|---|---|---|---|---|---|
| GPS | Navigation | USA | 24 | 6 | 4 | 20,180 |
| GLOSSNAS | Navigation | Russia | 24 | 3 | 8 | 19,100 |
| Iridium | Phone Communication | USA | 66 | 6 | 11 | 781 |
| Orbcomm | Store-Forward Comms | USA | 30 | 4 | 6-8 | 825 |
| Globalstar | Communication | EU | 24 | 8 | 3 | 1,400 |
| Galileo | Navigation | EU | 30 | 3 | 8+spares | 23,222 |
| Bei Dou | Regional Navigation | China | 4 | N/A | N/A | GEO |
| Quasi-Zenith | Regional Navigation | Japan | 4 | N/A | N/A | 42,164 |
| Regional Nav | Indian Regional Nav System | India | 4 | N/A | N/A | 36,000 |

**Fig. 2.26** Several satellite constellations

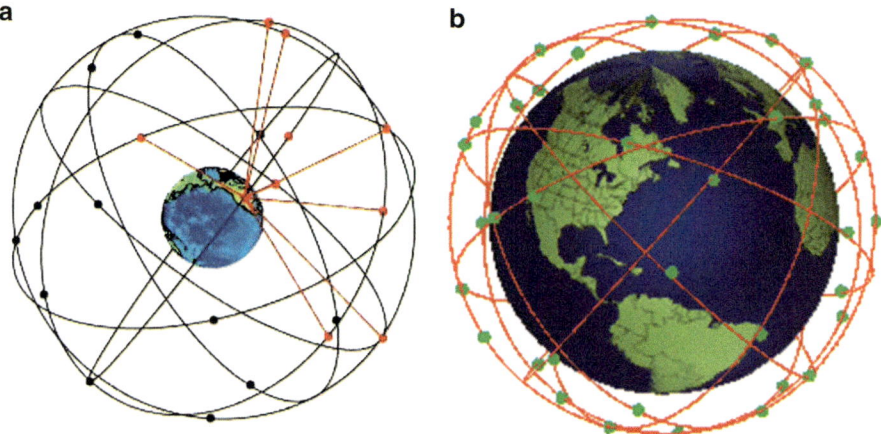

**Fig. 2.27** (**a**) GPS constellation of 24 spacecraft. (**b**) Globalstar constellation of 24 spacecraft

The main drivers in the design of a satellite constellation are (1) mission-imposed requirements, (2) degree of coverage (or permissible holidays of coverage), (3) number of planes and launches, (4) orbital altitude and (5) cost. Figure 2.28 illustrates a Polar plane of 8 spacecraft in a multi-plane constellation of satellites. In the illustration, each spacecraft is at an altitude of 550 km. The orange cones represent the coverage limits of the satellites. Each cone is tangent to the Earth surface at the edge of coverage. There are gaps in coverage between adjacent spacecraft.

In the following, the number of spacecraft in a plane, the number of planes and spacecraft altitudes will be considered to determine constellation coverage of the Earth. While the example is for Polar orbits, coverage considerations for differently inclined orbits are similar.

**Fig. 2.28** Illustration of earth coverage by a polar plane of 600-km altitude spacecraft

## 2.4.2 Coverage and Gaps

Figure 2.29 shows the ground range from a spacecraft at varying altitudes.

When a number of spacecraft are equally spaced in an orbit of a given altitude, their coverage may have gaps between adjacent spacecraft, shown in Fig. 2.28, or their coverage may overlap. Figure 2.30 shows the length of the gap (or overlapping coverage) as a function of spacecraft altitude and the number of spacecraft in the same plane of the orbit. If there were 9 spacecraft per plane, then, at an orbit altitude as low as 400 km, the coverage circles of adjacent spacecraft would just touch. At higher altitudes, spacecraft coverage would overlap, with the number of kilometers of the overlap indicated by the figure. With only 8 spacecraft per plane, orbit altitude would have to be raised to about 540 km to achieve the same result.

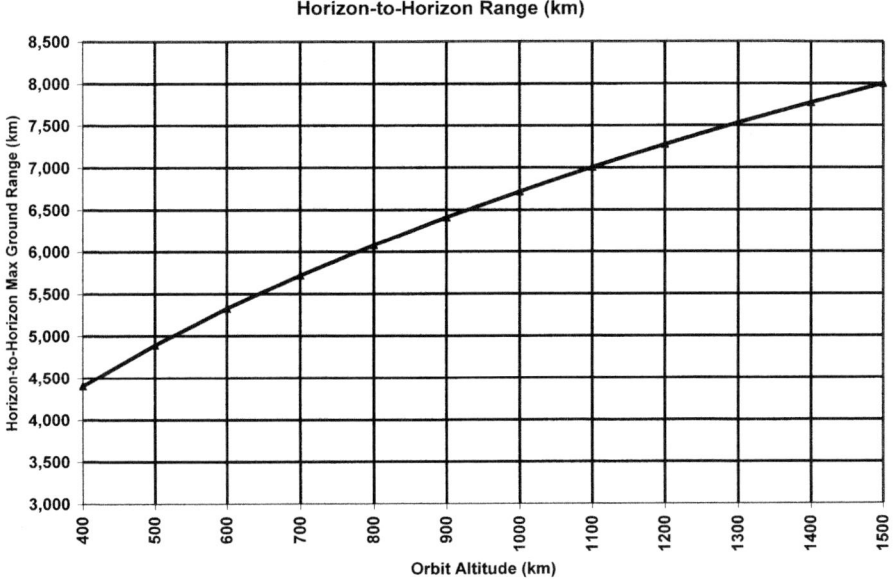

**Fig. 2.29** Ground range vs. altitude

The amount of overlap determines the swath width in which continuous coverage is achieved. In the Orbcomm constellation, for instance, at an orbit altitude of 825 km, overlapping coverage starts with about 6.5 spacecraft per plane. Orbcomm uses 6-8 spacecraft per plane. Iridium, flying at 781 km altitude, uses 11 spacecraft per plane, clearly achieving sufficient overlap for continuous coverage, which is required by the continuous voice communication mission. By contrast, Orbcomm is a store-forward system and does not require overlapping coverage. Nevertheless, in a plane with only 6 spacecraft, the coverage gap is a minimum of 300 km, or about 43 s.

The swath width as a function of the number of spacecraft per plane, and orbit altitude are shown in Fig. 2.31. It is seen that for 6 spacecraft per plane the swath is pretty small until an altitude of about 1100 km. For 8 spacecraft per plane, the orbit can be as low as 700 km for a reasonable swath width.

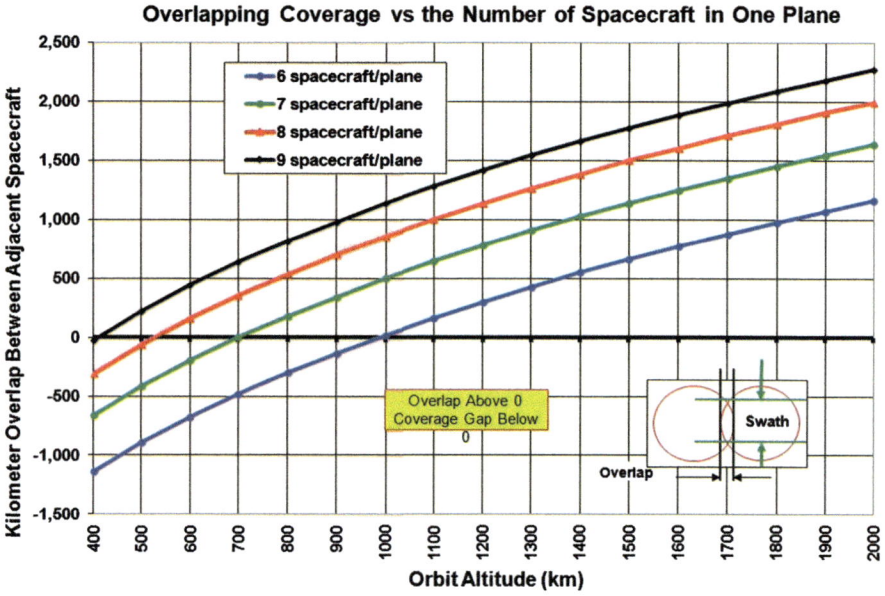

**Fig. 2.30** Overlapping coverage vs. the number of spacecraft per plane and altitude

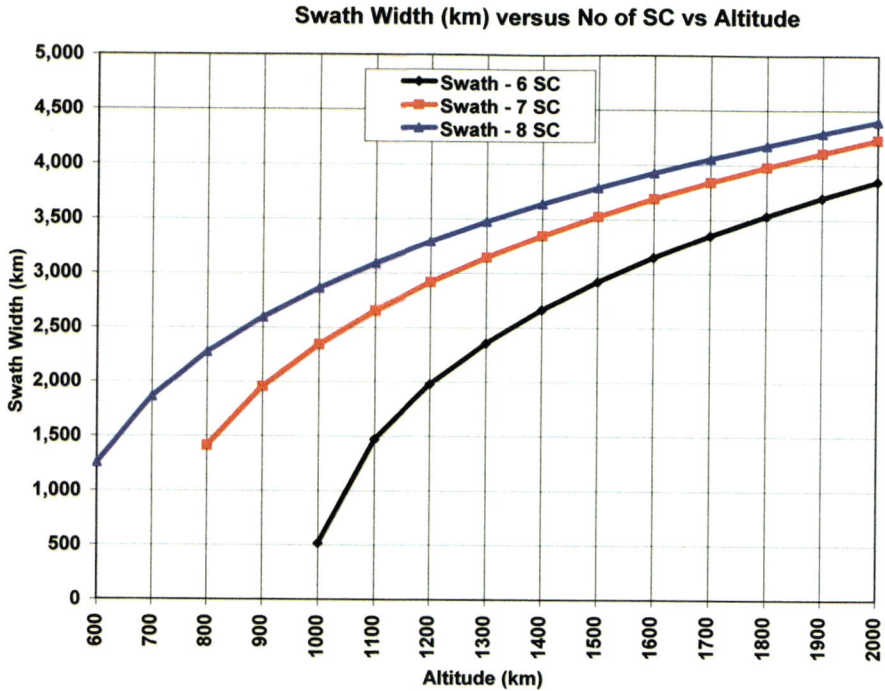

**Fig. 2.31** Swath width as a function of the number of spacecraft per plane and orbit altitude

## 2.4 Satellite Constellations

The overlap (or gap of coverage) between planes also depends on the latitude. For example, for a Polar orbiting constellation, each of the satellites in each of the planes covers the Polar region. Above a certain latitude, adjacent planes start to overlap. Below that same latitude there will be coverage gaps. Figure 2.32 illustrates above what latitude overlap starts, as a function of the number of equally spaced planes and swath width. The ordinate of the graph is the number of kilometers of gap or overlap.

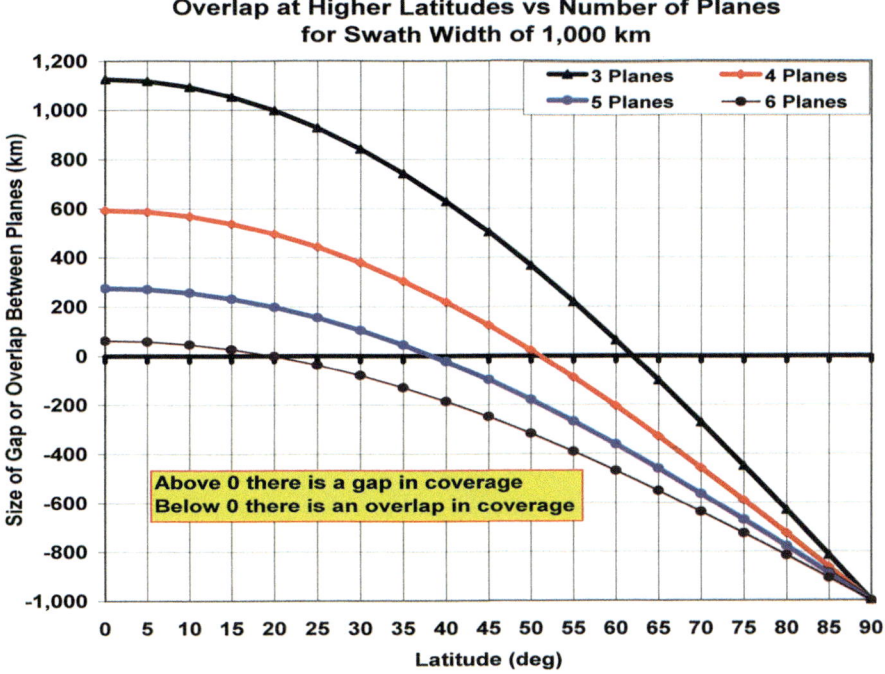

**Fig. 2.32** Planes overlap above a certain latitude, and there are gaps below that latitude. Positive numbers show the size of the gaps, negative numbers show the amount of overlap

An illustration, shown in Fig. 2.33, shows 8 spacecraft distributed uniformly on a 700-km, 65° orbit. There is a small amount of overlap in their footprints, making it possible to cover the swath width almost continuously.

When 3 planes of such spacecraft are on orbit, shown in Fig. 2.34, it is evident that every point on Earth is covered almost continuously. Thus, a 24-spacecraft constellation is almost large enough for continuous Earth coverage (from a communications point of view).

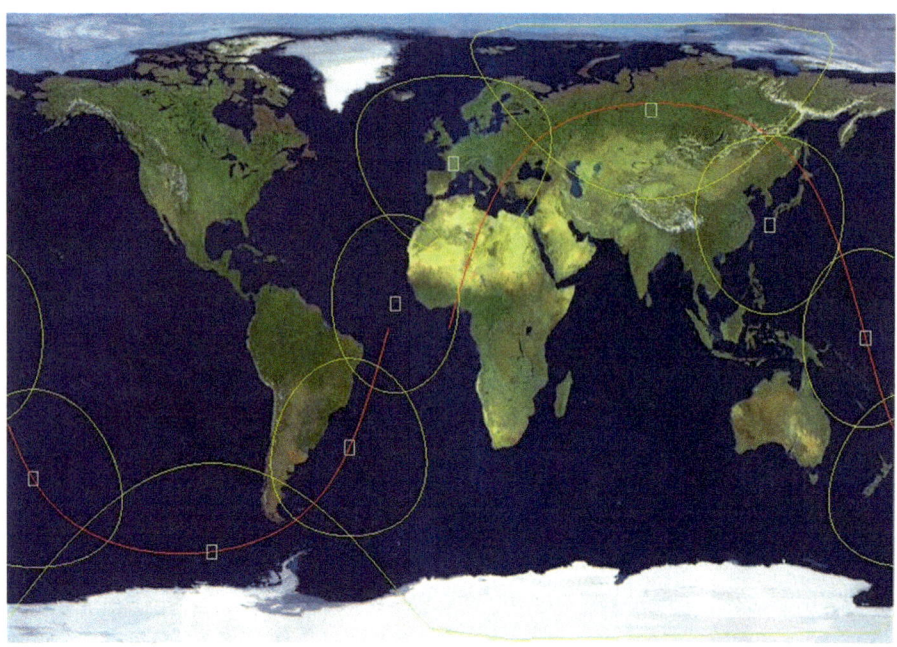

**Fig. 2.33** Single 65° inclination plane at 700 km with 8 spacecraft

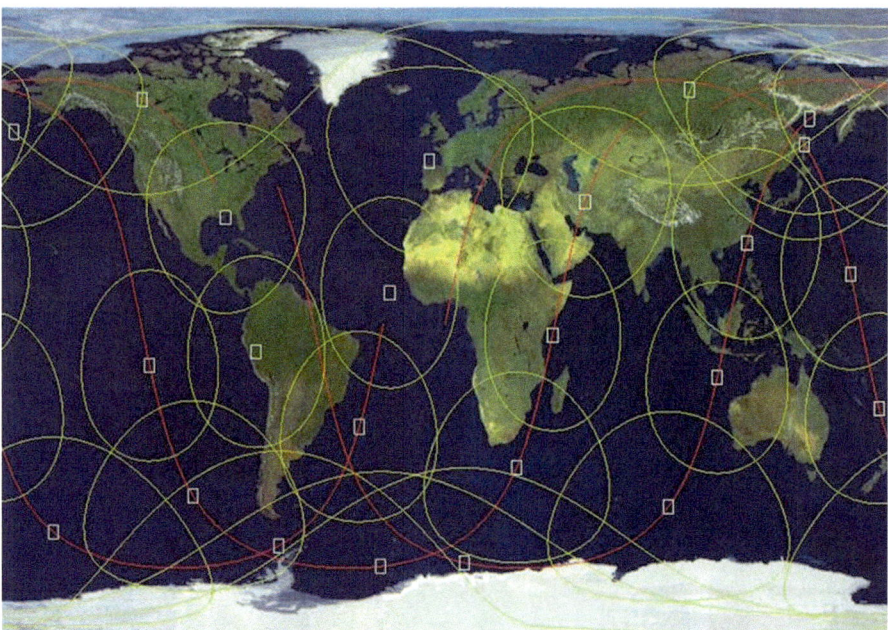

**Fig. 2.34** Three planes of 8 spacecraft per plane at 700-km altitude and 65° inclination provides almost continuous coverage

## 2.4.3 Other Satellite Constellation Considerations

*Launch Vehicles*  Constellations consist of a number of planes of spacecraft. It is important to have each of the spacecraft in a plane at the same altitude and inclination. For this reason, if possible, all of the spacecraft of a plane, including on-orbit spares, should be launched on the same launch vehicle.

Distributing the individual spacecraft of a plane launched by one launch vehicle can be done by having the launch vehicle deliver each spacecraft to its station, or by having the propulsion systems of the individual spacecraft move the spacecraft to its station. The former would require the launch vehicle to have restartable engines. For this reason, the latter approach is used most often.

*Station Keeping*  Station keeping can be a real manpower driver at the ground station. In a 24-spacecraft constellation, if a station keeping maneuver is performed once every 72 days on each spacecraft, then a station keeping maneuver has to be performed every 3 days. It is desirable that station keeping maneuvers be automated.

*Maximum Off-Nadir Angle*  Some missions, such as imaging, require that the Off-Nadir angle be limited. For example, if the target is to be imaged at a look angle greater than (say) $30°$ by a 600-km altitude constellation, the spacecraft off-nadir angle should not exceed $52.3°$. This corresponds to a maximum ground range of 843 km from the subsatellite point. While the horizon is 2631 km away, the useful maximum range for imaging can be substantially less. For this reason, imagery coverage requires more spacecraft per plane, or coverage gaps have to be accepted. Obviously, this is a major cost driver.

*Constellation Communications*  For some missions, intermittent spacecraft-to-ground communications is adequate. The Store-and-Forward digital communications system, Orbcomm, is an example. Imaging satellite systems can also operate with just spacecraft-to-ground communications of their images. However, a system designed for continuous access from and to the spacecraft must use either ground station communications relays, GEO satellite relays (like TDRS) or intersatellite links (like Iridium).

# Chapter 3
# Orbits and Spacecraft-Related Geometry

## 3.1 Acceleration of Gravity, Velocity, Period

Acceleration of gravity, g(H), at an altitude, H, varies inversely with the square of H. The equation for g(H) is given below. Its value at the surface of the Earth is $g_o$ and is approximately 32.2 ft./sec² or 9.8 m/sec². R is the radius of the Earth and H is the orbit altitude, both in km.

$$g_H = g_o \left[ R/(R+H) \right]^2 \tag{3.1}$$

A spacecraft traveling in a circular orbit with tangential velocity, V, experiences centrifugal radial acceleration of value $a_r = V^2/(R + H)$. If the spacecraft is to be in a stable orbit, the acceleration of gravity must be equal to the centrifugal acceleration. Solving for V, the equation below is obtained.

$$V = R^* (g_o)^{0.5} \left[ 1/(R+H) \right]^{0.5}$$
$$= 631.3363183^* (6378.137 + H)^{-0.5} \text{ in km/sec} \tag{3.2}$$

The period of the orbit, P, can be obtained by dividing the circumference of the orbit, $C = 2*\pi*(R + H)$ by the orbit velocity. P in minutes is given below.

$$P = 0.00016587^* (R+H)^{1.5} \tag{3.3}$$

These relationships are summarized in Fig. 3.1 for a 600-km altitude spacecraft.

| Earth Radius R | | 6378.137 | km | |
|---|---|---|---|---|
| Altitude H | | 600 | km | |
| g(0) | | 9.797919335 | m/sec² | |
| Acceleration of Gravity at H | g(H)=g(0)*(R/(R+H))² | | m/sec² | 8.18545 |
| Period = P (min) | P=0.00016587*(R+H)^1.5 | | min | 96.68900 |
| Orbit Circumference, C | C=2*π*(R+H) | | km | 43,844.93 |
| Orbital Velocity, $V_H$ | $V_H$=C/(P*60) | | km/sec | 7.55772 |
| Radial Acceleration $a_r$ | $a_r$=($V_H$)²/(R+H) | | m/sec² | 8.18545 |
| Acceleration of Gravity-Radial Acceleration | g(H)-$a_r$ | | m/sec² | 0 |

**Fig. 3.1** Altitude, velocity, period and radial acceleration example

If the orbit is elliptical, the same formulas can be used, except (R + H) should be the Semi-Major Axis of the elliptical orbit. If A is the Apogee and PE is the Perigee of the orbit, the Semi-Major Axis is given by:

$$\textbf{Semi} - \textbf{Major Axis} = \textbf{R} + (\textbf{A} + \textbf{PE})/\textbf{2} \qquad (3.4)$$

For example, the Period of a Molniya orbit with Perigee = 860 km and Apogee = 39,610 km has a Semi-Major Axis of 26,613 km and a period of 12 Hrs.

## 3.2 Position of Spacecraft as a Function of Time

Using the spacecraft orbital elements, the position of the spacecraft at a given instant can be determined.

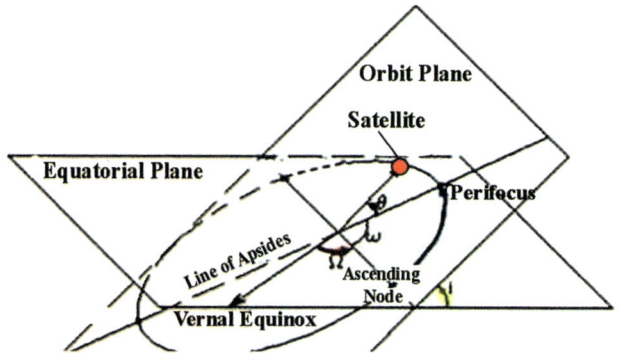

Ω = Right Ascension of Ascending Node (Angle from the Vernal Equinox to the Ascending Node)
I = Inclination of the orbit
a = Semi-Major Axis (Earth Radius + 0.5*(Apogee + Perigee))
E = Eccentricity = (Apogee-Perigee)/(Apogee+Perigee))
ω = Argument of Perigee (Angle in the Orbit Plane from the Ascending Node to the Perigee)
θ = True Anomaly (Angle in Orbit Plane from the Argument of Perigee to Present Spacecraft Position)

**Fig. 3.2** Illustrates and describes the orbital elements

## 3.2 Position of Spacecraft as a Function of Time

The spacecraft motion in its orbit can also be described from the orbital elements. Each of the elements, except True Anomaly, remain constant, and True Anomaly describes 360° in one orbit period. The orbit period in minutes is P = $2\pi*(a^3/\mu)^{0.5}$ where $\mu$ = GM, **a** is the semi-major axis in meters, **G** is the gravitational constant in $m^3/s^2$ and M is the mass of the Earth. Note that the period depends only on the Semi-Major Axis, SM. $R_E$ = 6378.1 km. Equation 3.5 is the Period in terms of the Semi-Major Axis:

$$\mathbf{P = 0.000106587^* (R_E + a)^{1.5}} \; \mathbf{P} \text{ in minutes}, R_E \text{ and SM in km} \quad (3.5)$$

For a 600-km altitude circular orbit of 60° inclination, and other orbital elements given below, the Ground Track of the satellite as a function of time can be computed and is shown in Fig. 3.3a. There are several good programs available commercially for computing the ground track. Programs such as the *Satellite Tool Kit* from Analytical Graphics, or *NOVA* from AMSAT are examples of such programs. Access times from Washington DC are given for a two-day period in Fig. 3.3b.

1. Apogee = 600 km
2. Inclination = 60°
3. Argument of Perigee = 0°
4. Lon of Ascending Node = 239.46°
5. Eccentricity = 0
6. True Anomaly = 0°

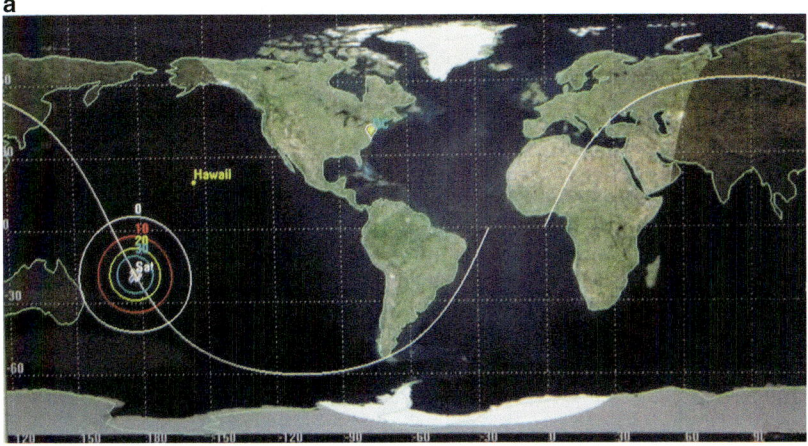

| Access | Start Time (UTCG) | Stop Time (UTCG) | Seconds |
|---|---|---|---|
| 1 | 31 Aug 2016 04:26:44.649 | 31 Aug 2016 04:40:06.258 | 801.610 |
| 2 | 31 Aug 2016 06:07:21.438 | 31 Aug 2016 06:18:33.846 | 672.408 |
| 3 | 31 Aug 2016 18:34:18.880 | 31 Aug 2016 18:43:05.994 | 527.113 |
| 4 | 31 Aug 2016 20:11:13.687 | 31 Aug 2016 20:24:34.033 | 800.346 |
| 5 | 31 Aug 2016 21:52:18.323 | 31 Aug 2016 22:04:21.323 | 723.000 |
| 6 | 31 Aug 2016 23:35:46.300 | 31 Aug 2016 23:44:24.396 | 518.095 |
| 7 | 01 Sep 2016 01:18:11.672 | 01 Sep 2016 01:26:49.920 | 518.248 |
| 8 | 01 Sep 2016 02:58:14.680 | 01 Sep 2016 03:10:17.845 | 723.165 |

**Fig. 3.3** (a) Ground track of a 600-km 60° inclination spacecraft. (b) Access times to Washington DC for 2 days

Note that the ground range from the subsatellite point to the circle of 0° elevation angle is 2631 km. However, because of local ground terrain obstructions, typically only a ground range to the 10° elevation range circle is used for communications purposes. This ground range is only 1740 km. Therefore, the satellite useful swath width (from a communications point of view) is 3480 km; and the useful ground station access times are shorter than those listed in Fig. 3.3b.

## 3.3 Spacecraft Elevation, Slant Range, CPA, Ground Range

The most often used geometrical relationships for circular orbit spacecraft will be described in this section. One of these is Slant Range vs. Elevation angle, needed for computing RF link margins. Figure 3.4 illustrates the geometrical situation in the orbit plane.

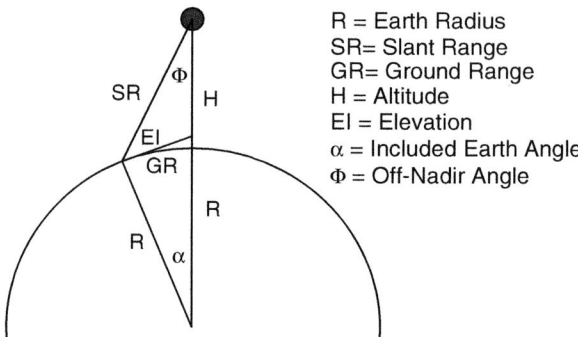

**Fig. 3.4** Spacecraft and earth geometrical situation

Equations for Slant and Ground Range are given below and are plotted for various orbit altitudes (Fig. 3.5).

$$\mathbf{SR} = \mathbf{R}^*\cos(90+\mathbf{El}) + \left[\mathbf{R}^*\left(\cos(90+\mathbf{El})\right)^2 + (\mathbf{R}+\mathbf{H})^2 - \mathbf{R}^2\right]^{0.5} \quad (3.6)$$

$$\mathbf{El} = \mathrm{Acos}\left((\mathbf{R}+\mathbf{H})/\mathbf{R}^* \sin(\varPhi)\right) \quad (3.7)$$

$$\mathbf{Gnd\ Rng} = (90 - \varPhi - \mathbf{E})^* \mathbf{R} \quad (3.8)$$

## 3.3 Spacecraft Elevation, Slant Range, CPA, Ground Range

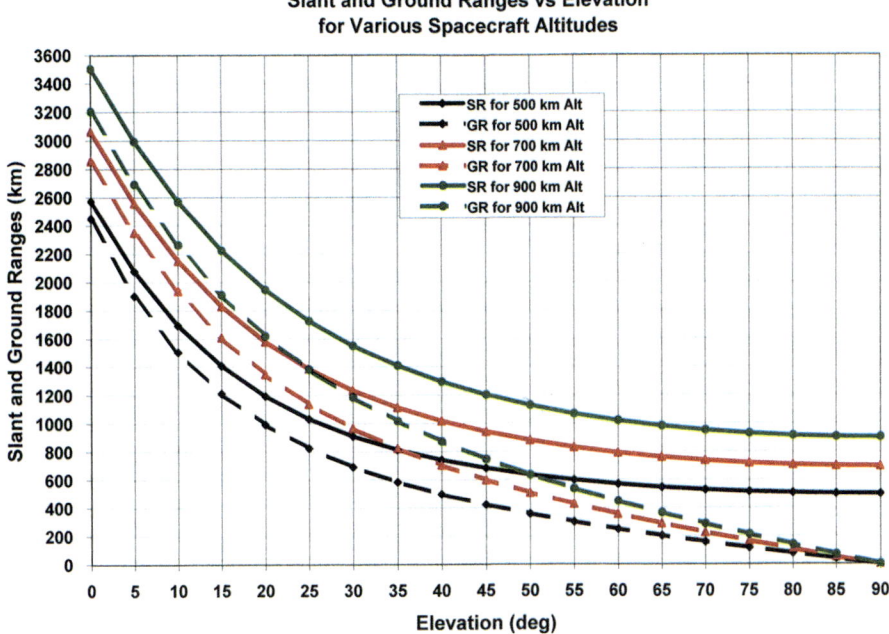

**Fig. 3.5** Slant and ground range (km) vs. elevation angle for various spacecraft altitudes

Figure 3.6 shows a spacecraft at altitude H and a ground station at CPA distance from the spacecraft ground track. The ground range between the spacecraft subsatellite point and the ground station is **GR**, while the ground range from the subsatellite point to the CPA point is **GR$_{CPA}$**. The central angle subtended by **GR** is $\alpha$, by GR$_{CPA}$ is $\omega$, and by the CPA distance is $\psi$.

For an altitude of 600 km, a CPA distance of 250 km and an elevation angle of 5°, the left-hand column gives the numerical results, while the right-hand column gives the equations from which the numerical results were computed.

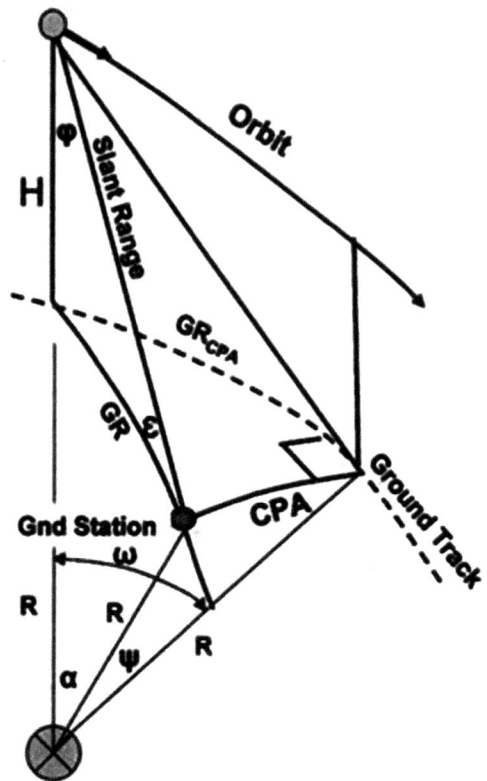

| | | | |
|---|---|---|---|
| R | 6,378.3 | km | |
| H | 600 | km | |
| El | 5 | deg | |
| CPA to Grnd Stat | 200 | km | |
| CPA Subtended Angle φ | 1.797 | deg | φ=CPA/(2*π*R)*360 |
| Slant Range=S | | | |
| Slant Range=S | 2,329 | km | SR=R*sin(-E*PI/180)+((R*sin((-E)*PI/180))^2+(R+H)^2-R^2)^0.5 |
| Ground Range (GR) | 2,162 | km | GR=R*asin(SR*sin(90+El)/(R+H) |
| GR Subtended Angle ψ | 19.420 | deg | GR/(2*π*R)*360 |
| SR Subtebded Angle on Gnd | 19.503 | deg | Ω=(φ^2+ψ^2)^0.5 |
| SR S | | km | |
| Depression Angle (d) | 65.815 | deg | SC Depression Angle=d=asin(R/(R+H))*cos(El) |
| Off-Nadir Angle (θ) | 24.185 | deg | Off-Nadir Angle = 90-d |
| Orbit Period P | 96.692 | min | P=0.00016587*(R+H)^1.5 |
| Pass Duration T | 9.535 | sec | Pass Duration=T=2*P/((2*π)*(R+H))*GR |
| Orbit Velocity $V_o$ | 7.558 | km/sec | Orbit Velocity=2*π*(R+H)/P/60 |
| Gnd Velocity $V_G$ | 6.908 | km/sec | Ground Velocity=R/(R+H)*Orbit Velocity |

**Fig. 3.6** Geometry and example for 600-km orbit and general equations, taking CPA into consideration

## 3.3 Spacecraft Elevation, Slant Range, CPA, Ground Range

The first thing to note is that as the CPA increases, the maximum elevation angle reached during a pass is reduced, as is the duration of the pass. This is illustrated in Fig. 3.7. For higher orbit altitudes, the peak elevation angle increases for the same CPA distance. Since this figure is plotted against the ground range to CPA, it also illustrates the elevation angle vs. time (when ground range is converted to time by the spacecraft ground velocity). The Azimuth vs. range to CPA is in Fig. 3.8.

The reduction of the pass duration as the CPA is increased is shown in Fig. 3.9. Here, pass duration is expressed as a function of the minimum elevation angle, which is considered to be the beginning of a pass. This is relevant because a pass in a mountainous region with low horizon obstructions may not be of value until a given minimum elevation angle (like 5° or 10°) is reached.

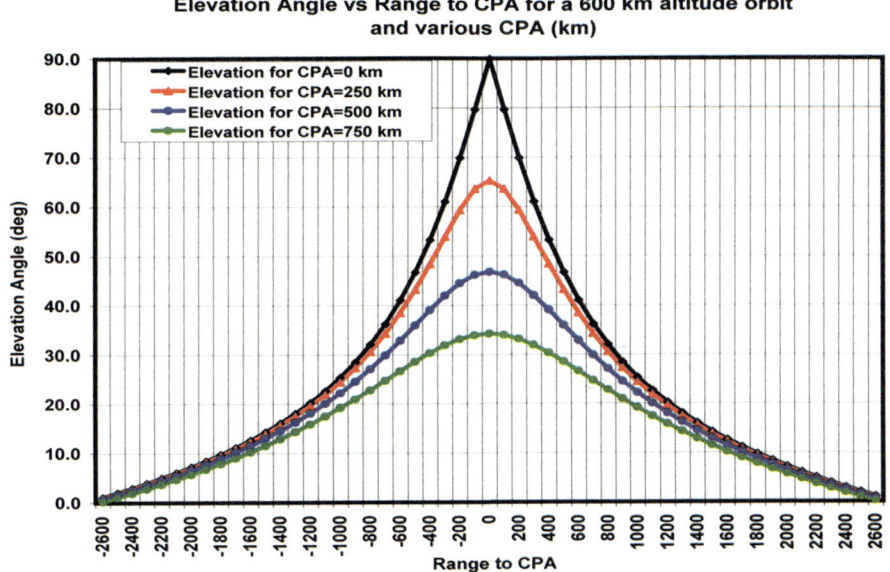

**Fig. 3.7** Elevation angle vs. ground range to CPA, various CPA

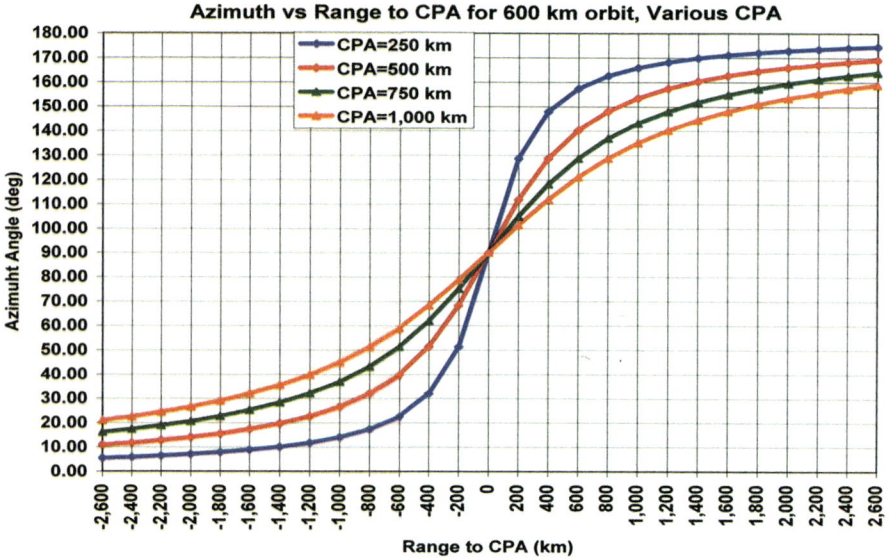

**Fig. 3.8** Azimuth or yaw to target vs. range to CPA

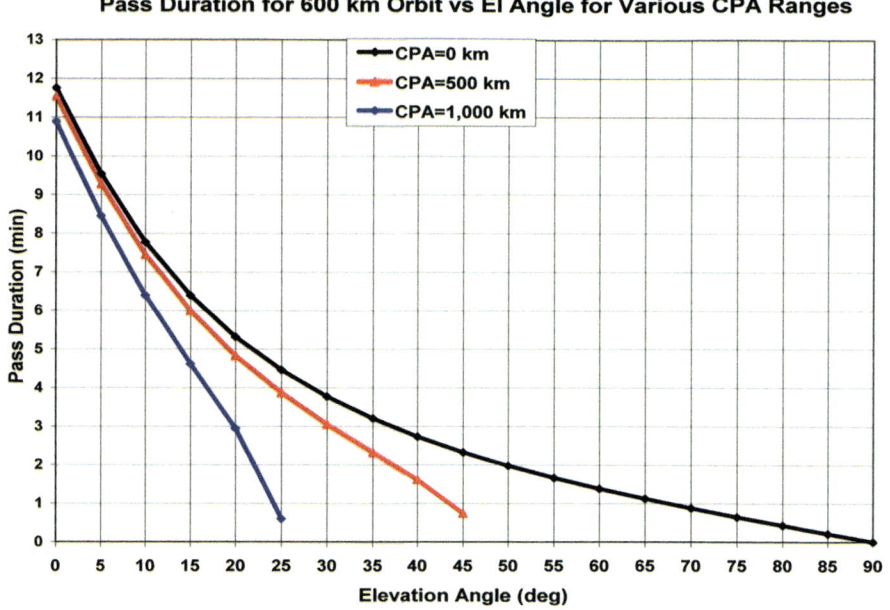

**Fig. 3.9** Pass duration vs. minimum elevation angle for various CPA

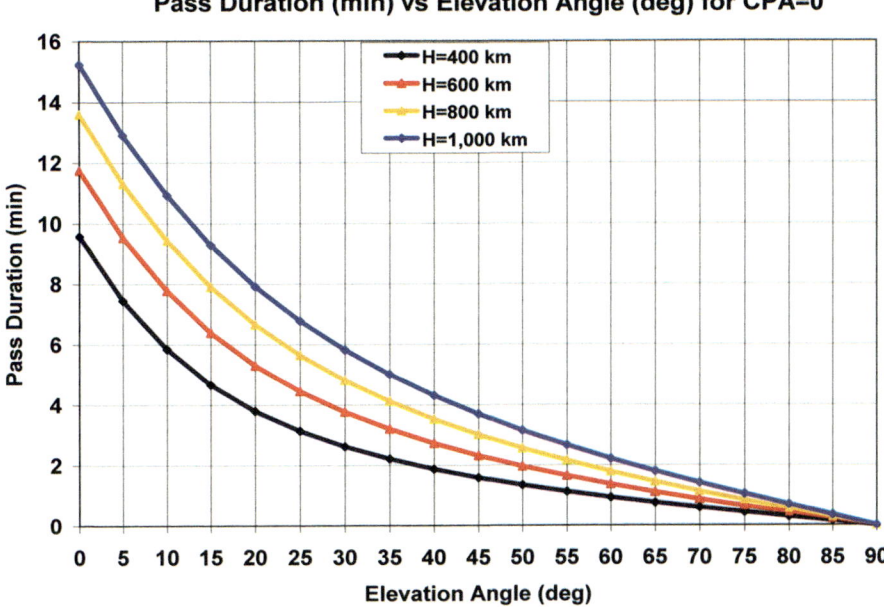

**Fig. 3.10** Pass duration a function of minimum elevation angle (for various orbit altitudes and for CPA = 0)

Figure 3.9 shows that, for example, for a CPA = 500 km and a minimum elevation angle of 15°, the pass is 6 minutes long. For a CPA = 1000 km, the pass is only 4.5 minutes long.

Figure 3.10 illustrates how pass duration increases with orbit altitude (for CPA = 0).

## 3.4 Pointing to a Target on the Ground From the Spacecraft

In many situations, such as in imaging a ground target from the spacecraft, the Pitch and Yaw angles from the spacecraft must be known to be able to point to the target. The target does not lie on the spacecraft ground track, but is off at a range of CPA km at the spacecraft Closest Point of Approach. The geometrical situation is shown on the left side of Fig. 3.11. The figure on the right is in the plane defined by the spacecraft, the center of the Earth and the Target. It is used only to assist in the derivation of the relevant equations. The equations are shown in Fig. 3.12 for a 600-km orbit and a CPA of 500 km.

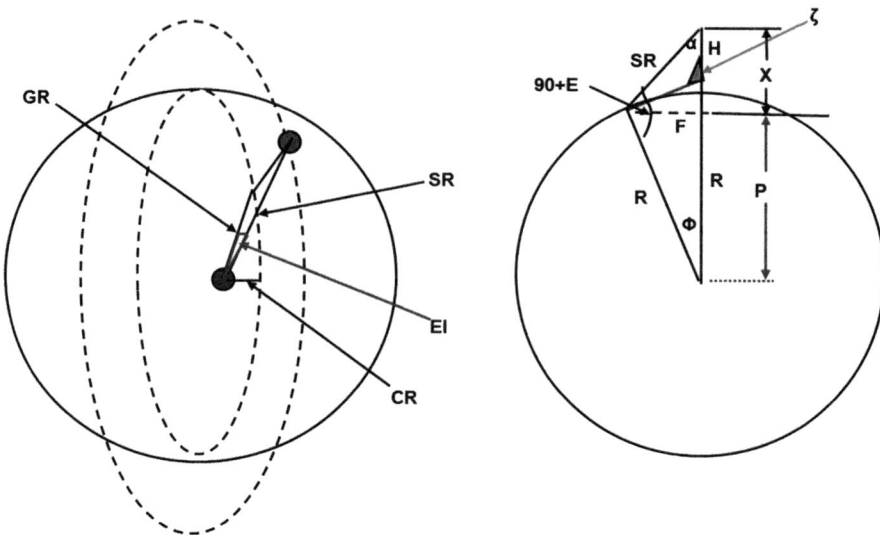

**Fig. 3.11** Spacecraft pointing to target at CPA km from ground track and geometry in plane of target, spacecraft and earth center

| Item | Symbol | Units | Quantity | Equation |
|---|---|---|---|---|
| Earth Radius | R | km | 6,378.3 | |
| Altitude | H | km | 600.0 | Input |
| CPA | CPA | km | 500.0 | Input |
| Grnd Range to CPA | $GR_{CPA}$ | km | 2,000.0 | Input |
| CPA Subtended Angle | c | deg | 4.491 | c=180/π*(CPA/R) |
| $GR_{CPA}$ Subtended Angl | φ | deg | 17.966 | φ=180/π*($GR_{CPA}$/R) |
| | P | km | 6,067.3 | P=R*cos(φ) |
| | F | km | 2,068.2 | F=R*tan(φ) |
| | X | km | 911.0 | X=H+R-P |
| Off-Nadir Angle | α | deg | 66.228 | α=atan(F/X) |
| | α | deg | | α=180/PI()*atan(R*tan(GRcpa/R)/(H+R-R*cos(GRcpa/R))) |
| Pitch Angle | p | deg | 23.772 | p=90-α |
| Azimuth to Target | az | deg | 14.036 | az=180/π*tan(CPA/$GR_{CPA}$) |
| Off-Nadir Angle | α | deg | 66.228 | 180/PI()*atan(R*tan(GRcpa/R)/(H+R-R*cos(GRcpa/R))) |
| Elevation Angle | ε | deg | 5.806 | ε=90-α-φ |

**Fig. 3.12** Spacecraft pitch, yaw to aim at ground target, various CPA

Above, the spacecraft look angles to the ground target are given as Pitch and Yaw. However, in many instances, it is more convenient to use spacecraft Roll and Pitch. In Sect. 2.3, the equations for that situation were derived. In Fig. 3.13, the Pitch, Yaw and Elevation are plotted for a 600-km orbit and a target CPA of 500 km.

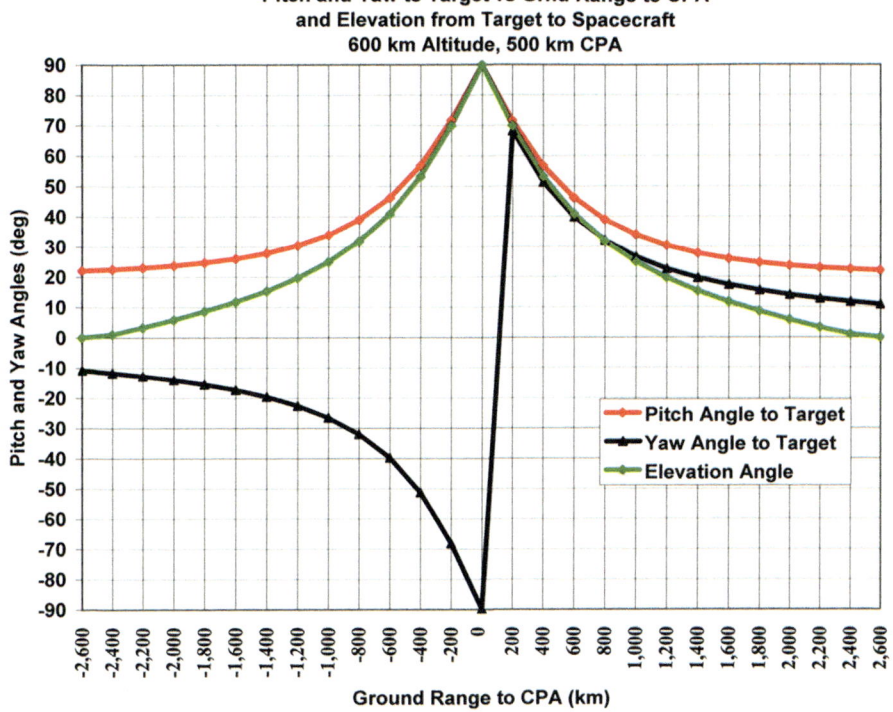

**Fig. 3.13** Pitch and yaw from spacecraft to target (600-km Altitude, 500 km CPA)

## 3.5 Ballistic Coefficient and On-Orbit Life

Ballistic Coefficient is the ratio of spacecraft mass to its drag. It is $\mathbf{B} = \mathbf{M}/\mathbf{C_D A}$, where $\mathbf{M}$ is the mass, $\mathbf{C_D}$ is the drag coefficient and $\mathbf{A}$ is the cross sectional area of the spacecraft. For example, a 200 lbs 24" × 24" × 30" nadir-pointing spacecraft has a cross sectional area of 720 sq. in (0.465 sq. m), a mass of 6.23 slugs (90.9 kg), a coefficient of drag of 2 and a Ballistic Coefficient of 97.7. Typical spacecraft ballistic coefficients vary between 50 and 200.

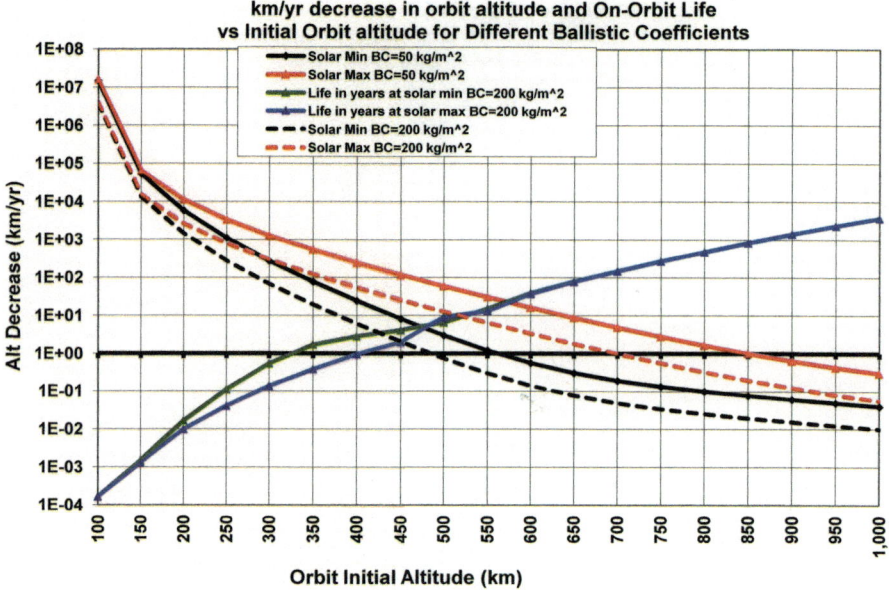

**Fig. 3.14** Orbit life and atmospheric density

| Alt km | Estimated Orbit Lifetime | | | |
|---|---|---|---|---|
| | Solar Min 50 kg/m² (days) | Solar Max 50 kg/m² (days) | Solar Min 200 kg/m² (days) | Solar Max 200 kg/m² (days) |
| 0 | 0 | 0 | 0 | 0 |
| 100 | 0.04 | 0.06 | 0.06 | 0.06 |
| 150 | 0.24 | 0.18 | 0.54 | 0.48 |
| 200 | 1.65 | 1.00 | 5.99 | 3.60 |
| 250 | 10.04 | 3.82 | 40.21 | 14.98 |
| 300 | 49.90 | 11.0 | 196.7 | 49.2 |
| 350 | 195.6 | 30.9 | 615.9 | 140.3 |
| 400 | 552.2 | 77.4 | 1,024 | 346.9 |
| 450 | 872 | 181 | 1,497 | 724 |
| 500 | 1,205 | 393 | 2,377 | 3,110 |
| | Years from this Row Down Years | | | |
| 550 | 4.5 | 2.2 | 15.0 | 13.1 |
| 600 | 9.0 | 7.0 | 38.6 | 36.7 |
| 650 | 15.2 | 12.5 | 78.1 | 76.4 |
| 700 | 36.7 | 34.5 | 146 | 144 |
| 750 | 67 | 66 | 270 | 268 |
| 800 | 115 | 112 | 480 | 477 |
| 850 | 209 | 208 | 842 | 840 |
| 900 | 348 | 342 | 1,427 | 1,425 |
| 950 | 605 | 575 | 2,337 | 2,334 |
| 1000 | 934 | 931 | 3,739 | 3,732 |

**Fig. 3.15** Orbit decay rate and orbit life time vs. altitude and ballistic coefficient (NASA)

## 3.6 Computing the Projection of the Sun on Planes on the Spacecraft

The website https://www.orbitaldebris.jsc.nasa.gov/mitigation/das.html provides downloadable NASA software for computing orbit life.

## 3.6 Computing the Projection of the Sun on Planes on the Spacecraft

To compute the power produced by a set of solar panels on the spacecraft, the projection of the sun on each panel must be computed. Using unit area panels, the projection of the Sun is a function of true anomaly and Beta. Ignoring eclipse, for the moment, the power produced by the 5 main panels (+X, −X, +Y, −Y and +Z) can be computed with the aid of the downloadable spreadsheet listed in the Chap. 21.

The Sun projection on panels in other orientations can be computed from the projections on these 5 principal panels. The bottom panel (−Z) is usually not available for power generation for it contains the separation system.

Next, eclipse must be factored in. Eclipse is a function of Beta and altitude. Figure 3.16 shows the eclipse duration vs. Beta for a 600-km and an 800-km orbit.

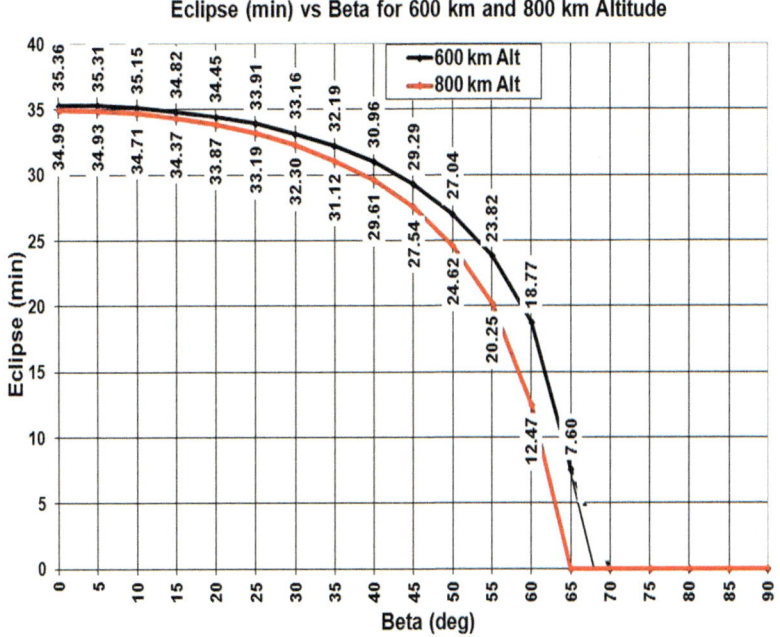

**Fig. 3.16** Eclipse (min) vs. Beta for 600- and 800-km Orbits

Note that (1) Beta does not vary much with altitude, and (2) for Beta > ≈65°, there is no Eclipse; the spacecraft is in sunshine all the time.

The eclipse duration corresponds to an anomaly angle range that straddles 0° anomaly (when the sun is in the equatorial plane). For instance, for ß = 0 the Eclipse is 35.36 minutes long, and, since the orbit period is 96.518 minutes, the eclipse lasts for an anomaly range of 131.89°. The anomaly ranges from 360 − 131.89/2 = 294.155° to 131.89/2 = 65.945°.

Figures 3.17, 3.18, 3.19 and 3.20 show the instantaneous powers of the 5 panels and the OAP of the sum. The spreadsheet from which these figures are obtained is among the downloadable spreadsheets listed in the Chap. 21.

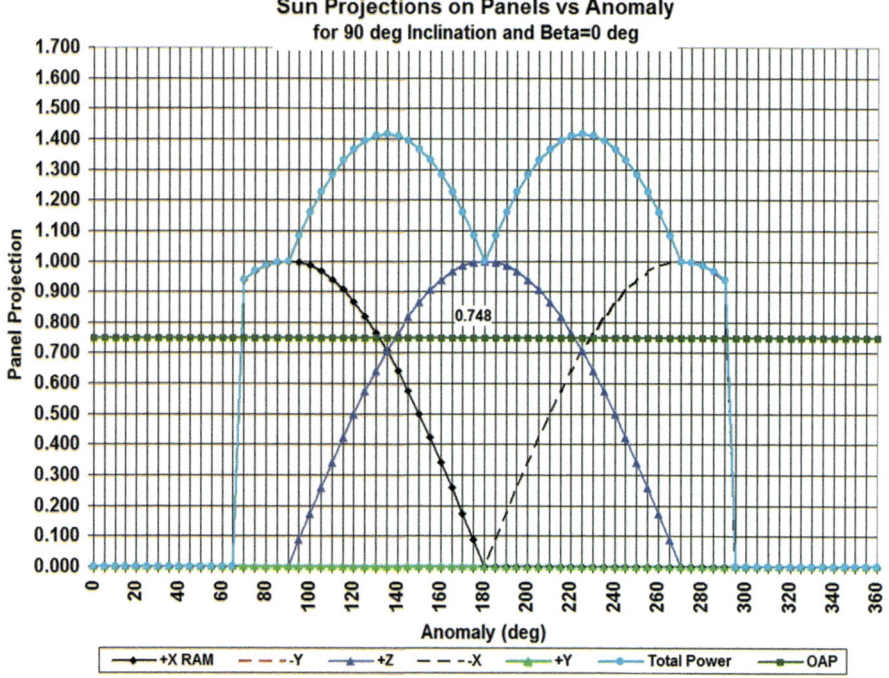

**Fig. 3.17** ß = 0°, Inc. = 90° Only +X, −X, +Z panels contribute to power

Note that the OAP (if each of the 5 panels have unit areas) is 0.748. Since there are 5 panels, the OAP produced is 14.96% of the solar cell power purchased and installed. Also note that the top (+Z) panel contributes almost half of the total OAP. If it were not used, the OAP would drop to 9.35% of the 4 remaining panel powers.

As ß is increased (to 45° and 90°) the OAP increases because more panels contribute and because the eclipse is shorter. See Fig. 3.18.

## 3.6 Computing the Projection of the Sun on Planes on the Spacecraft

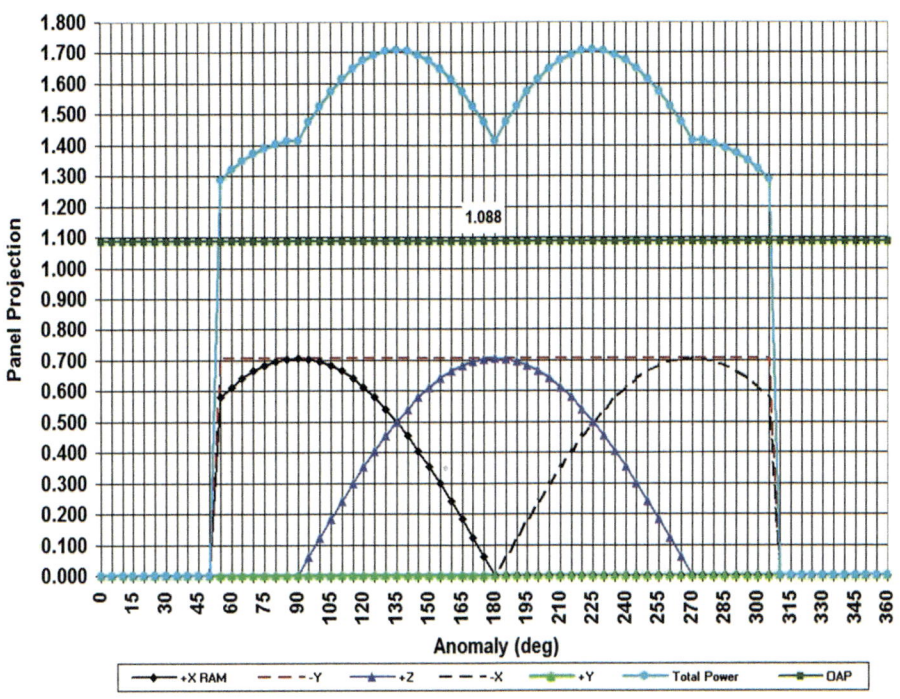

**Fig. 3.18** ß = 45° OAP increased to 1.088 or 21.76% of installed power

Note that now, the -Y panel starts to contribute, and the eclipse spans only 109.25°.

Increasing ß to 90°, Fig. 3.19, shows that there is no eclipse and that the OAP is 1.213 or 24.26% of the installed power.

Lowering the inclination to 65° reduces the OAP somewhat (to 1.107), resulting in an overall OAP efficiency of 22.14%.

If computation of the power produced by some other panel is desired, it can be constructed from the components of the 5 panels given here. For example, to compute the contribution of a panel tilted up 45°, we simply take 0.707 of the +Z and 0.707 of the (say) + X panel as a function of the anomaly, and compute the OAP as anomaly advances from 0 to 360°.

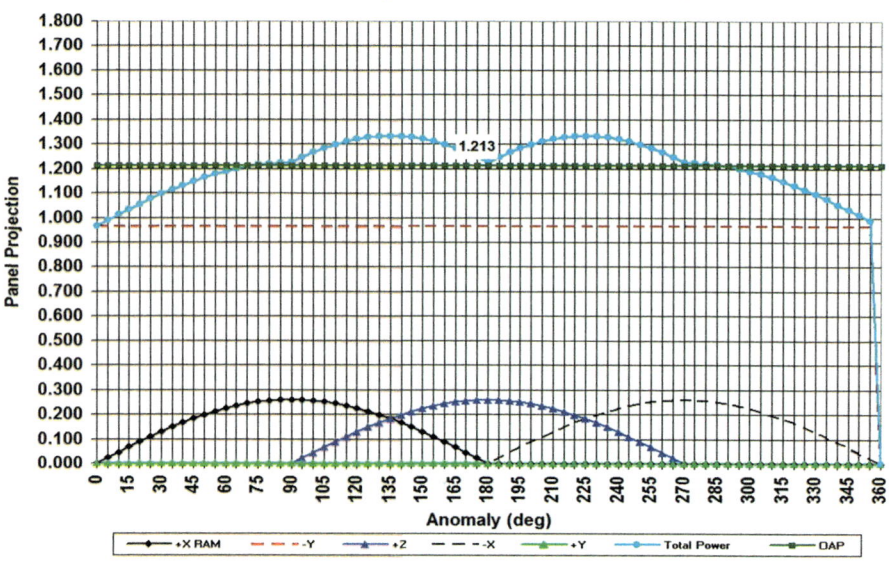

**Fig. 3.19** For ß = 75° There is no more eclipse

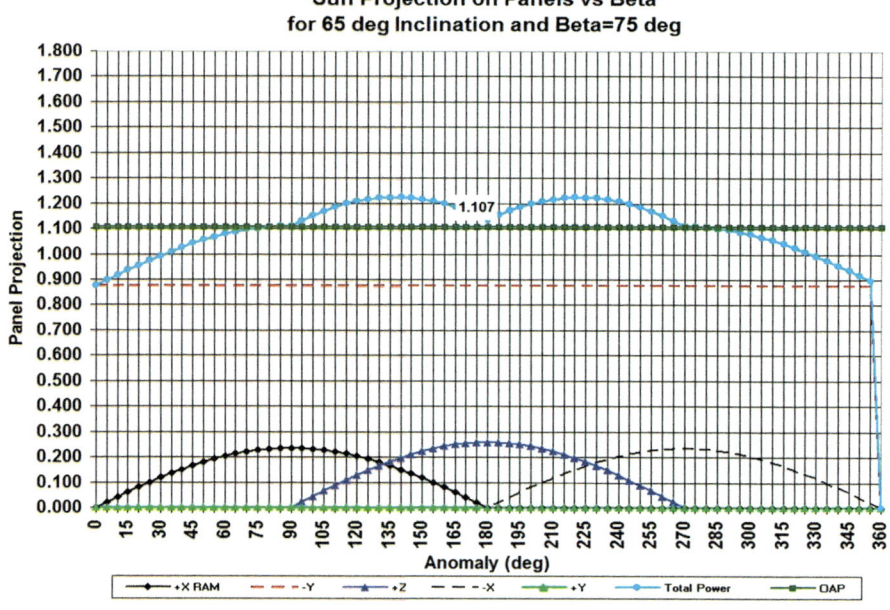

**Fig. 3.20** As inclination is reduced, OAP decreases slightly

# Chapter 4
# Electric Power Subsystem Design

The Electric Power Subsystem (EPS) of a satellite is a heavy and expensive subsystem. It is often about 25% of the weight and 25% of the cost of a spacecraft. Electric Power is also often underestimated, resulting in insufficient power to support the "mission creep" requirements of the spacecraft. The EPS design procedure is outlined below. Each of the steps will be discussed and illustrated in more detail later.

(A) **Determine the Required Spacecraft Orbit Average Power**

- List all of the electronic components of the satellite and the voltages and currents that each component requires
- Determine the power drawn by each component in each of the spacecraft operating modes. Augment these by the appropriate DC/DC conversion efficiency to obtain the OAP drawn from each voltage source in each spacecraft operating mode.
- Determine the peak OAP required

(B) **Determine the Battery Capacity Required and Choose the Battery Bus Voltage**

- Based on the power drawn during the eclipse (and the maximum eclipse duration), determine the battery WH requirements
- Select the battery cells that will be used
- Applying the battery output vs. input efficiency, determine the battery WH used during the eclipse
- Select the maximum Depth of Discharge below which the batteries should not be discharged. Apply this, and a large safety factor, to obtain the battery WH to be installed.
- Choose a battery bus voltage and divide by the cell voltage to determine the number of cells in series (in a string) of cells. Divide the total battery current by the current each parallel string will supply to determine the number of parallel battery strings.

(C) **Select a Solar Panel Configuration and Compute the OAP it can Supply**
- Select the solar panel configuration (the orientations and areas of each panel relative to the spacecraft axes). Also, determine how each panel will be stowed and released.
- Compute the instantaneous power generated by each panel as the spacecraft moves around an orbit. The total power vs. time is then computed, as is the OAP.
- Repeat this for all Beta angles (the angle between the sun line and the orbit plane) to determine what the minimum OAP is. Ensure that the minimum OAP generated is equal to or greater than the spacecraft OAP required.

(D) **Draw the EPS Block Diagram**
- Given the panel configuration, the various required voltages and the number of battery strings and cells per string, the EPS block diagram can now be drawn.
- Consider which groups of components should be turned ON/OFF on command, and whether the switch to turn these groups ON/OFF should be ahead or after the respective DC/DC converters that supply the voltage to the group.

(E) **Miscellaneous EPS Design Steps**
- Often, an EPS computer is included to collect telemetry regarding the state of health of the EPS, the battery capacity status, component temperatures and EPS status. This computer may also be used to turn ON/OFF power to the various electronic components.
- The Separation Switch that signals release from the launch vehicle and the start of spacecraft operations is also part of the EPS. The functions enabled or disabled to ensure that no electric power is drained from the spacecraft prior to launch are used to determine where in the spacecraft circuit the Separation Switch should be located.

## 4.1 Required Orbit Average Power (OAP)

The procedure above is illustrated for a 3-axis stabilized spacecraft that has two star trackers for attitude determination, a camera and image processor payload, an S-Band downlink communication transmitter and a UHF uplink command system. It also has a separate Command & Data Handling (C&DH) computer.

The three operating modes of the spacecraft are:

- **Idle Mode** - when no communications or imaging takes place, but all other systems are operating. In this mode, the CMD receiver is ON.
- **Imaging Mode** - when the camera and image processing computer are turned ON, but when there is no downlink communications
- **Communications Mode** - when the spacecraft downlinks telemetry and image data and receives commands from the ground station. In this example, it is assumed that downlink takes place for 5 min (out of a total orbit period of 90 min).

4.2 Battery Capacity and Battery System Design

The table in Fig. 4.1 shows the power consumption of each component and the voltage source from which it draws the power. The first group of three columns shows the peak powers required from each DC/DC converter. The DC/DC converter efficiencies (in %) are also shown.

The 2nd, 3rd and 4th groups of columns show the component percent utilizations in each of the three spacecraft operating modes. The OAP, after accounting for DC/DC converter efficiencies in each of the three modes, are shown in the *OAP in Each Op Mode* row. The peak OAP requirement for which the EPS must be designed is shown in the last row as the maximum of the OAP requirements in each of the three operating modes.

| Spacecraft OAP Requirements | Watts at Voltages | | | Idle Mode Watts | | | | Imaging Mode Watts | | | | Communications Mode Watts | | | |
|---|---|---|---|---|---|---|---|---|---|---|---|---|---|---|---|
|  | 5v | 12v | 28v | % | 5v | 12v | 28v | % | 5v | 12v | 28v | % | 5v | 12v | 28v |
| C&DH | 1.5 | | | 100 | 1.5 | | | 100 | 1.5 | | | 100 | 1.5 | | |
| EPS Processor | 0.2 | | | 100 | 0.2 | | | 100 | 0.2 | | | 100 | 0.2 | | |
| Imaging Payload | | | | | | | | | | | | | | | |
| Camera | | 4.0 | | | | | | 10 | | 0.4 | | | | | |
| Image Processor | 3.0 | | | | | | | 15 | 0.5 | | | | | | |
| ADACS | | | | | | | | | | | | | | | |
| Pitch Reaction Wheel | | | 3.5 | 100 | | | 3.5 | 100 | | | 3.5 | 100 | | | 3.5 |
| Roll Reaction Wheel | | | 3.5 | 100 | | | 3.5 | 100 | | | 3.5 | 100 | | | 3.5 |
| Yaw Reaction Wheel | | | 3.5 | 100 | | | 3.5 | 100 | | | 3.5 | 100 | | | 3.5 |
| Star Tracker #1 | 0.5 | | | 100 | 0.5 | | | 100 | 0.5 | | | 100 | 0.5 | | |
| Star Tracker $2 | 0.5 | | | 100 | 0.5 | | | 100 | 0.5 | | | 100 | 0.5 | | |
| Course Sun Sensors | 0.4 | | | 100 | 0.4 | | | 100 | 0.4 | | | 100 | 0.5 | | |
| 3 Torque Rods | | 0.8 | | 100 | | 0..8 | | 100 | | 0.8 | | 100 | | 0.8 | |
| Communication | | | | | | | | | | | | | | | |
| TTM & Image Xmitter | | 30.0 | | | | | | | | | | 5.5 | | 1.6 | |
| CMD Receiver | | 1.5 | | 100 | | 1.5 | | 100 | | | | 100 | | 1.5 | |
| Peak and Ave Power | 6.1 | 36.3 | 10.5 | | 3.1 | 2.3 | 10.5 | | 3.6 | 1.2 | 10.5 | | 3.2 | 3.9 | 10.5 |
| DC/DC Converter Eff. % | 87 | 85 | 85 | | 87 | 85 | 85 | | 87 | 85 | 85 | | 87 | 85 | 85 |
| OAP from Each Source | 7.0 | 42.7 | 12.4 | | 3.6 | 2.7 | 12.4 | | 4.1 | 1.4 | 12.4 | | 3.7 | 4.6 | 12.4 |
| OAP in Each Op Mode | | | | | 18.7 | | | | 17.9 | | | | 20.7 | | |
| Design for OAP of | | 20.7 | | | | | | | | | | | | | |

Fig. 4.1 Developing spacecraft orbit average power requirements

## 4.2 Battery Capacity and Battery System Design

### 4.2.1 Battery Capacity

Select the spacecraft operating mode when it will be in the eclipse. This is usually the *Idle Mode*. The spacecraft will spend a maximum of about 35 min in eclipse per orbit, during which it must be powered from the battery. In the above example, the *Idle Mode* power is 18.622 watts, and the battery energy used in eclipse is Eclipse Power × Max Eclipse Duration/60 min = 18.622*35/60 = 10.868 WH.

Typical battery output vs. input efficiencies are 85%. That is, 100/85 = 1.176 times the battery capacity is required to provide the required battery output WH. In the above example, this raises the battery capacity requirement to 12.780 WH.

Next, consider the expected mission and battery life that the EPS must support. In Low Earth Orbit (LEO), a satellite orbits the Earth typically 15 times per day. Therefore, the battery will be charged and discharged about 5000 times per year. A battery's useful life depends on the average Depth Of Discharge (DOD). The smaller the DOD, the longer the battery life. A typical Lithium Battery life vs. average DOD is shown in Fig. 4.2.

Figure 4.2 shows that if 5-year mission life is desired, the DOD should be less than 15%. For 10-year mission life, a DOD of 10% may be appropriate. Thus, the battery capacity that must be provided is many times that which the SC requires to support operations in the eclipse.

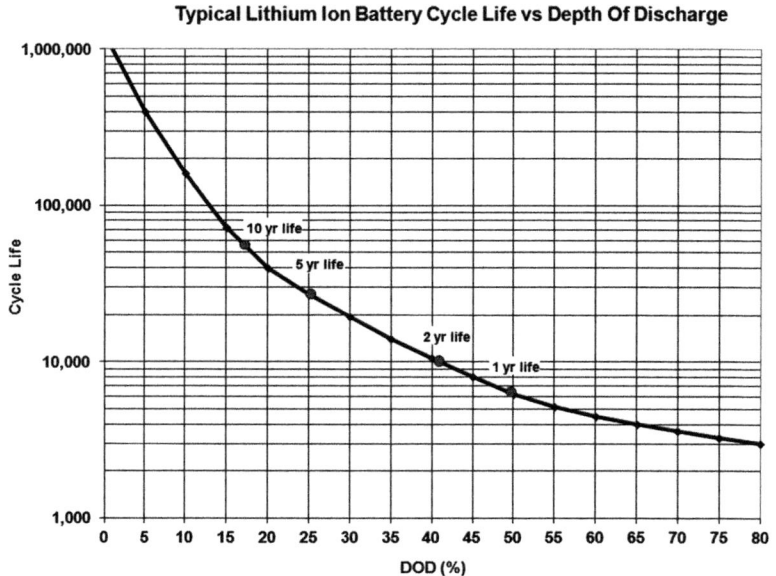

**Fig. 4.2** Typical lithium ion cycle life vs. depth of discharge (DOD)

In addition, a significant margin of battery capacity should be provided to compensate for temperature degradation of battery, solar cell capacity and for mission creep. These factors are reflected in the battery capacity table in Fig. 4.3.

**WH (installed)=(WH used in Eclipse)*(100/Efficiency)*(100/DOD)*(1+Margin/100)**

| Item | Quantity | Units |
|---|---|---|
| WH needed to operate in Eclipse | 10.868 | WH |
| Battery Efficiency | 85 | % |
| Depth of Discharge | 10 | % |
| Margin | 25 | % |
| Battery Capacity Installed | 159.8 | WH |

**Fig. 4.3** Computing installed battery capacity

This example shows that about 15 times the needed battery WH has to be provided to support a 10-year mission life.

To many, this may sound excessive. However, experience suggests that it is not. One can never have enough battery capacity!

## 4.2.2 Battery Choice

Legacy spacecraft batteries were NiCd until about the year 2000. Sealed lead acid batteries were used in some spacecraft that required battery charge and discharge at cold temperatures. More recently, Lithium Ion batteries have been used in spacecraft with good success. The table in Fig. 4.4 compares the characteristics of the different kinds of batteries used in spacecraft. The entries in this table are approximate. Specific manufacturers' product specifications should be used since there are large variations from battery to battery, and because the state of the battery art changes rapidly.

| Battery Types | Lead Acid | NiCd | NiMH | Li-Ion |
|---|---|---|---|---|
| Voltage/Cell | 2.0 | 1.2 | 1.2 | 3.6 |
| Approx No of Series Cells for 28 V bus | 14 | 24 | 24 | 8 |
| Density WH/kg | 30-50 | 45-80 | 60-120 | 150-250 |
| Charge Temp C° | -20 to +50 | 0 to +45 | 0 to +45 | 0 to +45 |
| Discharge Temp C° | -20 to +50 | -20 to +65 | -20 to +65 | -20 to +60 |
| Self Discharge in (months) | 3-6 | 1-2 | 2-3 | years |
| Notes | Do Not Discharge to 0 | OK to Discharge | OK to Discharge | Cell Protection Needed |
| Approx Weight of 150 WH | 3.75 kg | 2.5 kg | 2.1 kg | 0.8 kg |

**Fig. 4.4** Comparison of the characteristics of different battery types

There is a great deal of experience and data on the NiCd batteries that were used in spacecraft until recently. However, in the last several years, spacecraft have switched to Lithium Ion batteries because of their large energy density and excellent low-temperature performance. Lithium Ion batteries are recommended.

Since there are large differences between different Lithium Ion batteries depending on the manufacturer, or even between different models of the same manufacturer, the reader should research the cell specifications before making a selection. In addition, since battery capacity degrades with temperature, the installed battery capacity may have to be increased to allow for aging and temperature-dependent loss of capacity.

## 4.3 Solar Arrays Configuration

There are basically three types of spacecraft solar array configurations.

In the *Body Mounted* array, solar cells are mounted to a substrate to create a panel and then the panels are mounted to the spacecraft structure. This type of array configuration is simple, but the Orbit Average Power (OAP) generated is usually only a small fraction of the installed solar cell power.

In the *Deployable Panel Solar Array,* power generated as a fraction of the installed power is greater, for panel angles can be adjusted to be more nearly at right angles to the Sun. However, these arrays are more complex because deployment and panel release mechanisms are required.

The *Rotating (Solar Array Drive)* panels optimize the use of the installed solar cells by continuously rotating (either about one or two axes) to orient the panels to maintain normal incidence to the sun. Such panels are used when large amounts of power are required. Most geostationary spacecraft use such rotating panels.

The various typical spacecraft solar array configurations are shown in Fig. 4.5a–d. In the first group are 4, 6 or 8-sided spacecraft with solar cells mounted on the sides of the body. Cells may also be on top.

In the next group of two spacecraft, the panels are angled at 30° and 90° with respect to the body.

In Fig. 4.5c, the long panels are tilted at 30° (the upper half) and 45° (the lower half). Some designers have tilted panels this way to better match the OAP requirements over the range of Beta during the mission.

Some spacecraft are built with one-axis and others with two-axis solar array drives. The two-axis solar drive permits the panels to be normal to the Sun all the time, thus increasing array efficiency. However, solar array drives and deployment mechanism are complex and costly.

The SC with single-axis solar array drive can also point the solar panels to the Sun most of the time. If the SC is permitted to yaw, this array configuration is very effective.

The Electric Power generated as a function of time in one orbit can be computed by stepping the spacecraft around the Earth in true anomaly (stepping the spacecraft position around in its orbit one revolution). At each step, the vector from the Sun to the spacecraft is computed, and then the dot product between this vector and a panel normal vector is computed. The dot products are multiplied by the peak output of each panel and summed to get the total output power. This results in the spacecraft power generated as a function of true anomaly (time), The OAP can then be computed. Finally, the OAP computation for each Beta is repeated.

Figure 4.6 illustrates the results of these computations for a 4-sided satellite (like the one shown in Fig. 4.5a Type A) in a 560 km 90° inclined orbit. The Beta angle varies from 0° to 90°. For Beta = 0° the spacecraft generates only 11% of the purchased and installed power.

## 4.3 Solar Arrays Configuration

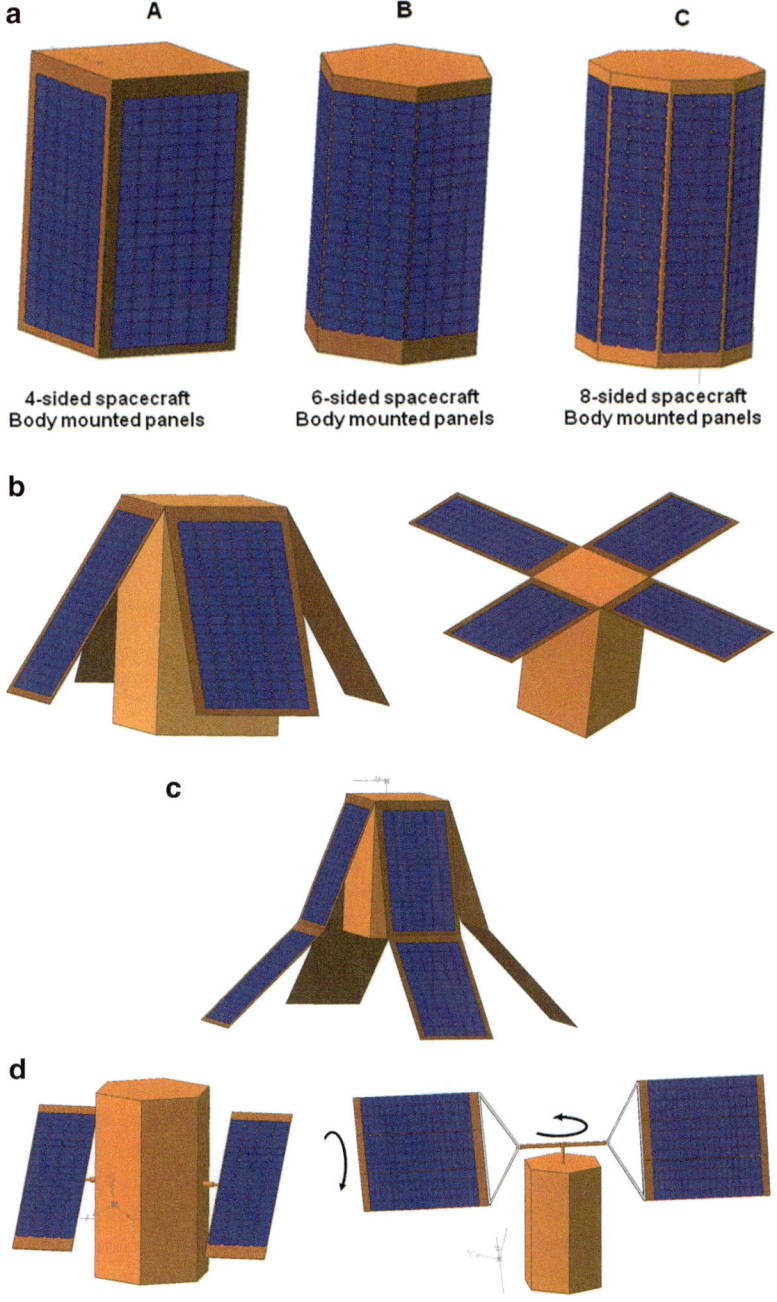

**Fig. 4.5** (**a**) Several body mounted solar array configurations. (**b**) Solar panels tilted 30° and 90° to the body. (**c**) Long panels are tilted 30° AND 45°. (**d**) Spacecraft with single or dual axis solar array drives

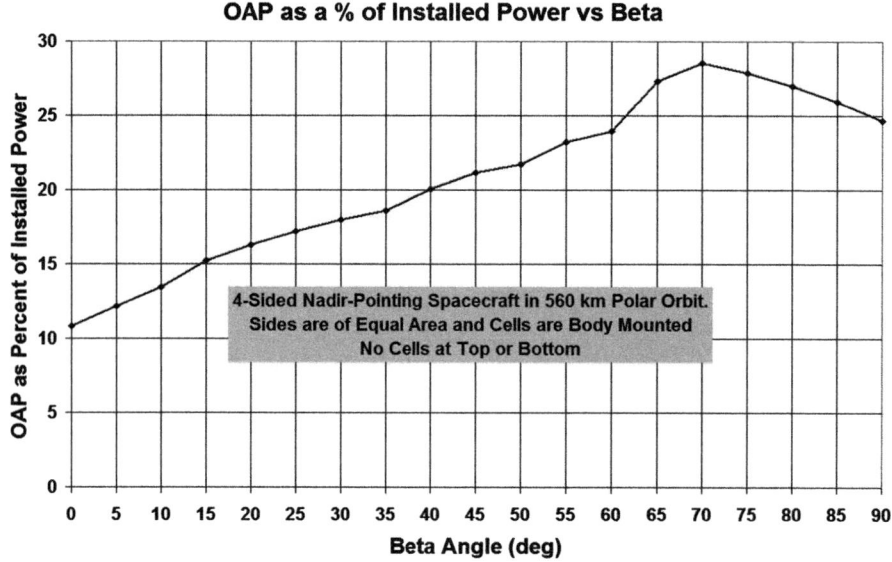

**Fig. 4.6** OAP as a percentage of installed power vs. Beta for polar orbit spacecraft

The OAP generated for all Beta angles can be increased significantly by tilting the panels, as shown in Fig. 4.5b. Tilting the panels 15° up increases the minimum OAP (for Beta = 0°) from 11% to 17% of the installed power. Increasing panel angles to 30° raises the minimum OAP to about 22%. While further increasing panel tilt increases the Beta = 0° OAP, the OAP at 90° Beta starts to decrease. The 30° panel tilt is nearly optimum if the spacecraft must operate at all Beta angles. At 30° panel tilt, the OAP is double that for body mounted panels. The OAP generated as a function of Beta for a 500 km Polar orbit spacecraft is shown in Fig. 4.7. Panel tilt is shown parametrically.

In an equatorial (0° inclination) orbit, the configuration with 90° panel angles, shown in Fig. 4.5b, can generate an OAP that is 33% of the installed power.

As the orbit inclination is reduced to 75°, 60°, 45°, 30° and 0°, OAP vs. Beta changes. Figure 4.8 shows the minimum OAP vs. Beta for each panel inclination. Panel tilt of 30° is near optimum for all inclinations until we approach the equatorial orbit, where a 90° tilt angle is best.

Note that in a geostationary orbit, the solar panels of a nadir-pointing satellite with a solar array drive can be aimed at all times to point to the Sun (except in the eclipse), resulting in the OAP becoming 95% of the installed power.

## 4.3 Solar Arrays Configuration

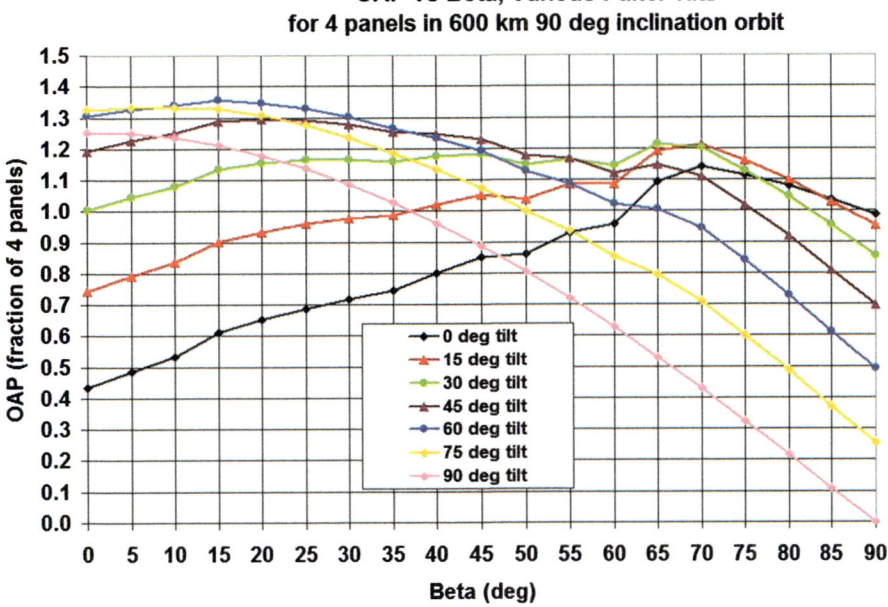

**Fig. 4.7** OAP vs. beta for a 4-sided spacecraft at 500 km altitude and 90° inclination with their solar panels tilted up by 0–90°

| Orbit Inclination (deg) | Panel Tilt (deg) | Min OAP vs. Beta | At Beta Equals (deg) |
|---|---|---|---|
| 90 | 30 | 22% of Total | 0 |
| 75 | 30 | 23% of Total | 0 |
| 60 | 30 | 23% of Total | 0 |
| 45 | 30 | 23% of Total | 0 |
| 30 | 45 | 25% of Total | 90 |
| 0 | 90 | 33% of Total | Any Beta |

**Fig. 4.8** Minimum OAP as a function of panel tilt, and the beta for which OAP is Minimum. 4 Solar Panels

The changes in Beta angle and inclination are illustrated in the pictures of Fig. 4.9.

In Sect. 3.6, the relationships were developed for a spreadsheet that provides the instantaneous and OAP for solar panels of arbitrary orientation. Employing those relationships, OAP for 4 configurations are shown in Fig. 4.10. These are (1) 4 body mounted panels, (2) 4 panels tilted 30° from the body, (3) 4 panels where half of the panels are at 15°, the other half at 30° and (4) where half the panels tilt 45°, the other half 60°.

The red line in Fig. 4.10 shows that tilting panels 30° increases the OAP significantly (relative to body mounted panels). Tilting the panels 45° and 60° further increases the OAP for low Beta, but at higher Beta the OAP is less. The bottom line is that tilting all the panels 30° is probably the best.

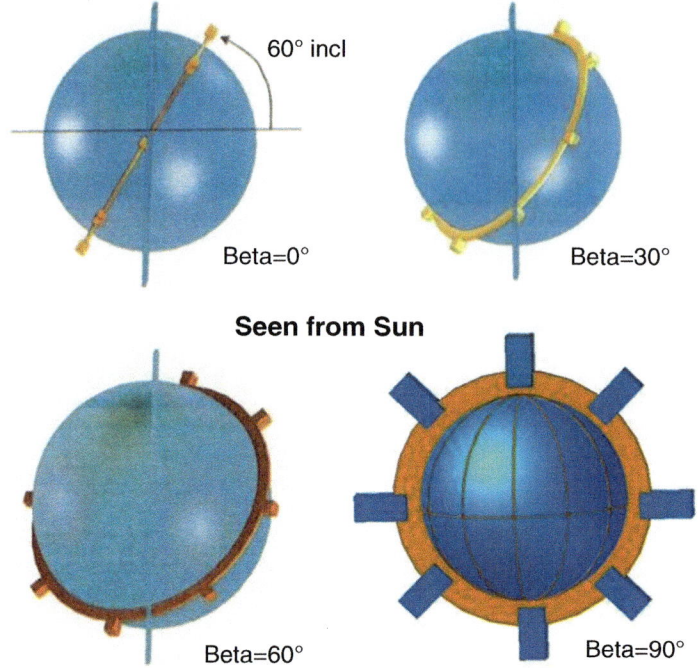

**Fig. 4.9** Pictorial illustration of changes in beta and inclination

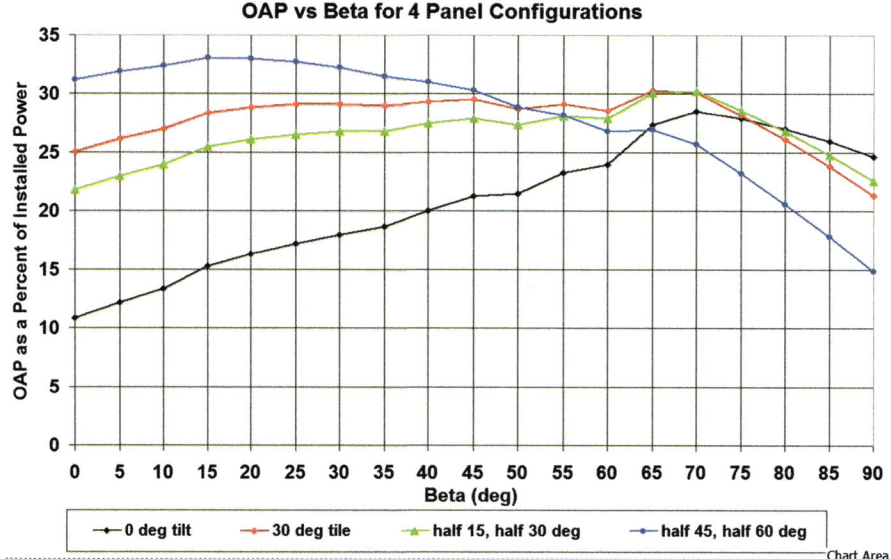

**Fig. 4.10** The OAP as % of installed power for four configurations

## 4.4 Beta Angle Vs. Time

As seen from the generated solar power's dependence on Beta, in planning a mission, the range of Beta angles encountered in that mission should be considered. The equation for Beta as a function of time, orbit inclination, launch date and hour is a complicated equation. It is given in the Appendix. Figure 4.11, however, illustrates how Beta varied in a specific one-year period.

Note that for the entire year, Beta does not exceed 80°. So, if this remains true for the entire mission life of the spacecraft, a solar panel configuration that maximizes OAP for Beta < 80° can be chosen. Significant savings in solar cell costs can then be realized.

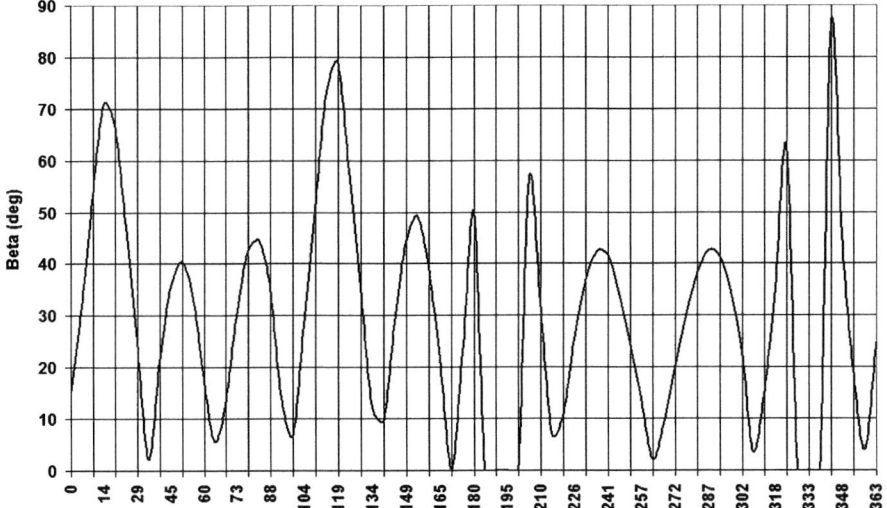

**Fig. 4.11** Beta as a function of time in a 1-year period for a 600 km, 60° orbit

## 4.5 Solar Cells and Cell Laydown

Solar cell technology is rapidly increasing the efficiency of solar cells. In the 1990s, Germanium cell efficiencies were only about 15%. Today, the Spectrolab Ultra Triple Junction Gallium Arsenide cells have efficiencies of about 28.3%. With Sun illumination density of 135.3 mW/cm$^2$, a typical 40 mm × 70 mm cell produces about 1 watt output at 28 °C. At maximum power, the cell voltage is 2.350 volts and the cell current is 425 mA. The bare cell weighs 84 mg/cm$^2$, and a little bit more with the cover glass.

The cell degrades a few (about 5) percent over 7–10 years due to radiation in space. The exact amount depends on the orbit. The cell output also degrades with temperature. The reader should consult the manufacturer's specifications for these degradations.

The number of cells in series is dependent on the bus voltage planned for the spacecraft. For a legacy 28 volt spacecraft bus, typically about 14–18 cells in series are used. The solar cell "string" output of 32.9–42.3 volts is sufficient to charge the batteries with an overvoltage to accommodate the voltage drop in the charge regulator. The strings are isolated from one another by diodes (see block diagram). In addition, wiring is typically built into the substrate in the reverse direction from the current in the string to eliminate the magnetic fields that the panel would generate otherwise.

Good manufacturing practice is to lay down these cells on a substrate with at least 1 mm gap between cells in series. The space between "strings" should be at least 2 mm. A typical cell laydown on two solar panels is shown in Fig. 4.12.

**Fig. 4.12** Two panels with 3 "strings" each and 9 cells per "string" produces 21 volts for a 14.4 volt battery bus. Each panel produces about 27 watts of power

## 4.6 EPS Block Diagram

A typical block diagram of the Electric Power Subsystem is shown in Fig. 4.13. Solar panel outputs are combined with blocking diodes. The current of each panel is monitored by the (optional) Electric Power Subsystem microprocessor and sent to the ground via the telemetry transmitter.

## 4.6 EPS Block Diagram

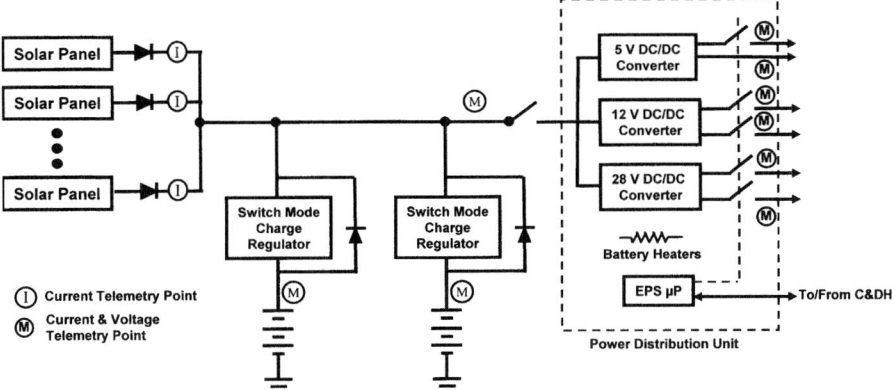

**Fig. 4.13** Typical electric power subsystem block diagram

Each of the multiple, series-connected string of batteries is charged by a charge regulator that assures that the (Lithium) cell voltages should not exceed 4.2 volts/cell. In a typical 28 volt bus, there are 8 series-connected battery cells. The string requires $8 \times 34.2 = 33.6$ volts of charging voltage, which requires solar array outputs of several volts more. The multiple batteries and charge regulators are connected in parallel. Blocking diodes from the battery strings to the main bus are used to ensure that no reverse currents flow into the batteries. Individual battery voltages and currents are monitored.

The Power Distribution Unit (PDU) is powered from the battery bus. It contains the various DC/DC converters required by the spacecraft electronics components. Each component can be switched ON/OFF as needed. Note that some DC/DC converter outputs may not be switched to ensure that they could not be turned OFF. One of the 5 V DC/DC converter outputs that powers the C&DH may not be switched to make sure it is never turned off (which could end the spacecraft mission).

The EPS micro-processor controls the switches, collects EPS telemetry and communicates with the spacecraft C&DH computer from which it receives commands and to which it sends EPS telemetry.

The spacecraft separation switch signals that the spacecraft and the launch vehicle have separated. This switch is often placed in series between the battery bus and the PDU.

In addition, if there are battery heaters, they are turned ON/OFF by either temperature sensors mounted to the batteries or by the EPS processor.

This diagram shows only one DC/DC converter for each different voltage required by the spacecraft components. However, a more distributed architecture would use multiple DC/DC converters, often one associated with each of the spacecraft components.

Also, this diagram shows the power switches at the outputs of the DC/DC converters. Some prefer to switch the converter inputs to avoid DC/DC converter power drain from the batteries when components are turned OFF.

This diagram shows that each Charge Regulator charges an entire string of series-connected battery cells. The cells should be matched in this design. There are charge regulators, however, that can charge each cell in a string separately.

The EPS design procedure was described in the foregoing. There are, of course, other considerations, such as the thermal design of the EPS, which are not described here. Some of these EPS considerations are discussed elsewhere in this book.

# Chapter 5
# Spacecraft Communications

RF communication is used to command the spacecraft from the ground, to have the spacecraft send Health and Status (Telemetry) about the condition of the spacecraft and to send payload data to the ground station(s). This chapter discusses spacecraft frequencies, communication link margins, Bit error rates and RF hardware.

## 5.1 Frequency Allocation

Satellites are only permitted to transmit and receive in allocated bands. The Abbreviated Table in Fig. 5.1 and the detailed Table in Fig. 5.2 list the frequencies from within allocations must be obtained to operate satellites by amateurs, researchers and commercial organizations. This Figure was extracted from the FCC Table of Frequency Allocations (July 28, 2016). The FCC Table enumerates further restrictions that will not be discussed here.

The VHF frequencies are suitable for low data rate applications. The spacecraft antennas, because of their size, are usually omnidirectional (or Earth Coverage) and are usually fixed (not steerable). Antennas must often be stowed, because of their deployed size, and released when on orbit. Since radio frequency interference on the ground raises the ground station RF noise above the KTB noise level by more than 10 dB, successful communication requires either sufficient spacecraft transmitter power or high gain ground station antennas.

The UHF frequencies are also used for low data rate applications. However, the spacecraft antennas are smaller (or they have more gain). Terrestrial noise, while greater than KTB, is less than at VHF, and ground stations with larger antenna gains are smaller in size. All of these factors permit operating at higher data rates (typically several hundred Kbps).

| Frequency Range (MHz) | Typical Uses or users |
|---|---|
| 137-138 VHF | Low Data Rate Satellite-to-Earth |
| 145-146 VHF | Amateur Satellites |
| 148-150 VHF | Uplinks to Low Data Rate Satellites |
| 240-270 UHF | Military Satellites |
| 400-403 UHF | Mobile Communications |
| 432-438 UHF | Amateur Satellites |
| 2,025-2,300 S-Band | Space-to-Earth and Earth-to-Space |
| 8,000-9,000 X-Band | X-Band High Data Rate Satellites |
| Above 20,000 | Increasingly used despite weather degradation |

**Fig. 5.1** Abbreviated list of frequencies used by spacecraft

| From (MHz) | To (MHz) | Allocated to |
|---|---|---|
| 7.000 | 7.1000 | Amateur Satellite |
| 14.000 | 14.250 | Amateur Satellite |
| 19.068 | 18.168 | Amateur Satellite |
| 21.000 | 21.450 | Amateur Satellite |
| 24,890 | 24.990 | Amateur Satellite |
| 137.000 | 138.000 | Space-to-Earth |
| 144.000 | 146.000 | Amateur Satellite |
| 148.000 | 149.900 | Earth-to Space |
| 149.900 | 150.050 | Earth-to Space |
| 161.9625 | 161.9875 | Earth (Mobil)-to-Space |
| 162.0125 | 162.0375 | Earth (Mobil)-to-Space |
| 399.900 | 400.050 | Earth-to Space |
| 400.150 | 401.000 | Space-to-Earth |
| 401.000 | 402.000 | Space-to-Earth and Earth-to-Space |
| 402.000 | 403.000 | Earth-to Space |
| 1,164.000 | 1,215.000 | Radio-Navigation Sat Space-to-Earth, Space-to-Space |
| 1,215.000 | 1,240,000 | Radio-Navigation Sat Space-to-Earth, Space-to-Space |
| 1,240.000 | 1,300.000 | |
| 1,300.000 | 1,350.000 | Radio-Navigation |
| 1,400.000 | 1,427.000 | Earth Exploration and Satellite Research |
| 1,525.000 | 1,535.000 | Space-to-Earth |
| 1,535.000 | 1,559.000 | Space-to-Earth |
| 1,559.000 | 1,610.000 | Radio-Navigation Sat Space-to-Earth, Space-to-Space |
| 1,610.600 | 1,613.800 | Earth-to-Space (Mobil, Aeronautical Navigation) |
| 1,626.500 | 1,660.000 | Mobile Earth-to-Space |
| 1,660.000 | 1,675.000 | Mobile Earth-to-Space |
| 1,675.000 | 1,695.000 | Meteorological Space-to-Earth |
| 1,695.000 | 1,710.000 | Meteorological Space-to-Earth |
| 1,761.000 | 1,850.000 | Space Operation Earth-to-Space |
| 2,000.000 | 2,020.000 | Mobile-to-Satellite Earth-to-Space |
| 2,025.000 | 2,110.000 | Space Research, Earth-to-Space, Space-to-Space |
| 2,200.000 | 2,290.000 | Space-to-Earth, Space-to-Space |
| 2,310.000 | 2,345.000 | Broadcasting Satellite |
| 2,483.500 | 2,500.000 | Space-to-Earth |
| 2,655.000 | 2,700.000 | Space Research, Passive |
| 3,100.000 | 3,300.000 | Earth Exploration and Space Research |
| 3,600.000 | 4,200.000 | Fixed Satellite Space-to-Earth |
| 4,500.000 | 4,800.000 | Fixed Satellite Space-to-Earth |
| 5,000.000 | 5,010.000 | Radio Navigation Earth-to-Space |

**Fig. 5.2** Frequencies used by satellites (up to 20 GHz)

| | | |
|---|---|---|
| 5,010.000 | 5,030.000 | Radio Navigation Space-to-Earth |
| 5,150.000 | 5,250.000 | Fixed Satellite Earth-to-Space |
| 5,250.000 | 5,460.000 | Earth Exploration, Space Research |
| 5,460.000 | 5,570.000 | Earth Exploration, Space Research |
| 5,839.000 | 5,850.000 | Amateur Satellite Space-to-Earth |
| 5,850.000 | 7,075.000 | Fixed Satellite Earth-to-Space |
| 7,145.000 | 7,235.000 | Space Research Earth-to-Space |
| 7,250.000 | 7,850.000 | Fixed Satellite Space-to-Earth |
| 7,900.000 | 8,215.000 | Earth-to-Space |
| 8,215.000 | 8,400.000 | Earth Exploration Space-to-Earth, Mobile Earth-to-Space |
| 8,400.000 | 8,450.000 | Deep Space Space-to-Earth |
| 8,450.000 | 8,500.000 | Space Research Space-to-Earth |
| 8,550.000 | 8,650.000 | Earth Exploration and Space Research |
| 9,300.000 | 9,500.000 | Earth Exploration and Space Research |
| 10,600.000 | 10.700.000 | Earth Exploration and Space Research, Passive |
| 10,700.000 | 12,200.000 | Fixed Satellite Space-to-Earth |
| 12,200.000 | 12,700.000 | Broadcasting Satellite |
| 12,700.000 | 13,250.000 | Fixed Satellite Earth-to-Space |
| 13,250.000 | 13,750.000 | |
| 13,750.000 | 14,470.000 | Earth-to-Space |
| 14,500.000 | 15,350.000 | Space Research |
| 16,600.000 | 17,100.000 | Deep Space Research Earth-to-Space |
| 17,200.000 | 17,300.000 | Earth Exploration, Space Research |
| 17,300.000 | 17,700.000 | Broadcasting Satellite |
| 17,700.000 | 17,800.000 | Fixed Satellite Earth-to-Space |
| 17,800.000 | 20,200.000 | Fixed Satellite Space-to-Earth |

**Fig. 5.2** (continued)

At S-Band, ground noise interference is small, and high gain spacecraft antennas are easier to build and aim. For this reason, S-Band operations at data rates of a few MHz are possible.

At X-Band, all of the properties described for S-Band also apply. In addition, more bandwidth is available. For this reason, data rates of tens of MHz can be accommodated. However, rain attenuation must be considered.

At all of these frequencies the FCC requires that the ground illumination density be less than 1–5 mW/cm². This effectively limits the maximum usable spacecraft transmitter power and antenna gain.

## 5.2 Modulation Types

Communications from and to a spacecraft use digital data and one of several modulation types. The most frequently used modulation types and their general characteristics are given in Fig. 5.3. Modulation types are listed in the order of decreasing S/N ratio required for the same Bit Error Rate (BER).

| Modulation Type | Description | Comments |
|---|---|---|
| FSK (Incoherent) | Carrier frequency is toggled between 2 values | Heritage modulation, simple, doppler insensitive, Modulation Index is 0.3 |
| FSK (coherent) | Carrier frequency is changed with continuous phase shift | Requires less S/N for the same BER |
| GMSK (Gaussian Minimum Shift Keying) | Binary data is first Gaussian rounded before applying to FSK | Similar to FSK except the spectrum is more contained |
| PSK (Phase Shift Keying) | Carrier phase is changed ±90° | Improved BER performance |
| BPSK (Bi-Phase Shift Keying) | Carrier is modulated by a signal that is +1 or −1 | Efficient but doppler sensitive |
| QPSK (Quadrature Phase Shift Keying) | Carrier phase is changed to one of four phases | Improved spectral efficiency. Two bits per step. Phase changes occur for every two bits of information |
| O-QPSK (Offset QPSK) | One of the two phase changes is delayed by one bit | Improved spectral efficiency. Phase changes occur every bit |

**Fig. 5.3** Most frequently used modulation types and their characteristics

## 5.3 Bit Error Rate (BER) and Forward Error Correction (FEC)

Before the communication links can be designed and the required received S/N can be determined, properties of the different modulation techniques must be assessed.

Depending on the type of modulation chosen, there are significant differences in the Signal-to-Noise ratio needed for reliable communication. This is expressed by the Bit Error Rate as a function of signal bit energy per noise power density ($E_b/N_o$). Figure 5.4 shows the BER vs $E_b/N_o$ for the most often used modulation types.

Forward Error Correction (FEC) is often used to reduce the required $E_b/N_o$. Error correction increases the number of bits in the message to provide a way of making the message less susceptible to noise or other interference. There are a large number of different error correction codes with different properties. Here, we will only use the most frequently used Viterbi Rate 1/2, k = 7 convolutional code.

Applying FEC either reduces the effective message data rate (if the bit rate is held constant), or it requires increasing the bit rate and the bandwidth if the data rate is held constant. Thus, while using FEC has significant advantages, it also has disadvantages.

Figure 5.4 shows the BER vs. Eb/No for FSK, BPSK, QPSK, FSK with FEC and QPSK with FEC. It is seen that introducing FEC has a significant impact on reducing the Eb/No required for a given BER. This is best seen from the summary given in Fig. 5.5, illustrating the Eb/No required for BERs ranging from $10^{-5}$ to $10^{-7}$.

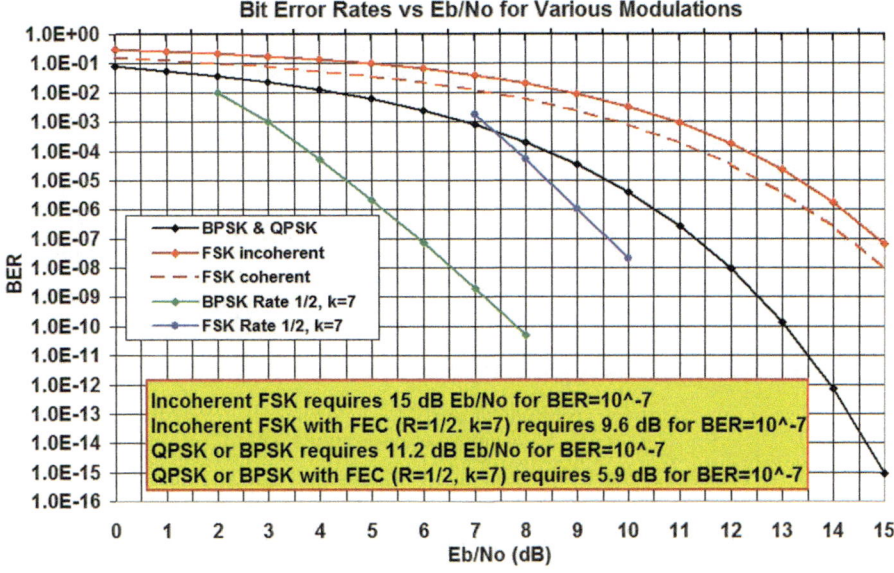

**Fig. 5.4** Bit error rate vs. Eb/No for various modulations and FEC

| Modulation | Approximate $E_b/N_o$(dB) Required for | | |
|---|---|---|---|
| | $BER=10^{-5}$ | $BER=10^{-6}$ | $BER=10^{-7}$ |
| FSK (incoherent) | 13.2 | 14.2 | 14.8 |
| BPSK or QPSK | 9.8 | 10.6 | 11.7 |
| FSK with FEC* | 7.8 | 9.0 | 9.6 |
| QPSK with FEC* | 4.5 | 5.6 | 5.9 |
| * FEC is Viterbi Rate 1/2 with k=7 | | | |

**Fig. 5.5** Summary of Eb/No requirements for different modulations and BER

## 5.4 Link Equations

The downlink RF Link Equations express the signal and noise received at the ground station receiver in terms of the spacecraft transmitter power, spacecraft antenna gain, frequency-dependent path loss, data rate, received antenna gain, pointing loss, ground receiver noise figure, ground station receiver noise and FEC for a given modulation type. Similarly, the uplink Link Equation expresses the signal and noise received at the spacecraft receiver. Figure 5.6 illustrates the link margin as a function of spacecraft ground elevation angle. The spacecraft has a 10 watt FSK transmitter and a quarter wave Quadrifilar (Earth Coverage) antenna (see black line in the figure). The spacecraft operates at S-Band (2.250 GHz) and the downlink data

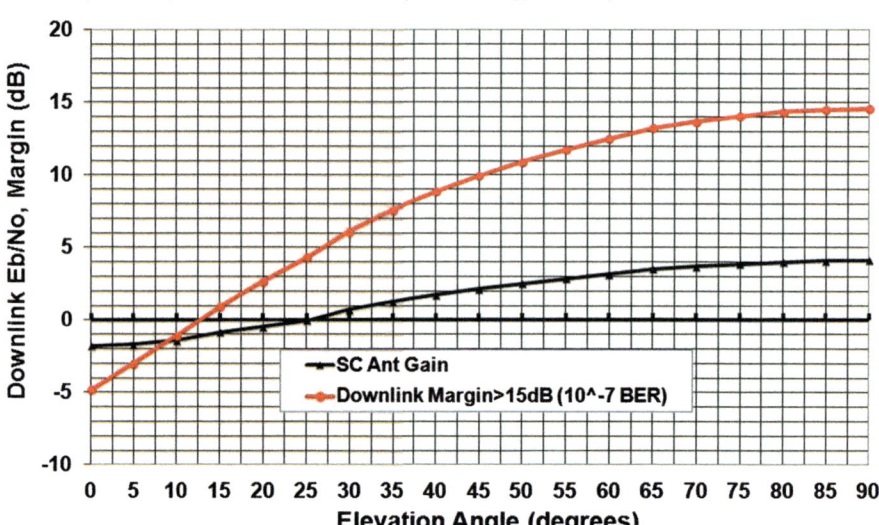

**Fig. 5.6** S-Band downlink margin example

rate is 1 Mbps. No FEC is used. The ground station has a 2.4 meter antenna of 3.9° beamwidth. The ground station receiver noise figure (NF) is 3 dB. In addition, a 3 dB implementation loss at the receiver is assumed. To operate at a BER of $10^{-7}$, the $E_b/N_o$ should exceed 15 dB. The minimum required $E_b/N_o$ of +15 dB is included in the graph plotted in Fig. 5.6. The Link Equations are shown under Fig. 5.6.

It is seen that the satellite can close the Space-to-Earth link at ground elevation angles greater than ≈12°, and it has excess $E_b/N_o$ at higher elevation angles. If Rate 1/2, k = 7 FEC were used, link performance would improve by 5.2 dB (see Fig. 5.5), and the spacecraft could close the downlink at the horizon.

The downloadable software contains a detailed Link Margin as a function of Orbit Altitude, Ground Elevation angle from the ground observer to the spacecraft and all of the other variables already described (Fig. 5.7).

$S_{GR}$ = **Received Signal Strength at Gnd Rcvr** = $P_T + AG_{SC} - ML - SL + AG_{GS} - PL$

$S_{Required}$ = **KTB noise** + **GN** + **NF** + **NBW** + **IL** + $(Eb/No\ R)$, where

$SL = 37.8 + 20^* LOG(Freq - MHz) + 20^* LOG(Slant\ Range)$

and **KTB** noise = $-174$ **dBm / Hz**

**Link Margin** = $S_{GS} - S_{Required}$ in dB and where :

5.4 Link Equations

$P_T$ = Spacecraft Transmitter Power in dBm
$AG_{SC}$ = Spacecraft Antenna Gain in the direction of the Ground Station in dBic
SL = Space Loss (varies inversely with the square of the slant range to the Ground Station) in dB
$AG_{GS}$ = Ground Station Antenna Gain in dBic
ML = Modulation Loss in dB
PL = Ground Station Antenna Pointing Loss in dB
$S_{GR}$ = Ground Receiver Input Signal in dBm
NF = Ground Receiver Noise Figure in dB
NBW = Ground Station Receiver Noise Bandwidth in dB/Hz
KTB = KTB noise in dBm
GN = Ground Noise in excess of KTB noise in dB
IL = Implementation Loss in dB
$E_b/N_o$ R = Required $E_b/N_o$ taking into account modulation, BER requirements and FEC in dB
$S_{Required}$ = Signal Strength Required at the Ground Receiver in dBm

| Quantity | Description | Values for Example in Figure 7.6 |
|---|---|---|
| $P_T$ | Spacecraft Transmitter Power | 40.00 dBm |
| $AG_{SC}$ | Spacecraft Antenna Gain in the direction of the Ground Station | -1.45 dBm |
| ML | Modulation Loss | -0.50 dB |
| SL | Space Loss (from 600 km altitude spacecraft to Ground at 15° Elevation) | -163.70 dB |
| $AG_{GS}$ | Ground Station Antenna Gain | 32.04 dB |
| PL | Ground Station Antenna Pointing Loss | 0.00 dB |
| $S_{GS}$ | **Received Signal Strength at the Ground Station Receiver Input** | **- 93.11 dBm** |
| NF | Ground Receiver Noise Figure | 3.00 dB |
| NBW | Grnd Station Receiver Noise Bandwidth (for 1 Mbps Data Rate) | 60.79 dB |
| KTB | KTB Noise Density (per Hz) | -174.00 dBm |
| GN | Ground Noise in Excess of KTB noise | 0.00 dB |
| IL | Implementation Loss | 2.00 dB |
| $E_b/N_o$ R | Required $E_b/N_o$ (for FSK, no FEC, BER=$10^{-7}$) | 15.00 dB |
| $S_{Required}$ | **Required Signal Strength at Grnd Rcv Input to close the link** | -93.21 dBm |
| | Link Margin | 0.10 dB |

**Fig. 5.7** Link margin example

If FEC were employed, the Link margin would increase to 5.3 dB; and that would enable closing the link at the horizon with $10^{-7}$ BER.

Using QPSK with FEC would increase the link margin by about 8.9 dB. This excess margin could be used to close the link at the horizon (using up 4.8 dB) AND increasing the data rate to 2 Mbps.

Additional improvements would have to be achieved mostly by increasing the spacecraft antenna gain.

## 5.5 Spacecraft Antennas

The most frequently used antennas on LEO spacecraft are:

- N-Turn Helix
- 1/2 wave Quadrifilar Helix
- Full wave Quadrifilar Helix
- Patch
- Horn (at microwaves)
- Dish (mostly at microwaves)

### 5.5.1 The N-Turn Helix Antenna

Helix antennas are very popular for spacecraft use, and they can achieve gains up to about 15 dB. The antenna can also be pressed flat (like a bedspring) so that in the stowed configuration it occupies little space. When on orbit, the turns of the antenna are released to let the helix deploy to its design length. Flexible lines tied between consecutive turns ensure that the deployed antenna has its design length.

Different references provide different values of antenna gain for a given numbers of turns. The Table in Fig. 5.8 uses conservative values. This author has used 2.5 and 5-turn helixes on several spacecraft. The antenna gain for the 2.5 turn helix was about 5.5 dB.

| Helix Antenna Characteristics | | | | | |
|---|---|---|---|---|---|
| Frequency | | f | 2.250 | GHz | |
| Lambda | | 30/f(GHz) | 13.333 | cm | |
| Space Between Turns | | 0.22169*$\lambda$ | 2.960 | cm | |
| Pitch | | 12.5° | 12.5 | degree | |
| Diameter | | $\lambda/\pi$ | 4.24 | cm | |
| Ground Plane Diameter | | 0.8*$\lambda$ | 10.66 | cm | |
| Turns | Gain (dBic) | 3 dB BW (°) | Ant Length (cm) | Grnd Plane Dia (cm) | Diameter (cm) |
| 1 | 3.76 | 110 | 2.96 | 10.66 | 4.244 |
| 2 | 6.77 | 78 | 5.91 | 10.66 | 4.244 |
| 3 | 8.53 | 64 | 8.87 | 10.66 | 4.244 |
| 4 | 9.78 | 55 | 11.82 | 10.66 | 4.244 |
| 5 | 10.75 | 49 | 14.78 | 10.66 | 4.244 |
| 6 | 11.54 | 45 | 17.74 | 10.66 | 4.244 |
| 7 | 12.21 | 42 | 20.69 | 10.66 | 4.244 |
| 8 | 12.79 | 39 | 23.65 | 10.66 | 4.244 |
| 9 | 13.30 | 37 | 26.60 | 10.66 | 4.244 |
| 10 | 13.76 | 35 | 29.56 | 10.66 | 4.244 |

**Fig. 5.8** N-Turn Helix antenna gain, beamwidth and length

## 5.5.2 Half Wave Quadrifilar Helix Antenna

Quadrifilar antennas were first developed in the 1970's. The half wave antenna shown in Fig. 5.9 has a 1 GHz bandwidth and a peak antenna gain of about 4 dB. It provides Earth Coverage from LEO.

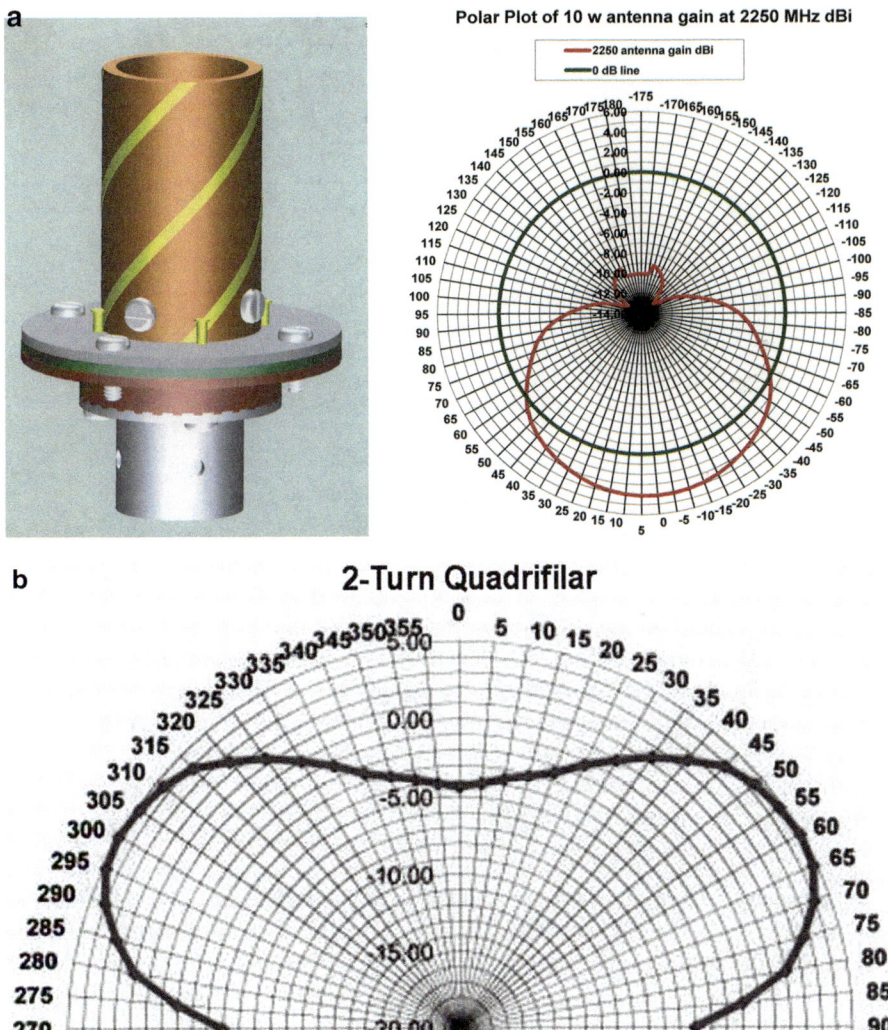

**Fig. 5.9** (a) 0.75″ Diameter half wave quadrifilar antenna at 2.250 GHz. (b) 2-turn quadrifilar pattern a better match to path loss vs. angle

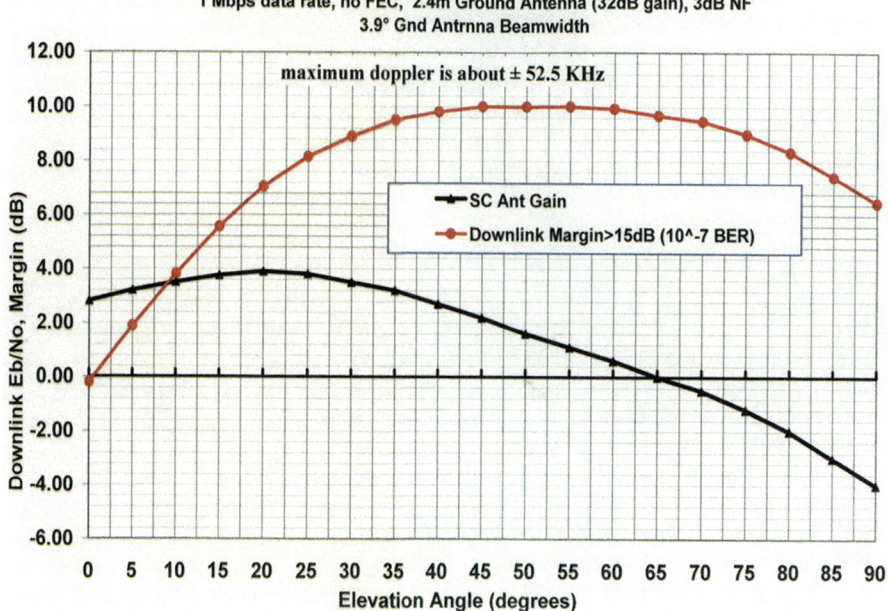

Fig. 5.10 Improving the low elevation link margin through use of full wave quadrifilar antenna

A full wave quadrifilar has improved gain, but it is used mainly because its antenna pattern peaks at 60° (30° down from the spacecraft horizontal). This is the angle of the horizon from about 600 km. In addition, the antenna gain is reduced at nadir, better matching the slant range path loss. The antenna pattern of a full wave Quadrifilar antenna is shown in Fig. 5.9b. The link margin vs elevation angle for the same conditions as in Fig. 5.6, except for substituting the full wave quadrifilar, is shown in Fig. 5.10.

### 5.5.3 The Turnstile Antenna

This type antenna is well-suited to provide Circularly Polarized Earth Coverage for spacecraft below about 800 km. The antenna is small (at S-band, about $2'' \times 2''$) and rigid. It easily withstands launch vehicle loads (Fig. 5.11).

### 5.5.4 The Patch Antenna

For higher gains, a circularly polarized Patch Antenna can be used. This is a rectangular, flat antenna that can be built with gains ranging from about 3 dB to in excess of 15 dB. A commercial 2.4 GHz patch antenna of 8 dBic gain is typically 11.4 cm × 11.4 cm × 2.3 cm.

## 5.5 Spacecraft Antennas

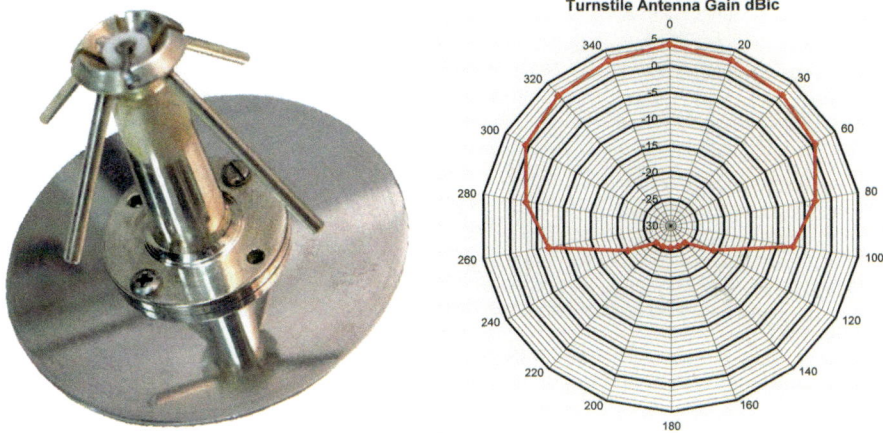

**Fig. 5.11** Turnstile antenna (courtesy SpaceQuest)

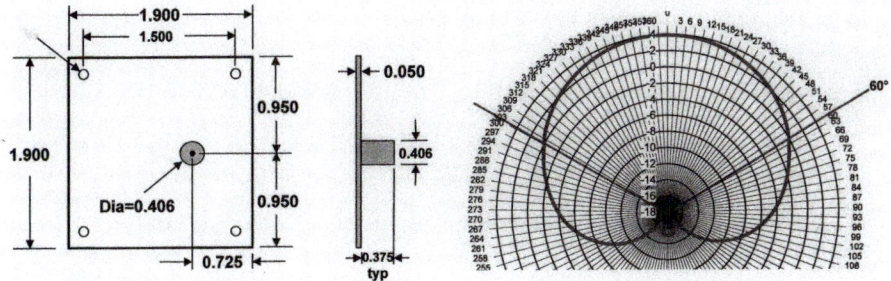

**Fig. 5.12** Low gain earth coverage patch antenna

Another patch antenna, designed to provide telemetry from a satellite at S-Band (2.2–2.3 GHz), had to provide Earth Coverage so the Ground Station could be accessed any time the spacecraft was in the Field Of View (FOV) of the ground station. This antenna was 1.9 in × 1.9 in × 0.050 in and had the antenna gain, shown in Fig. 5.12. Note that the Earth Horizon is about 30° down from the spacecraft. The antenna has 0 dBic gain at the horizon, and reaches 4 dB gain toward nadir.

### 5.5.5 Horn Antennas

At microwave frequencies, a horn antenna can provide medium-to-high antenna gain. At X-Band stock horns are readily available with 10 dB, 15 dB and 20 dB gains. A typical 10 dB (linearly polarized) horn is shown in Fig. 5.13. It also shows the physical characteristics of a 15 dB horn.

**Fig. 5.13** X-Band horn antennas

## 5.5.6 Dish Antennas

For higher gains dish antennas are used. The gain of a parabolic dish of diameter D and its beamwidth are given below (Fig. 5.14):

$$\mathbf{Gain} = 10^* \log\left(\mathbf{k}(\pi/\lambda)^2\right) \qquad\qquad \mathbf{BW}\deg) = 70^* \lambda / \mathbf{D} \qquad (5.1)$$

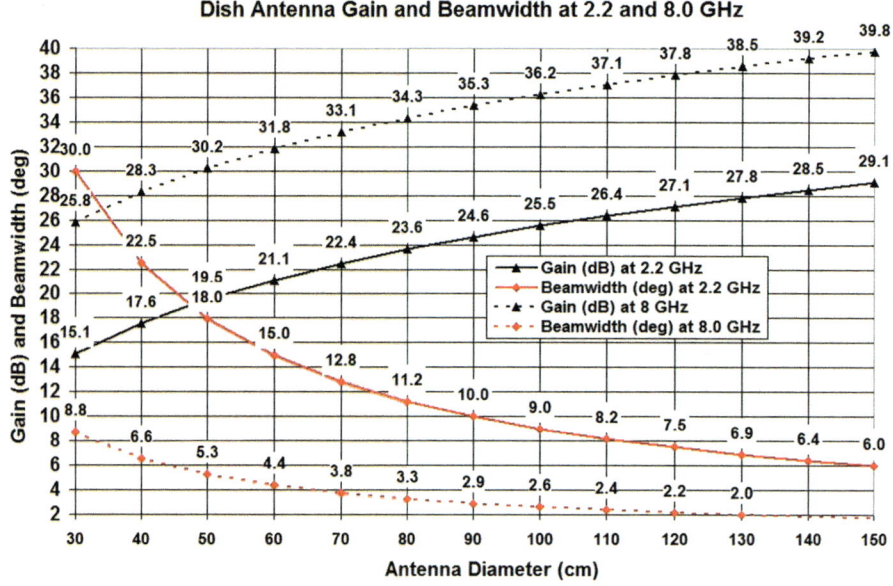

**Fig. 5.14** Dish antenna gains and beamwidths at 2.2GHz and 8 GHz

## 5.5.7 Intersatellite Links and Steerable Antennas

There are some spacecraft applications that require very high gain antennas. For example, an intersatellite link at 23.2 GHz, if it is to communicate with another satellite at 3500 km distance at 25 Mbps bit rate, requires a spacecraft antenna gain of about 30 dBic. Such an antenna has a beamwidth of about 5.3°. This presents a difficult problem of acquiring the other satellite and to keep the antenna pointing to it during a rapidly changing geometrical environment. Typically, such antennas are mechanically pointed. However, the angular momentum and CG changes of the antenna during steering must be taken into account by the spacecraft attitude control system, for both of these will affect spacecraft attitude.

Two examples of mechanically steered microwave antennas (both from Surrey Satellite Technology) are shown in Fig. 5.15. The one on the left is a 19 dB 8 GHz antenna, while the one on the right is a steerable dish antenna. These antennas are typically used to communicate from spacecraft to a ground station, and are used to track the ground station during the satellite pass.

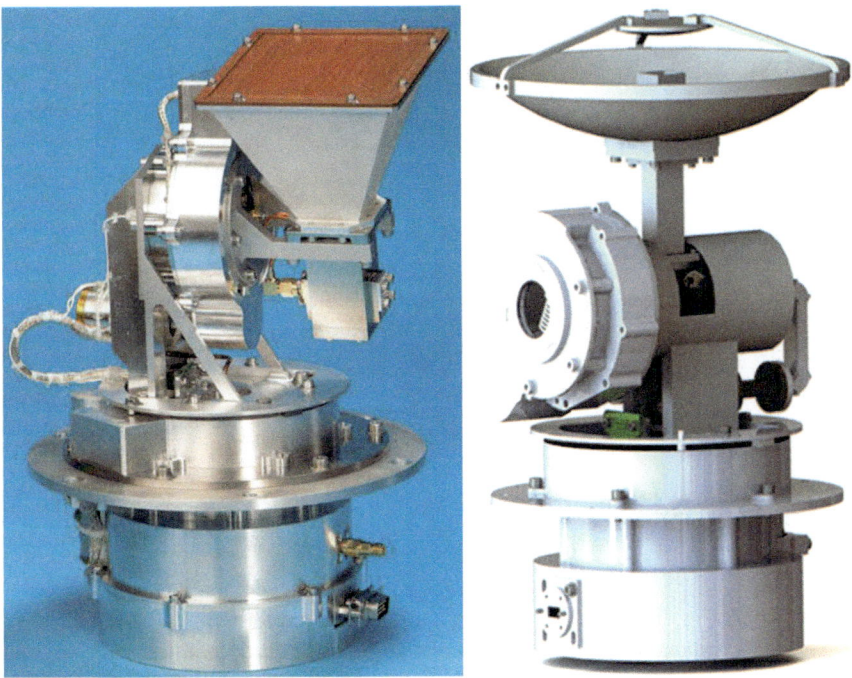

**Fig. 5.15** Mechanically steered microwave antennas (Surrey Satellite Technology)

## 5.5.8 Phased Arrays

Phased array antennas can be steered rapidly and can exhibit moderate gains. Steering is usually restricted to about ±45° in elevation and azimuth. The array of elements that comprise the antenna operate at low power and are phase shifted digitally. The main advantage of such antennas is their ability to switch antenna direction very rapidly and to serve as a multibeam antenna. The main disadvantage, from a spacecraft point of view, is that the phase shifters draw a lot of power. In small spacecraft, this power requirement may prohibit the use of phased arrays.

## 5.5.9 Deployable Antennas

In most larger spacecraft where very large antennas are needed, the antennas are deployable. Examples of deployable antennas are shown in Fig. 5.16 (ATS-6 and the 2-meter APL Hybrid Inflatable Reflector).

**Fig. 5.16** Pictures of deployable antennas

## 5.6 Increasing Throughput by Varying Bit Rate or Switching Antennas

It was shown in Fig. 5.10 that a link margin of 0 dB limited the data rate to 1 Mbps. However, $E_b/N_o$ reached 10 dB for higher elevation angles before it dropped to about 6 dB at Nadir. The data rate could have been increased to take advantage of the excessive link margin at higher elevation angles. For example, after 10° elevation, the data rate could have been increased to 2 Mbps, and after about 17° the rate could have been increased to 4 Mbps. While the majority of time is spent at low elevation angles, varying the data rate during the pass can substantially increase the throughput.

Another approach to increasing the throughput is to use (in addition to the Nadir pointing Quadrifilar antenna) two high gain antennas pointing in the horizontal direc-

## 5.6 Increasing Throughput by Varying Bit Rate or Switching Antennas

tion fore and aft. The maximum gains that such antennas could have are plotted in Fig. 5.17. If 6-turn Helix antennas of 11.54 dB peak gains were used, the link equation would change, shown in Fig. 5.18, and 4 Mbps could be used for the entire pass.

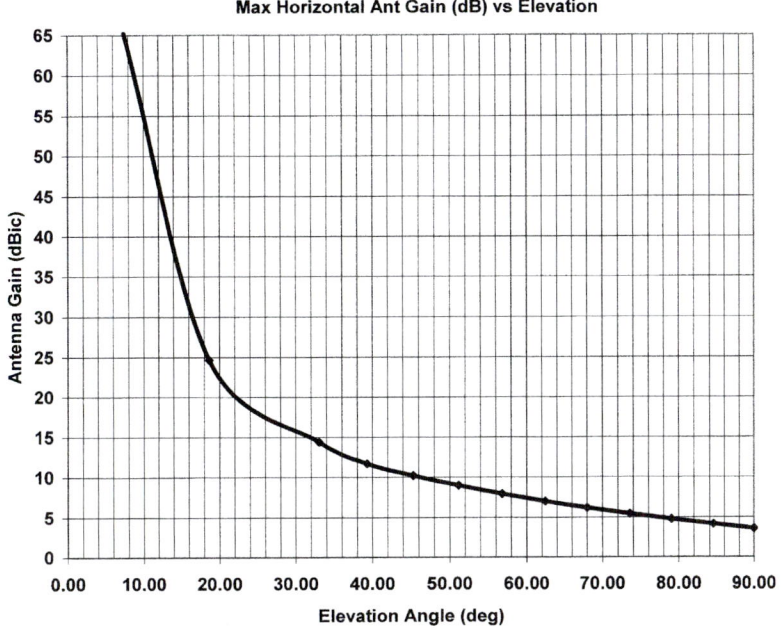

**Fig. 5.17** Maximum gain of horizontally aimed antennas to increase link margin at low elevations

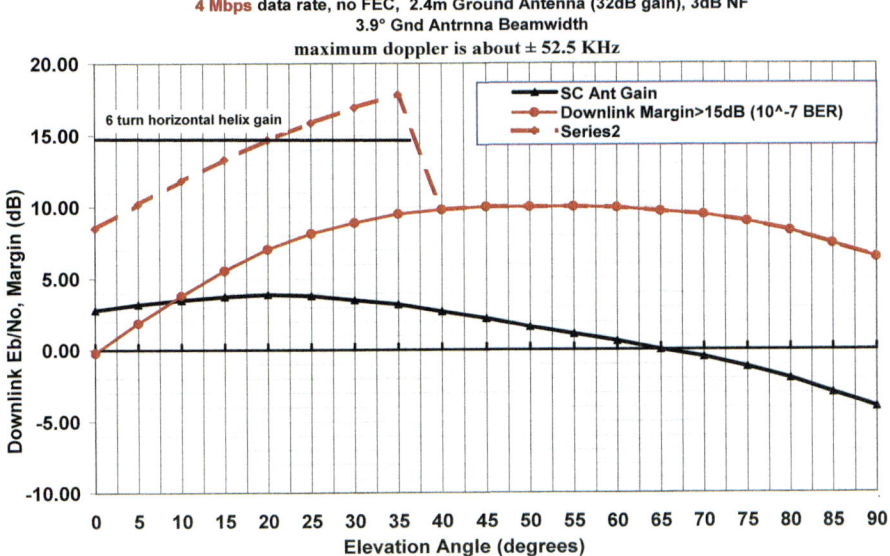

**Fig. 5.18** Switching antennas during the pass could increase the data rate to 4 Mbps

## 5.7 Geometrical Constraints on Space-to-Ground Communication

The relatively short duration of a pass of a LEO spacecraft over a ground station, and the low maximum antenna gains for Earth coverage severely limit the amount of data that can be downlinked from a spacecraft during a pass. Figure 5.19 illustrates the duration of a pass from a 600 km orbit altitude spacecraft to a groundstation at various CPA distances from the ground trace of the spacecraft.

As the Spacecraft Ground Trace CPA to the Ground Station increases, pass durations get shorter, and the maximum elevation angles become lower. At a 500 km CPA and an elevation angle above 20°, for example, the pass is only about 5 min long.

With an Earth Coverage spacecraft antenna and a large (2.4 meter) ground station dish antenna operating at 1 Mbps data rate, the maximum amount of data that can be downlinked in a pass is about 300 Mbits or 37.5 MBytes. At greater CPA distances, the amount of data that can be downlinked is even less. This severely limits how LEO satellites can be used for large data dumps.

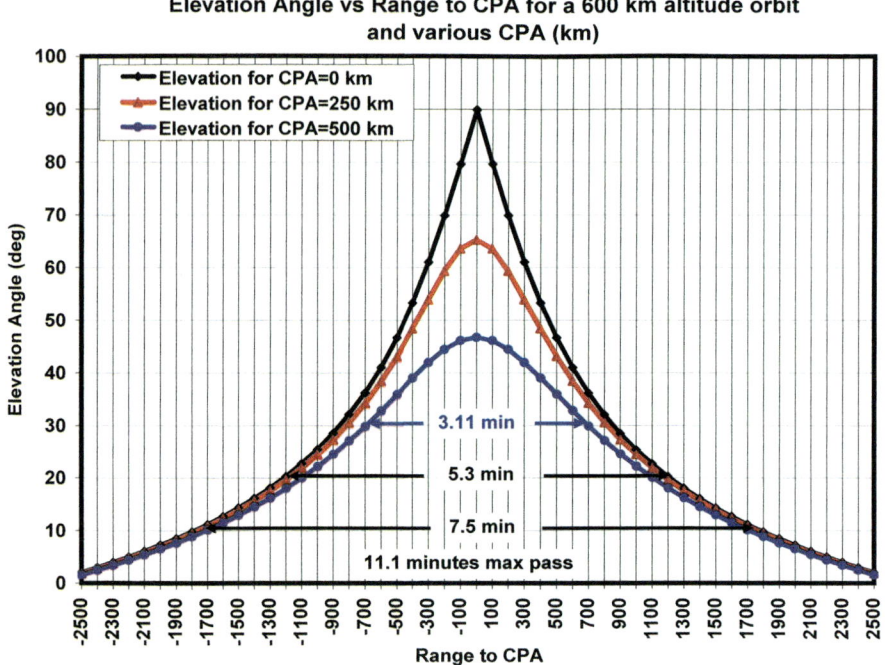

**Fig. 5.19** Elevation angle vs. Ground range and pass durations in communicating to a ground station at a given CPA from the satellite ground trace

A steerable spacecraft antenna that tracks the ground station can provide a large enough antenna gain to increase the amount of data that can be downlinked to 10 times more than an Earth coverage spacecraft antenna.

A spacecraft orbits the Earth about 15 times a day, but it can access a given ground station only about 5 times a day. Therefore, the total amount of data a spacecraft can downlink to a given ground station per day is about 1.5 Gbits. With steerable antennas, this number can increase 10 fold or even more.

Other methods of increasing a spacecraft's ability to downlink large amounts of data are (a) use multiple ground stations, (b) use ground stations at high latitude where the number of orbits visible per day is larger - at least for highly inclined orbits and (c) use a geostationary relay satellite.

## 5.8 RF Subsystem Block Diagram

A typical LEO spacecraft RF subsystem is shown in Fig. 5.20. It is assumed that packetization, randomization to balance the data stream for 0 DC component, Forward Error Correction and Encryption are performed either in digital hardware or software in the C&DH computer. The data stream ready for transmission enters the S-Band 10 watt transmitter where it is QPSK modulated. The data may be telemetry (TTM) or payload data.

The design employs a nadir looking main antenna and an anti-nadir looking secondary antenna to assure that TTM could be sent to the ground even if the spacecraft suffered an attitude control anomaly, or if the spacecraft has just been released from

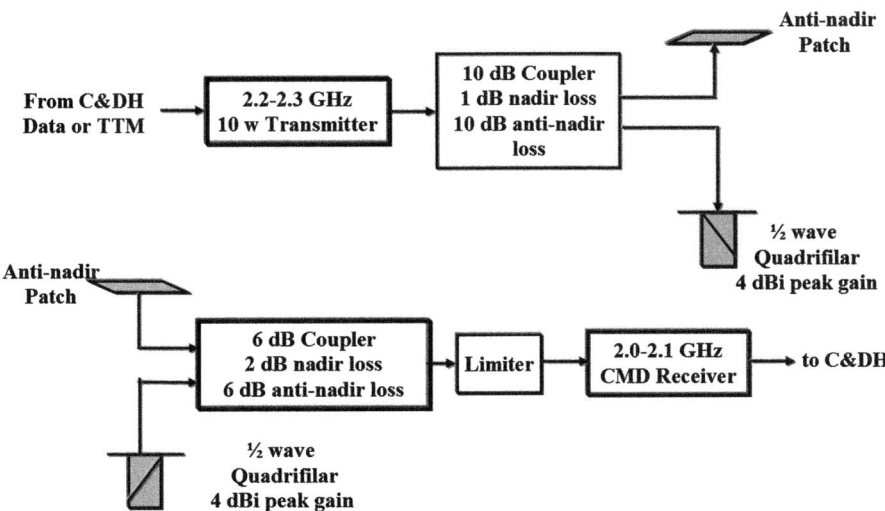

**Fig. 5.20** Typical LEO spacecraft RF subsystem

the Launch Vehicle and is tumbling. Since the data rate of TTM is very small compared to the data rate of the main mission payload, a power splitter is used to dump most of the power into the main antenna, and only a small amount of power is fed to the anti-nadir patch antenna. In this example, 1 Mbps data is fed to the main antenna and the TTM is only 9.6 kbps. Therefore, when the spacecraft is inverted, TTM transmissions reach the ground with the same power density as the main data when the spacecraft is nadir pointing.

The same procedure is used to receive the CMD uplink transmissions. Since the ground station is assumed to have a powerful transmitter, and since the command uplink is at a low data rate, the spacecraft received signal is strong even if the spacecraft is upside down.

# Chapter 6
# Spacecraft Digital Hardware

Digital computer selection for spacecraft applications is not as big of a problem as it used to be, because computers have increased in speed and in the size of their memories. Thus, these are no longer major factors in the selection process. However, radiation hardness, sensitivity to SEU, power consumption and I/O capabilities are still properties that influence the choice of digital processors. A list of factors that should be taken into consideration is given below:

- Architecture (one central or a distributed set of computers)
- Software Operating System
- Instruction Set
- Number of Bits Per Word
- I/O Capability
- Rad Hardness
- Reliability
- Redundancy
- Memory Size
- Speed
- Cost
- Weight
- Power Consumption

## 6.1 Computer Architecture

The space community is divided between a single, centralized processor performing most of the functions versus several small, low power consumption processors (associated with the spacecraft subsystems) in a distributed architecture. The community is also divided between those who like to use the many high technology, low cost COTS processors of low RAD hardness versus the very expensive RAD hard processors, usually lagging in technology and speed behind the COTS processors. The trend is toward the use of a distributed computer architecture.

Advantages of the decentralized, distributed computer architecture are:

- Less capable computers can be used to command, handle and receive information from a single subsystem than from all the subsystems of a spacecraft
- Computer speed requirements are reduced
- Fewer I/O channels are needed in each computer
- Power consumption is usually less
- There is inherent redundancy in a distributed system (one computer malfunction may not take down the entire spacecraft)
- It is easier to integrate all spacecraft software (written by multiple people) if each person only writes for one computer
- A distributed system is often much less expensive

An example of a distributed digital architecture is shown in Fig. 6.1 for a complex spacecraft with a main mission of imaging selected targets on Earth. This spacecraft has four computers. The C&DH is mostly concerned with housekeeping and communication functions. It requires a large memory. It also controls other payloads that may exist in the spacecraft. The ADACS computer performs the

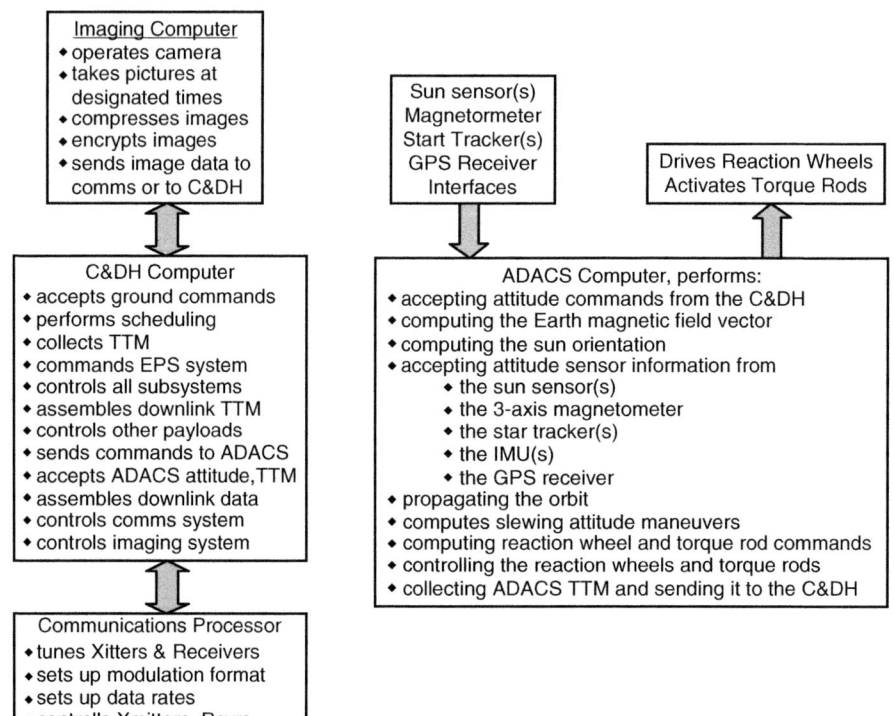

**Fig. 6.1** Example of a distributed spacecraft processor architecture

hardest tasks and may require the highest speed. The image processor is also a high speed processor and requires a large memory. The communications processor (if separate from the C&DH computer) is of modest speed and requires little memory. Interconnections between computers are not shown.

## 6.2 Computer Characteristics and Selection

The requirements for each of the multiple computers of the spacecraft should be determined from mission and cost considerations. The RAD hardness, speed, amount of memory, the number of analog and digital I/O channels, power consumption, weight and cost are the most important characteristics on which a selection is based.

In spacecraft weighing under 150 lbs, low power consumption is more important than in larger spacecraft. Single board computers for small spacecraft typically consume 1–3 watts, and that power consumption can be reduced further by operating the computer at lower clock speeds when it is not needed. Large concentrated RAD hard computers typically consume 30–40 or more watts, an amount a small spacecraft cannot afford.

The I/O capability is another major consideration. A small spacecraft may need to access about 30–40 points to collect TTM data. In a larger spacecraft, this number may be 100 or more. If a distributed architecture is used, any one computer may only be required to access less than 20 points. This not only reduces I/O requirements, but it also reduces the required speed.

Memory size used to be a big issue, but in recent years, memories have gotten larger and cheaper. However, SEU sensitivity and RAD hardness are still big issues.

Single board computers typically weigh about 0.5–1.0 lbs, while larger central computers can weigh as much as about 20 lbs.

The form factor ranges from the PC/104 ($3.6'' \times 5''$) to much larger sizes.

COTS for Single Board Computers range from about $10 K, while larger RAD hard machines cost several hundred thousands of dollars.

So, while it seems that selecting a set of computers for a spacecraft would be a challenging task, in reality it is much easier when one lists the main features and compares different machines in terms of those features.

Cost is most often the deciding factor, as long as radiation and SEU characteristics can be managed to provide the required mission life.

## 6.3 Spacecraft Computers Available Today

It must be recognized that COTS computer state-of-the-art is changing rapidly, and any set of choices would be obsolete in a year or two. The space computer state-of-the-art is changing much more slowly, but even here, specific recommendations will become obsolete in a few years.

Nevertheless, it is worthwhile to list some of today's space computers. The list below is but a small fraction of space computers available today.

Space Micro manufactures a set of radiation and SEU tolerant single board space computers. They are the Proton 200 k, the Proton 300 k and the Proton 400 k. They weigh about 0.5 lbs, consume about 1.5 watts and are PC-104 size. Each is modestly rad hard and SEU insensitive. Their costs are at the low end of the scale.

Surrey Satellite Technology (UK) also manufactures a very capable single board 5 k Rad Hard computer. It weighs less than 3.3 lbs and consumes less than 10 watts. It is more expensive.

Synova, Inc. manufactures the Mongoose-V 32-bit radiation-hardened spacecraft computer. It is packaged in a 256-pin ceramic flatpack.

# Chapter 7
# Attitude Determination and Control System (ADACS)

The ADACS is one of the most expensive subsystems of a spacecraft. Over-specifying ADACS performance requirements can easily "bust the budget." A common mistake is to call for more stringent accuracy than is required by the payload. For example, if the mission is to simply scan the atmospheric spectrum from the Northern Hemisphere, then pointing much more than 5° is hardly required. Specifying pointing accuracy 1 or 2 orders of magnitude better rapidly leads to expensive equipment, such as star trackers, when a simple momentum-biased system with Earth sensors would suffice.

This chapter describes ADACS development beginning with the process of flowing down requirements from mission to system and component level requirements with the added objective of minimizing cost while providing adequate performance. Next, different types of attitude control system configurations and their accuracies and suitability to various missions are described. Finally, the processes of on-orbit checkout, operations and anomaly resolution are addressed.

## 7.1 ADACS Performance Requirements Flowdown

The first step is to evaluate mission pointing accuracy requirements. How accurately does the payload need to point to a target on the Earth or in space? Are there any maneuvering requirements (that is, how far and how fast)? The simplest lowest cost and most reliable solution is nearly always that which yields the lowest performance acceptable to the mission. Following are some representative examples.

If the mission is space-to-ground two-way communications, attitude accuracy of 0.2 times the antenna beamwidth is usually sufficient. Since Earth Coverage antennas typically have beamwidths of about 100°, attitude control can be as poor as 20°. This is readily achieved by a gravity gradient stabilization system.

If spacecraft-to-spacecraft communications is employed, the spacecraft-to-spacecraft antenna beamwidth is quite narrow, but pointing accuracy probably does not have to be better than about 2°. This can readily be achieved with a Pitch Bias Momentum stabilization system.

If the pointing accuracy required by the mission is to point a camera to the ground with an accuracy of 60 m (at nadir) from an altitude of 600 km, then the pointing accuracy needs to be 0.0001 radians (0.00573°), and attitude sensing capability needs to be twice as good, 0.002865° or 10.25 arc-sec. This results in a very expensive ADACS and requires a 3-axis zero momentum stabilization system with 3 reaction wheels and two star trackers. Because of the cost involved, one must be sure that 60 meter pointing accuracy on the ground is really required.

Another factor that drives the cost and choice of the ADACS is the agility with which a spacecraft must be able to change attitude. The more agility is required, the larger the torque the ADACS must be able to impart to the spacecraft.

The lowest cost, most reliable system usually results by considering the minimum performance required to fulfill mission requirements. The challenge is knowing what the simplest configuration is that can meet the requirements of the payload.

Most missions are either Earth or inertial-pointing. Earth-pointing missions point a payload at, or offset from, the subsatellite point, requiring the spacecraft to rotate at or near the orbit rate. Inertial-pointing missions point the payload at the Sun or other target fixed in position with respect to the stars. Either kind of mission can have various agility requirements.

Many of the Earth or inertial-pointing missions can be satisfied by a single momentum wheel or even an all magnetic spinning system. Such systems have demonstrated performance of about 1° in roll and pitch and 3° in yaw. They are generally inertially fixed or have only limited mobility about the momentum bias axis.

Another class of missions have low (5° or greater) Earth-pointing requirements. This accuracy can sometimes be achieved by an all passive system, consisting of a gravity gradient boom and magnetic hysteresis rods. Because the gravity vector is used to orient the spacecraft, this solution is only for nadir pointing.

The highest performance category involves use of 3 orthogonal reaction wheels operated with little or no bias speed, termed a 3-axis zero momentum system. Spacecraft attitude can be controlled with excellent accuracy (0.01° or better), accommodating payloads with stringent pointing requirements. One or more star trackers are generally required to sense such small attitude errors. This is the most costly ADACS system.

Mission requirements must flow down to specify the minimum pointing accuracy and spacecraft agility. The table shown in Fig. 7.1 summarizes ADACS requirements for various types of missions

| Mission | Pointing Requirement | Maneuver | ADACS Type | Equipment | Relative Cost |
|---|---|---|---|---|---|
| Ram Pointing | 3° into Ram | None | Aerodynamic at Altitudes Below ~300km Only. | Tail Feathers, Optional Pitch Bias Wheel and Torque Rods with Magnetometer | Low |
| Nadir Pointing | >3° to 10° | None | Gravity Gradient | Gravity Gradient Boom, Hysteresis Rods, No Attitude Sensing | Low |
| Nadir or Single Axis Inertial Pointing | 1° | 0 to 1° per day | Magnetic Spinner | Torque Rods, Magnetometer, Optional Earth or Sun Sensors | Low |
| Nadir Pointing | 1° | Track Sub-satellite Point to Slight Pitch Axis Excursions | Pitch Momentum Bias | Single Reaction Wheel, Torque Rods, Magnetometer Earth Sensor, Optional Sun sensor | Medium |
| 3-axis Stabilized Earth or Inertial Pointing | < 0.1° | None to Highly Agile | 3-axis zero Momentum | 3 Reaction Wheels, Torque Rods, Star Tracker(s), Magnetometer | High |

**Fig. 7.1** Stabilization approaches for various missions

## 7.2 Description of the Most Common ADACS Systems

### 7.2.1 Gravity Gradient Stabilization

The simplest method of stabilizing a spacecraft so it would be nadir-pointing is by gravity gradient stabilization, shown in Fig. 7.2. This is a completely passive method, requiring no electric power. Here, a Tip Mass of weight $W_T$ is placed at the end of a boom of length, L, protruding from the spacecraft of weight $W_{SC}$. Hysteresis rods at the tip mass damp the oscillations. Nadir-pointing accuracy of 5–10° peak swing can be achieved in pitch and roll, but not in yaw.

Since the acceleration of gravity is inversely proportional to the square of the distance from the center of the Earth to the spacecraft altitude, there will be a gravity

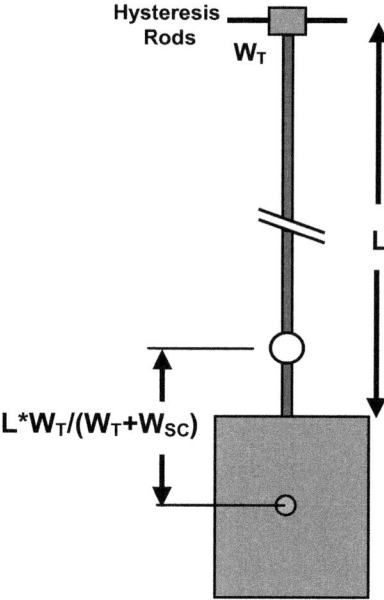

**Fig. 7.2** Gravity gradient stabilization

gradient between the tip mass and the spacecraft that will cause the mechanical system to act like a pendulum and oscillate. The thermal energy generated by the hysteresis rods cutting the Earth magnetic field is used to dissipate the pendulum oscillation energy and damp the magnitude of oscillations.

The acceleration of gravity at an altitude H km is given by Eq. 7.1, the gravity gradient by Eq. 7.2, the restoring torque at θ° deflection by Eq. 7.3 and the frequency of oscillation by Eq. 7.4. R is the Earth radius.

$$g(H) = g_o * (R/(R+H))^2 \text{ where } g_o = 9.8 \text{ N/s}^2 \text{ or } 32.2 \text{ ft/s}^2 \qquad (7.1)$$

$$g(H)/dH = -2*g_o*R^2/(R+H)^3 \qquad \text{m/s}^2/\text{m} \qquad \text{or ft/s}^2 \qquad (7.2)$$

$$T(\theta) = -2*W_T*L*W_T/(W_T+W_{SC})*\sin(\Theta)*g_o*R^2*(R+H)^{-3} \qquad (7.3)$$

$$f_o = -2*\pi*(L/dg(H)/dH)^{0.5} \qquad (7.4)$$

To get a "feel" for the stabilization achieved, numbers are put into these equations. These are illustrated in Fig. 7.3 for a 600 km altitude spacecraft.

Since a gravity gradient stabilized spacecraft has two stable states (right side up and upside down), it is possible that when first achieving a stable orientation, the spacecraft will be upside down. Employing the Z-coil against the Earth magnetic field in the Polar region can be used to flip the spacecraft so it becomes stable in the right side up orientation.

## 7.2 Description of the Most Common ADACS Systems

| SC Weight (lbs) | Tip Weight (lbs) | Boom Length (ft) | Period (min)=1/$f_o$ | Restoring T(10°) |
|---|---|---|---|---|
| 200 | 5 | 20 | 47.732 | $3.98*10^{-5}$ ft-lbs |
| 200 | 10 | 20 | 58.281 | $7.77*10^{-5}$ ft-lbs |
| 400 | 10 | 50 | 91.874 | $2.95*10^{-4}$ ft-lbs |
| 10 | 1 | 10 | 65.161 | $3.71*10^{-6}$ ft-lbs |
| 10 | 1 | 20 | 92.152 | $7.41*10^{-6}$ ft-lbs |

**Fig. 7.3** Different spacecraft and tip mass weights and periods of oscillation, restoring force

Oscillations are damped out usually by hysteresis rods that dissipate some of the energy due to oscillation by converting oscillatory energy to heat. "Fat" hysteresis loop soft iron rods at right angles to the orbit plane are typically used in a passive damping system.

### 7.2.2 Pitch Bias Momentum Stabilization

Pitch bias momentum stabilization, shown in Fig. 7.4, provides excellent nadir-pointing stability in Roll, Pitch and Yaw at the cost of one reaction wheel, 3-axis torque rods, a magnetometer, and Earth horizon sensor(s).

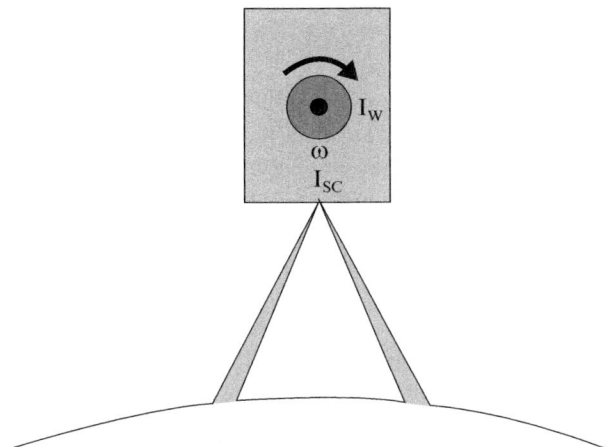

**Fig. 7.4** Pitch bias momentum stabilization

The reaction wheel is aligned with the spacecraft pitch axis and turns with an angular velocity of $\omega$. The wheel has a moment of inertia of $I_W$ while that of the spacecraft is $I_{SC}$. Earth horizon sensor(s) (EHS) determine spacecraft attitude in pitch in a narrow range (where the spacecraft is nearly nadir-pointing). Magnetic torque rods interact with the secular magnetic field to provide damping and precession torques. A magnetometer senses the phase angle of the torque rods with respect to the secular field.

The reaction wheel serves two functions. It provides gyroscopic stiffness to allow the spacecraft to retain attitude when the magnetic field is not favorably aligned, and it facilitates a high bandwidth loop in pitch. By conservation of angular momentum, when the reaction wheel speed increases, spacecraft pitch angular velocity decreases in the ratio of $I_W/I_{SC}$. Thus, the pitch attitude of the spacecraft can be adjusted by speeding up or slowing down the reaction wheel in response to changes in the angular position of the Earth horizon.

The gyroscopic stiffness of the spinning wheel provides inertial stability. When initially deployed, simple –B dot rate damping (turning the torque rods ON, out of phase with the sensed magnetic field to remove kinetic energy from the body) will drive the momentum bias vector to the negative orbit normal (the minimum energy state) where pitch can be captured by the EHS.

Subsequently, continuously damping nutation with the torque rods and serving to null pitch error with the reaction wheel will drive the Z axis to nadir. Quarter orbit coupling through gyroscopic stiffness will keep yaw error small also. Typical errors are 1° in roll and pitch and 3° in yaw.

It is recommended by GSFC that the reaction wheel angular momentum be about 10 times that of the spacecraft due to orbital rate. When the spacecraft is nadir-pointing it rotates at $P/360/60 = P*2.773*10^{-5}$ degrees/s, where P is the orbit period in minutes. Expressing the nadir-pointing spacecraft angular velocity in terms of its altitude, H, Eq. 7.5 is obtained. The reaction wheel angular velocity is given in Eq. 7.6.

$$\omega_{SC}(\text{deg/sec}) = 7.68*10^{-9*}(R+H)^{1.5} \text{ where R and H are in km} \quad (7.5)$$

$$\omega_W(\text{RPM}) = 0.16589^* I_{SC} / I_W \quad (7.6)$$

Numerical examples will help to get a "feel" for the important parameters of wheel speed and inertia. The spacecraft in this example weighs 200 lbs. and is a rectangular prism with 24 inches square base and 28 inches height. The Reaction Wheel has diameter of $D_W$ inches and is $T_W$ inch thick. It is made of steel of 0.270 lbs./cu in density. Orbit altitude is 600 km or 800 km. From Fig. 7.5, it is seen that the reaction wheel speed is in a reasonable range although the wheel is quite large.

The horizon sensor can be a horizon scanner where a rotating beam shines the incident energy onto a detector. When the rotating beam intersects the Earth horizon, the IR radiation-induced pulse indicates the angular span of the Earth pulse. The bisector of this pulse indicates the direction to nadir.

Alternatively, a linear array of static IR sensors spread out in the pitch direction can be used to look at the Earth limb. The specific sensors at the boundary of the high and low IR radiation indicate the angular position of the Earth limb. From this, the nadir angle can be computed. Earth horizon sensor accuracies are typically a fraction of a degree.

| SC Weight (lbs) | Altitude (km) | $I_{SC}$ (slug-ft²) | $D_W$ (in) | $T_W$ (in) | $I_W$ (slug-ft²) | $\omega_W$ (RPM) |
|---|---|---|---|---|---|---|
| 200 | 600 | 3.004 | 10.0 | 1.5 | 0.051418 | 2,175 |
| 200 | 800 | 3.004 | 10.0 | 1.5 | 0.051418 | 2,085 |

**Fig. 7.5** Reaction wheel speed (RPM) for example

## 7.2.3 3-Axis Zero Momentum Stabilization

A generalized functional diagram of a typical 3-Axis Zero Momentum spacecraft ADACS is shown in Fig. 7.6.

A block diagram of the ADACS for a 3-axis Zero Momentum system is shown in Fig. 7.7

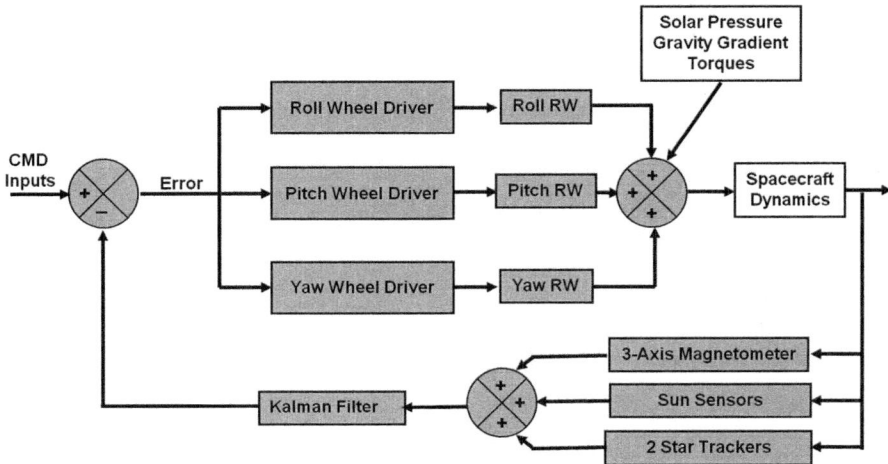

**Fig. 7.6** Functional diagram of a 3-axis zero momentum ADACS

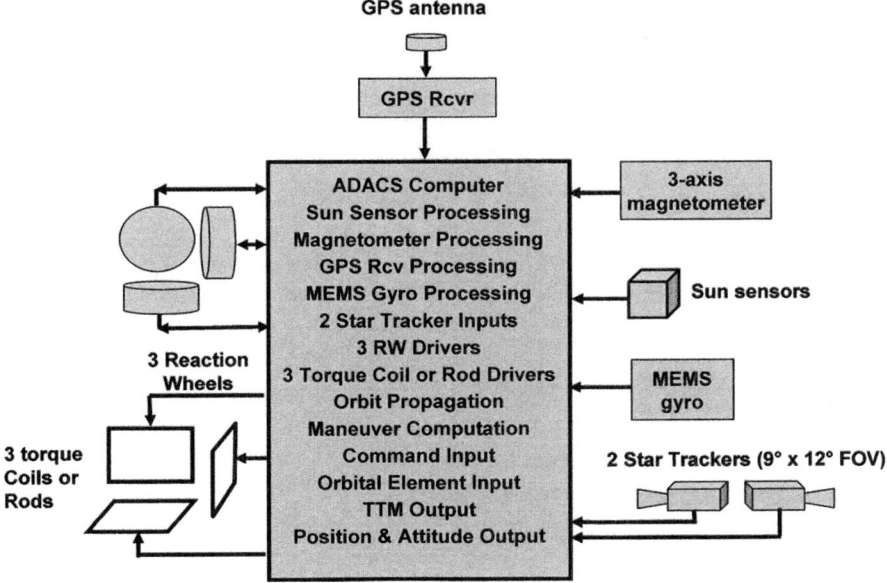

**Fig. 7.7** Block diagram of a 3-axis zero momentum ADACS

## 7.2.4 Magnetic Spin Stabilization

Another method of stabilizing a spacecraft is to cause it to spin about an axis. A property of a gyroscope is that the direction of the spin axis will remain fixed with respect to inertial space. If the satellite mass distribution is designed with a diagonal inertia matrix (cross products are zero or negligible) with respect to the spacecraft principal axes, then stable spin can be maintained about the major or minor inertia axis (but not the intermediate). Spinning can be initiated either by interaction between a torque rod on the spacecraft with the Earth magnetic field, or by thrusters.

The early Hughes communication satellites were spin stabilized. Another example of spin stabilized satellites is a three-plane, seven spacecraft per plane constellation performing continuous "bent-pipe" analog voice communication. Only one plane of these were built and flown. These spacecraft were spinning at 3 s/revolution normal to the 823 km orbit of 82° inclination. Spinning was accomplished by a single torque rod in the orbit plane, spinning with the spacecraft, and activated in response to a radial Earth horizon scanner. Images of one of the spacecraft and the footprints of the spacecraft in the constellation are shown in Fig. 7.8.

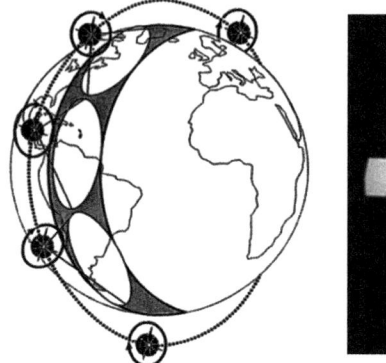

**Fig. 7.8** Three planes of 7 spin stabilized "Bent Pipe" satellites, each spinning about an axis normal to the orbit plane, provide nearly continuous communications

The spin RPM should be selected so that the spacecraft angular momentum should be 10 or more times the angular momentum due to the spacecraft rotation around the orbit. Since the orbit rate is 360° per Orbit Period (96.518 min), or 0.01 RPM, spinning the spacecraft faster than about 0.1 RPM (36°/sec) should result in a stable spin with the spin axis normal to the orbit plane. This orientation is the minimum energy state.

## 7.3 The ADACS Components

Selecting and sizing components of an ADACS system are described next.

### 7.3.1 Reaction Wheels and Sizing the Wheels

Reaction wheels consist of a mass rotated by a brushless DC motor. The motor applies torque to the wheel inertia and, hence, an equal and opposite torque to the spacecraft body; this is used for attitude maneuvers and for pointing the payload. In addition to providing torques for maneuvering spacecraft attitude, reaction wheels also provide momentum storage. The wheel parameters of interest are the available torque and momentum storage (product of wheel inertia and maximum speed).

Non-conservative external environmental moments (e.g. aerodynamic, gravity gradient, residual magnetic dipole, etc.) cause the control system to command corrective torques that accumulate in the wheels as stored momentum $J_W\omega$. If not unloaded by the electromagnets, $\omega$ would eventually reach a saturation limit (typically 5000 – 10,000 rpm) and not allow the wheel to apply any more torque in that direction.

Unbalances in the rotors (on the order of mg-mm for CubeSat wheels to gm-cm for SmallSat wheels) can create jitter torques that disturb pointing the payload and can also excite structural modes that further disturb pointing.

The torque necessary to provide a given maneuvering bandwidth determines maximum torque requirements. The dynamics of rigid body tracking a sinusoidal command of frequency, $\omega$, and amplitude, A, are:

$$\ddot{\theta} = \omega^2 A \sin(\omega t) \tag{7.7}$$

Substituting: $\ddot{\theta} = T/I_{SC}$ the peak torque required is:

$$\tau = I_{SC} A \omega^2 \tag{7.8}$$

For illustrative purposes, a typical small spacecraft has an inertia of 2.1 kg-m² and tracks a target having a position uncertainty of 0.1 deg. with a bandwidth of 0.25 Hz. The torque required is 9 mNm. Typical small spacecraft wheels can provide about 25 mNm maximum torque.

The maximum momentum that has to be stored influences the required wheel size. Due to external torques, momentum accumulates and is transferred to and from the body during maneuvers. Momentum storage during maneuvering is usually much greater than that due to momentum management. As already stated, wheels saturate around 5000–10,000 rpm, placing additional constraints on the system. If the same spacecraft is to slew at a rate of 5°/sec, then the required momentum storage is $I_{SC} \omega$ or 183 mNms. If the wheel inertia is 0.001 kgm² (typical for a small spacecraft wheel) then the peak speed is 1750 rpm.

The table in Fig. 7.9 shows torque and momentum capacities of representative wheels

| Manufacturer | Model | Maximum Available Torque (mNm) | Max. Momentum Storage @ rpm |
|---|---|---|---|
| Honeywell | HR-0610 | ±55 mNm | 4 Nms @ 6,000 rpm |
| Millenium | RWA-1000 | ±100 mNm | 1 Nms |
| Vectronics | VRW-1 | ±25 mNm | 1 Nms @ 5,000 rpm |
| MAI | MAI-300 | ±100 mNm | 1 Nms |

**Fig. 7.9** Representative small reaction wheels

In most 3-axis stabilized spacecraft, maneuvering requirements determine the maximum wheel torque needed. For example, if the spacecraft weighs 120 lbs. and is 20″ × 20″ × 30″, it has a pitch/roll moment of inertia about its CG of ≈2.812 slug-ft$^2$. If the goal is to have the agility of slewing the spacecraft 20° in 5s, a reaction wheel with maximum torque capability of 0.157 ft.-lbs. (0.13558 Nm) is required. The Millenium reaction wheel in Fig. 7.9 comes close to this maximum torque capability. With the Millenium wheel, the above spacecraft could be slewed 20 degrees in 7.296s. The slew maneuver consists of constant maximum torque for 3.648s, at which instant the spacecraft will have slewed 10°. This is followed by a constant negative torque of equal duration to decelerate the spacecraft to a halt at 20°.

## 7.3.2 Torque Coils or Rods: Momentum Unloading

Torque Rods are used to unload accumulated momentum of the reaction wheels and to detumble the spacecraft after separation from the launch vehicle. To determine the amount of built-up momentum that can be unloaded by a torque rod (or coil) involves a simulation that runs the spacecraft around the Earth and integrates the external torques as a function of time. The total integrated momentum is the maximum that can be dumped by the magnetic torquers per orbit. This needs to be less than the expected accumulation due to all external disturbance torques.

When the spacecraft is on orbit, external disturbance torques will cause the reaction wheels to spin up over time. To prevent this, magnetic coils are used to continuously desaturate the wheels by applying a torque to the body that causes the wheels to despin when they correct for it.

It is easy to run a simulation in Matlab to model the spacecraft on orbit and model the geomagnetic field. Consider a nadir-pointing spacecraft flying in a circular orbit at 450 km altitude and 45° inclination. The secular magnetic field in the body frame is shown in Fig. 7.10. It was generated by a 10th order IGRF model. In this figure, Blue is X, Green is Y, and Red is the Z axis.

A CubeSat ADACS system, the MAI-100, has magnetic coils of 0.037 Am$^2$ maximum dipole strength. The magnetic moment **r,** produced by energizing the coils, is the cross product of the dipole m and the secular magnetic field **B**.

## 7.3 The ADACS Components

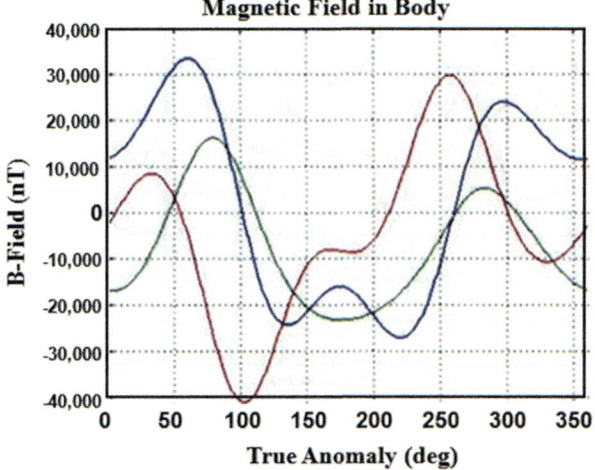

**Fig. 7.10** The earth magnetic field vector over one orbit

Next, take the cross product of the maximum magnetic field dipole, m = [0.037 0.037 0.037] with the magnetic field to obtain the available magnetic torque. The result is shown in Fig. 7.11. Again, Blue is X, Green is Y and Red is Z.

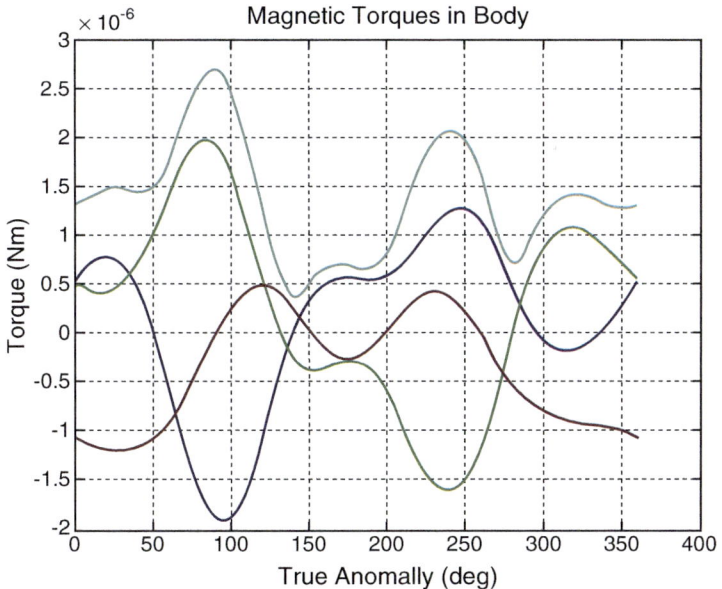

**Fig. 7.11** Available magnetic torque during one orbit

Finally, integrate the magnetic torques over an orbit to get the total momentum that can be unloaded by torque coils. This is shown in Fig. 7.12.

A total of almost 8 mNms can be dumped per orbit. To correct for inefficiencies in the secular field model and for other reasons, the actual momentum that can be

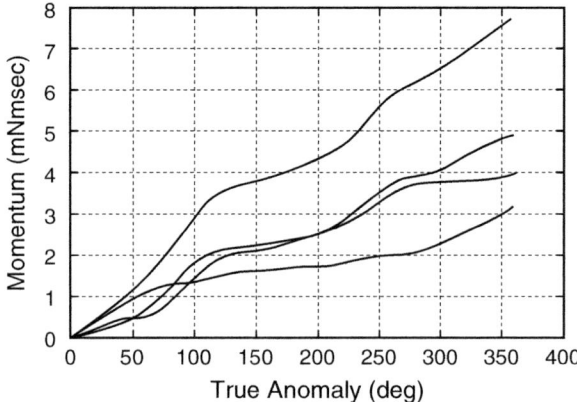

**Fig. 7.12** Total momentum that can be unloaded by the torquers

dumped per orbit is about 0.5–0.3 times that which the model calculated. The spacecraft designer should be sure that the total momentum accumulated from external torques should not exceed this.

Some spacecraft use torque rods while others use torque coils. Torque coils have an advantage over torque rods in that coils do not produce residual magnetic fields (when no current is applied), while even the best torque rods with the thinnest magnetic hysteresis curve iron core retain some residual magnetic field. If the spacecraft is relatively small, for a given B field, torque rods are lighter than coils; but when the spacecraft is large and has a large area around which a coil can be wound, coils are lighter.

One other feature of torque coils or rods that the designer must take into account is the inductance of the torquer. Torquers are most often used in a bang-bang manner (the torquer is either energized at full strength or not energized at all). Varying the duty cycle (or the duration of pulsed operation of the torquers) produces the required average torque.

Since the spacecraft has a magnetometer to measure the Earth magnetic field, care must be taken that the use of torquers should not affect magnetic field measurements. For this reason, the magnetometer readings are taken when the torquers are OFF, and when the magnetic field created by them has decayed enough to be below the Earth magnetic field intensity at the location of the magnetometer.

### 7.3.3 Star Trackers

If spacecraft attitude knowledge better than $0.1°$ is required, Star Trackers must be used. A star tracker is a (typically) $15° \times 15°$ Field of View CCD camera with computer to match the observed star field to a star catalog to determine spacecraft position and attitude.

Star tracker accuracies, weights and costs vary considerably. The Ball Aerospace High Accuracy Star tracker (HAST), for example, achieves 0.2 arc-second ($5.5 \times 10^{-5}$ degree) accuracy. The Sinclair ST-16 accuracy is about 7 arc-sec ($0.0002°$).

## 7.3 The ADACS Components

Star Trackers image fixed stars, measuring their vectors in the body frame and calculating the spacecraft attitude with respect to inertial space. Star Trackers have been historically large and expensive; however, recently a new low cost generation has become available, based on technology used in Digital Cameras and Smart Phones. The Maryland Aerospace Inc. Star Tracker, shown in Fig. 7.13, for example, weighs less than 200 gm, and features automatic star identification and attitude determination to 0.013° at a 4 Hz update rate.

**Fig. 7.13** Miniature star tracker in a 5 cm$^3$ form factor (Courtesy of Maryland Aerospace)

A unique feature of contemporary units is their ability to autonomously recognize visible stars in real time with no apriori information, known in the industry as Lost-In-Space attitude determination. Generally, the implementation is based on the observation that the angular separation between any 2 bright stars is a unique signature, and by comparing the measured separation of stars seen on an image with a catalog of separations, the two components may be identified. A third star is required to resolve the ambiguity and, in practice, a 4th star is required for positive identification, Ref. 37.

To give adequate all-sky coverage with a compact star catalog, a lens of short focal length and a wide field of view is required. The lens of the above unit is f 1.2 with a FOV of 14°×19°; the required catalog has only 1825 stars. This is large enough to insure that at least 4 stars of the catalog are visible in 99.6% of the sky.

Most star trackers can tolerate the Moon in the FOV but saturate when the Sun or Earth are in the FOV. Large baffles can reject bright objects outside the FOV. Star trackers are usually mounted on the satellite tilted up and away from the Earth and on a side facing away from the Sun.

Because of spacecraft maneuvering requirements, and because of the possibility that either the Sun or the Moon may enter the star tracker FOV, many spacecraft fly two star trackers (usually oriented 90° apart).

Figure 7.14 shows the fraction of passes on which a Star Tracker, pointing in the -X (anti-RAM) direction, is unavailable because either the sun or the moon are in its FOV.

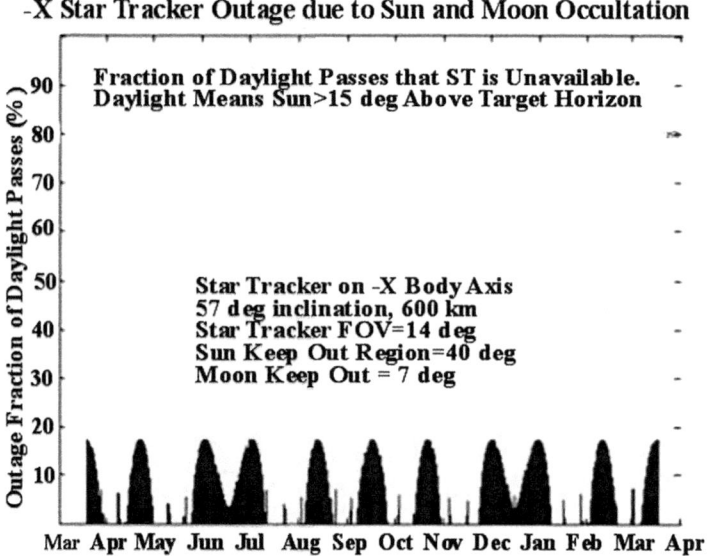

**Fig. 7.14** Star tracker (pointing in -X direction) unavailable because either the sun or the moon is in its FOV

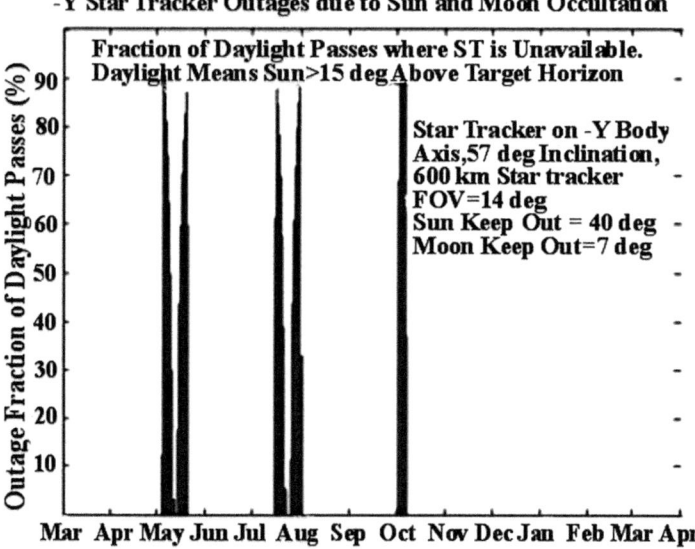

**Fig. 7.15** Star Tracker (pointing in -Y direction) unavailable

It is seen from Fig. 7.15 that pointing the Star Tracker to the side reduces the fraction of the time the Sun or Moon are in the FOV of the star tracker. If the spacecraft is permitted to yaw a little bit on days the star tracker would be otherwise unavailable, yawing would make it useful again. The conclusion is that the designer must calculate the times the star tracker would be unavailable, and decide whether to use one or two trackers and also decide in which direction the trackers should point.

### 7.3.4 GPS Receivers

GPS data are most useful when pointing a payload at a terrestrial target relative to the orbit frame, when it is required to know precise relative position. Imaging payloads may have FOVs of 0.5° or less. From an altitude of 600 km, the FOV may be on the order of 3 km. Thus, the SC position should be known to about 300 m. The permissible time error is on the order of 50 msec. This is well within the capability of inexpensive GPS receivers. An example of a low cost GPS receiver that was used on several CubeSat spacecraft is the NovaTel OEM7 receiver, shown in Fig. 7.16.

**Fig. 7.16** A small GPS receiver (NovaTel)

Generally, these receivers are intended for terrestrial use and contain software that limit the maximum velocity and altitude. These limitations need to be removed for space flight. In addition, they output position data in the ECEF frame that needs to be transformed to the J2000 (or other inertial) frame. This needs to be done precisely since nutation and precession terms can account for large position errors. The mathematics for ECEF to ECI transformations are in Reference 37, and Matlab, C++ implementation in http://www.celestrak.com/software/vallado-sw.asp

Accuracy of GPS receivers is generally on the order of 10 m or so; however, depending on the geometry of the orbit relative to the NAVSTAR GPS constellation, periodic dropouts of up to 20 min may occur. This can be mitigated by the use of data from the GLONASS and Galileo constellations. The aggregate GPS availability is still not 100%.

To propagate across the dropouts and to refine the position and velocity data, a Kalman filter is used to estimate the orbit state. The filter compares the residuals between successive GPS observations and an analytical propagation model to refine a state estimate that becomes more precise with time. The model may include the effects of perturbations also, such as J2 and atmospheric drag. During dropouts, the model is integrated forward in time to produce a state estimate in the absence of data. An estimate of the error in the state, called the covariance, is also calculated. Reference 67 is a treatise on the mathematics and methods of orbit determination by Kalman filtering.

GPS receivers can also be used to provide spacecraft attitude (by use of multiple GPS receive antennas). For example, for an antenna baseline of 20″ (RADCAL spacecraft), attitude accuracy of $0.3°$ was achieved.

As stated above, from GPS data the spacecraft can also propagate the equations of motion and obtain its own spacecraft orbital elements. It is also a very accurate time standard. In addition, a GPS receiver is relatively inexpensive, small and has low power consumption. It is a very cost-effective instrument to have on a spacecraft.

## 7.3.5 Other ADACS Components

In addition to the reaction wheels, torquers, Star Trackers and GPS receivers already discussed, a 3-axis zero momentum spacecraft stabilization system often has Sun sensors, a 3-Axis Magnetometer and an Inertial Measuring Unit (IMU) for tracking during a maneuver.

## 7.3.6 The ADACS Computer and Algorithms

In a pitch bias momentum stabilized spacecraft, the ADACS computer functions are quite simple. The computer determines the nadir direction (by bisecting the Earth IR pulse from a scanning earth sensor). It then compares the nadir direction with the Z axis of the spacecraft, and it commands increasing or decreasing reaction wheel speeds to minimize the error between the measured nadir and the Z axis of the spacecraft. If the wheel speed approaches saturation, the torquers are used to unload wheel momentum.

7.3 The ADACS Components

In a 3-axis zero momentum stabilized spacecraft, the ADACS computer performs the following functions:

- Accepts attitude commands from the C&DH
- Accepts sensor outputs from the magnetometer, Sun sensor, GPS receiver, star trackers
- Determines present spacecraft attitude and the Error from the commanded attitude
- Computes commanded functions (slewing, changing attitude, pointing to targets)
- Computes reaction wheel and torque rod drives
- Propagates the orbit
- Computes the orbital elements
- Outputs ADACS telemetry and Time to the C&DH

In the Maryland Aerospace Inc. ADACS system, all this is performed in a single board computer. This PCB also contains the reaction wheel drivers and the other I/O required to interface with all of the ADACS hardware components.

## 7.3.7 ADACS Modes

The ADACS can function in one of several attitude determination or attitude control modes. These modes can be selected by either ground command or autonomously, e.g. if a fault were detected. Modes are specific to every spacecraft, since they depend upon the mission and equipment complement. For robustness and operating simplicity, the number of modes should be minimized. Three or four modes are usually all that are required. For a representative 3-axis zero momentum spacecraft with reaction wheels, electromagnets, sun sensors, star tracker, gyro and magnetometer, the modes are as follows:

**Attitude Control Modes**

- **Acquisition** – After deployment from the launch vehicle, the spacecraft may tumble with unknown body rates. The torque rods are used in a Bdot mode where they are turned on and off out of phase with the secular magnetic field rate as sensed by the magnetometer. This has the effect of removing energy from the body and damping the rates. The end of the acquisition mode is when body rates are below $\approx 0.1°$/sec. The acquisition mode can also be used as a safe mode if an anomaly is detected.

- **Nadir Pointing** – The wheels are used to capture the attitude and stabilize the spacecraft with the Z body axis pointing toward the subsatellite point and the X body axis pointing into the ram direction. The nadir direction is determined from knowledge of the spacecraft position on orbit. Wheel momentum is managed with electromagnets. Offsets from the nadir attitude can be commanded to point to targets displaced from the ground track. By continuously updating the pointing command it is possible to keep the payload line of sight stabilized at a constant latitude and longitude on the Earth geoid.

- **Inertial Pointing** – This mode is similar to nadir pointing except that the payload line of sight is pointing to a fixed target in the sky such as the Sun or a distant star. Moving targets can be tracked by continuously updating the pointing quaternion as in a laser communication or space surveillance mission.

**Attitude Determination Modes**

- **Bdot** – Attitude is not determined, but the rate of change of the secular magnetic field in the body frame is measured by the backward difference of consecutive magnetometer measurements. The magnetic field rate is known as Bdot and is multiplied by a gain to specify the commands to the torque rods to effect body rate damping in acquisition mode.
- **Sun Sensor/Magnetomet**er – Sun Sensor and Magnetometer are used to measure the Sun and secular magnetic field vectors in the body frame. These are compared with the inertial vectors calculated from the Sun and IGRF magnetic field models, respectively, using Cross Product or Triad attitude determination. Sun sensor measurements can only be performed during the daylight portion of the orbit and require an initialization of the ephemeris. Sun sensors and magnetometers provide an all-aspect measurement of attitude that allows the ADACS to capture attitude for nadir pointing.
- **Star Tracker/Gyro** – A very precise quaternion is provided by star vector measurements. Between updates, attitude is propagated using the gyro. A Kalman filter is generally used to estimate gyro bias and to reduce measurement noise. Once the filter has converged, large angle slews can be made at rates that exceed the maximum tracking rate of the Star Tracker, generally 3–5°/sec.

## 7.4 Attitude Control System Design Methodologies

A methodology to develop ADACS control laws is to design them using analytical techniques, and then verify, test and fine tune them using time domain simulations. Algorithmic formulations are adequately covered in the literature, so they will not be repeated here. The authors have found References 68 (Wertz), 14 (Flatley) and 71 (Wie) to be very useful, and have had success designing attitude controllers with them.

As a first step in developing control laws, preliminary gains are calculated using linear analytical techniques. Methodologies for doing this are well covered in the literature, see for example Reference 71. In addition, the Matlab control system toolbox contains many utilities that can be employed, e.g. place.m, lqr.m, rlocus.m, etc. Given the inertia properties of the spacecraft and the desired dynamic performance characteristics, gains for magnetic, reaction wheel and thruster actuators may be determined along with their stability margins.

Once a gain set has been determined, it is essential to verify and test it in a high fidelity time domain software simulation. A dynamic simulator provides an all

## 7.4 Attitude Control System Design Methodologies

software standalone test set, and can also be used later in a Hardware-In-The-Loop (HITL) environment. The concept is to simulate the dynamics of the satellite and the space environment in real time on a PC. A screenshot of the simulation is shown below (Fig. 7.17).

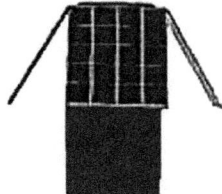

**Dynamic Simulator Display Contains:**

- **Commanded and Actual Reaction Wheel Speeds (3) vs Time**
- **Commanded and Actual Torquer Activities vs Time**
- **Earth Magnetic Field Components in Inertial Coordinates vs Time**
- **Magnetic Field Components in Spacecraft Coordinates**
- **Reaction Wheel Torque Commands (3) vs Time**
- **Sun Vector Components vs Time**
- **Spacecraft Velocity vs Time**
- **Spacecraft Position vs Time**
- **Spacecraft Attitude vs Time**
- **Spacecraft Attitude Picture vs Time (in RAM and Side View vs Time**

**Fig. 7.17** KEDS3D dynamic simulator GUI (Maryland Aerospace)

On the simulator computer is the spacecraft simulation that can be tailored for the dynamics of any arbitrary spacecraft through an input file. High fidelity math models of the spacecraft and the on-orbit environment are incorporated including:

- Rigid body dynamics
- Orbital mechanics model
- IGRF magnetic field model
- Atmospheric density model
- Aerodynamic torques
- Residual Dipole Torques
- Gravity gradient torques
- Sun position
- Solar pressure torques

Many of the math models used in KEDS3d were adapted from Refs. 72, 16 and 74. They are highly recommended for anyone doing work in this field. An example of one of the math models is the magnetic field shown below Fig. 7.18.

In addition, the progress of the mission can be followed in real time on the GUI which features an animated display of spacecraft attitude.

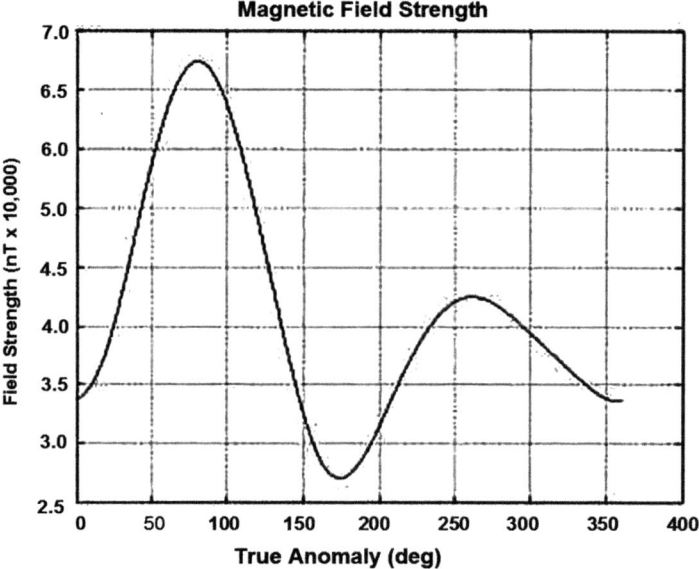

**Fig. 7.18** Magnetic field math model

Many different mission scenarios of spacecraft maneuvering can be constructed and run in order to validate the ADACS in all of its various pointing modes. The dynamic simulator thus forms the cornerstone of a comprehensive system validation and verification test program.

Simulations may be controlled and monitored in real time using a GSE GUI that sends and receives ADACS commands and telemetry. The GUI is identical in function to the display that would be seen on a ground control station, Fig. 7.21. By clicking buttons and entering numbers in the text boxes, commands can be sent to the simulation to change modes and point to specified target attitudes. Telemetry displays the output states of various sensors and actuators Fig. 7.19.

Please refer to the red numbers in the figure

1. **Receive Buffer** - A scrolling display of the raw telemetry bytes as they are received. The TLM valid light is green if the data are being received at 4 Hz rate with a valid checksum.
2. **Command History** - A scrolling display of the raw command bytes sent.
3. **General Information** includes:

   Version Number of the flight software
   TLM subframe counter
   Last Command Received by MAI-200
   Temperature of Motor Block
   Voltage at bus
   Pressure inside the hermetic enclosure

## 7.4 Attitude Control System Design Methodologies

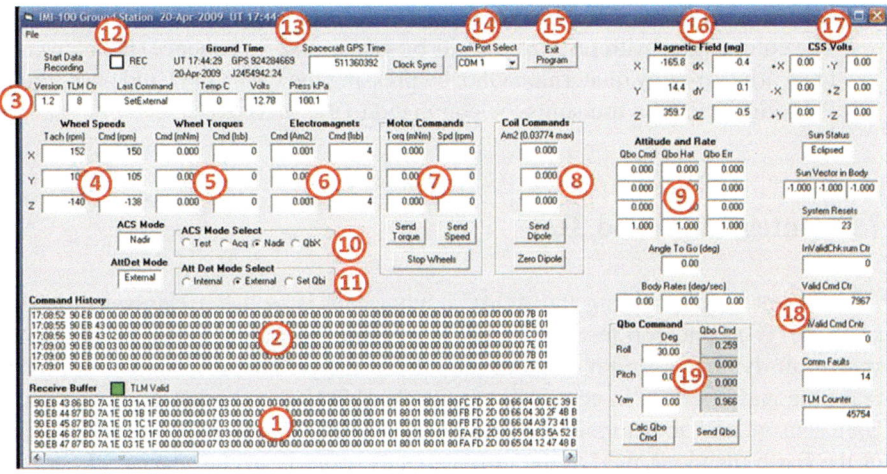

**Fig. 7.19** ADACS simulation control GSE

4. **Wheel Speeds** - Commanded and Tachometer
5. **Wheel Torques** - Commands in mNm and least significant bits (lsb)
6. **Electromagnet Commands** - in $Am^2$ and lsb
7. **Test Mode Motor Commands** – Open loop commands can be sent in Test Mode only. Either Torques or Speeds can be commanded. Type a torque or speed command in the text box and click on Send. Stop Wheels sends a zero speed command.
8. **Test Mode Electromagnet Commands** – Open loop commands can be sent in Test Mode only. Type a dipole command in the box and click Send. Zero Dipole sends a zero dipole command.
9. **Attitude and Rate Displays** – Commanded, Estimated and Error body-to-orbit quaternions from the ADACS. Angle-To-Go is the eigen angle of the Error quaternion. Body rates are the rate of change of the Estimated quaternion.
10. **ACS Mode Select** – Radio button selects the ACS mode, see Sect. 7.3.7. The text box displays the actual mode returned from TLM.
11. **Attitude Determination Mode Select** – Radio Button selects the Attitude Determination Mode, see Sect. 7.3.7. The text box displays the actual mode returned from TLM.
12. **Record Data** – Clicking the Button creates a data file (text, space delimited) of the raw TLM data.
13. **Time** – The Current GPS Time, Julian Date and Spacecraft GPS Time are displayed. The Clock Sync pushbutton send the ground GPS time to the MAI-200.
14. **ComPort** – Select Com Ports 1–10. Contact us if you need a different port.
15. **Exit Program** – The program can only be exited by this pushbutton.
16. **Magnetic Field** – The processed magnetometer outputs and bdot rates
17. **CSS Outputs** – The processed Coarse Sun Sensor outputs
18. **Statistics** – Counters for various normal and anomalous events

19. **Qbo Command** – Allows input of a quaternion command in Roll, Pitch, Yaw. The euler angles with respect to the orbit frame (3-2-1 sequence) are converted to an orbit-to-body quaternion Qbo, by pressing the Calc Qbo Cmd pushbutton. Pressing Send Qbo transmits the command to the MAI-200.

## 7.5 Integration and Test

An efficient way of testing the ADACS subsystem is to use a Hardware-In-The-Loop (HITL) benchtop test set. A HITL test employs a computer model of the spacecraft dynamics, and it contains a model of the orbital environment. It runs in real time and operates in conjunction with the ADACS computer and other hardware sensors and actuators of the spacecraft. The HITL generates the components of the Earth magnetic field at the instantaneous location of the spacecraft, and the components of the Sun vector. It also produces the disturbance torques the spacecraft is likely to see. Feeding these quantities back to the spacecraft, the spacecraft responds by generating the reaction wheel and torquer commands. The HITL then "flies" the spacecraft by executing the changes in the spacecraft attitude (and computing spacecraft position changes).

In this way, the performance of the ADACS can be tested. The use of a HILP test is invaluable in testing and debugging the software and characterizing on-orbit performance and behavior. Simulations several days long can be run, lending confidence that the system will perform reliably on-orbit.

A block diagram of a typical HITL test set is shown in Fig. 7.20. This diagram was drawn for the QbX 4 U CubeSat. Attitude determination and control is performed by the MAI-100 integrated ADACS. The 3D software simulator computes the spacecraft dynamic responses to reaction wheel, electromagnet and orbital environmental torques in exactly real time. Electrical interfaces between the MAI-100 ADACS and the simulation computer are identical to the actual QbX spacecraft, so that the MAI-100 in the loop believes that it is flying the real spacecraft on-orbit. The HITL system can provide an End-to-End performance validation of ADACS pointing accuracy. In addition, other related spacecraft systems that impact pointing performance are simulated, including performances of the Electrical Power System, the Command and Data Handling System, and the Communications System.

To visually see the performance of the ADACS subsystem, the point to which the spacecraft Z axis points as a function of time can be displayed together with the ground track of the spacecraft, as shown in Fig. 7.21. In this figure, one of the outputs of the Dynamic Simulator, the spacecraft ground track is shown as the purple line, while the "aim point" of the spacecraft as a function of time is shown by the red line.

The scenario shown in this figure is one where three targets must be imaged. The spacecraft enters the area covered by the display at the bottom of the figure and

## 7.5 Integration and Test

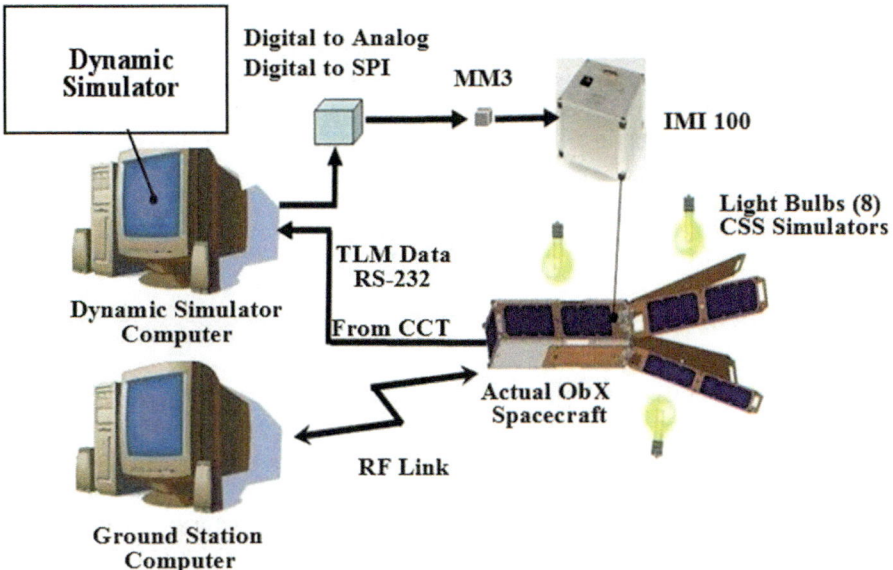

**Fig. 7.20** Flatsat dynamic simulator for QbX, Earth-Pointing Nanosat

**Fig. 7.21** The dynamic simulator "flies" the spacecraft in a hardware-in-the-loop to permit assessing ADACS performance

advances initially along the yellow arrow. Attitude commands then cause the spacecraft to roll to reach the first target roll angle and then settle, awaiting the instance when the FOV (shown as the white rectangles) reaches the target. The first image is then taken. The spacecraft then slews to the roll angle of the second target and settles at that roll angle until it is time to take the second image. Imaging the third target is accomplished in the same way. Finally, the spacecraft is caused to slew back to nadir, awaiting further commands.

## 7.6 On Orbit Checkout

Once the spacecraft is on orbit, checkout of the ADACS can begin. First, proper functioning of each of the ADACS components is verified by causing the spacecraft to perform simple functions. An example is to roll the spacecraft (say) $10°$. The roll acceleration and deceleration commands are executed (open loop), and the change in attitude, as measured by the Star Tracker(s), is noted. This verifies that the reaction wheel torques are as specified on the ground.

Next, ADACS system level tests are performed. The spacecraft is caused to roll and pitch to point to a known target on the ground. Repeated performance of this test can establish component misalignments and other bias errors, and it can also be used to verify that the loop gains are correct. ADACS software uploads can be used to correct for these biases.

# Chapter 8
# Spacecraft Software

Spacecraft software is discussed here in terms of (1) the functions the software has to perform, (2) the software architecture and (3) the manner in which the functions are performed. While the functions the software has to perform are generic, there can be different architectures and different methods of implementing the functions. Here, a point design is described, a design with a lot of flight heritage over many different spacecraft. The spacecraft missions in this example are:

- Store and Forward communications
- Collecting data from and commanding unattended remote sensors
- Imaging designated targets and downlinking image data from the spacecraft to ground stations

The spacecraft software utilizes a distributed processing environment. It has C&DH and ADACS computers. They operate independently; only ADACS housekeeping telemetry data flows from ADACS to the C&DH and only changes in ADACS requirements or orbital parameters (or target location) data and time flow from the C&DH to the ADACS computer. Communication functions are performed by the C&DH, but there may be a separate, dedicated image data processor (not discussed here).

The software can be thought of as consisting of three distinct software modules, each consisting of many different submodules. The three main modules are:

1. Command and Data Handling System (C&DH) software
2. Attitude Determination and Control System (ADACS) software
3. Communications Processor Subsystem software

The C&DH software contains the spacecraft bus operating software and the mission payload specific software. Spacecraft communications software may reside in the C&DH or it may be in a separate communications processor.

## 8.1 Functions and Software Architecture

The spacecraft bus C&DH software performs the following functions:

1. Initializing the C&DH processor, hardware, and operating system
2. Executing scheduled events, housekeeping, communications, telemetry
3. Communicating with ground stations or mobile users (when in the vicinity of these users)
4. Management of the on-board electric power system
5. Management of the on-board thermal control system
6. Collecting, computing and formatting spacecraft telemetry
7. Storing and retrieving data, messages and statistics
8. Managing the message memory

Expanding on the Communication functions listed above, the functions include:

- Communications transmitter frequency selection,
- Transmission of messages and data,
- Selection of CMD or uplink message frequency
- Reception and processing uplink commands
- Storing messages
- Accepting uplinked software to augment or change spacecraft software
- Collection, storage and processing of user access authorization data,
- Message collection, routing and acknowledgment

The ADACS Processor performs:

- Propagating the orbit from uplinked orbital elements or from on-board GPS
- Reading the data from all attitude and position sensors
- Kalman filtering sensor data to obtain optimum spacecraft attitude
- Receiving and interpreting commands to the ADACS (from the C&DH computer)
- Performing attitude control computations
- Controlling reaction wheel speeds and torque rod activities
- Controlling propulsion subsystem activities (if applicable)
- Controlling thruster activities
- Collecting ADACS telemetry and sending it to the C&DH
- Sending accurate orbital elements to the C&DH
- Sending accurate time to the C&DH

The Payload Processor performs the management of payload functions and collecting payload data.

The software hierarchy diagram is shown in Fig. 8.1. The Executive controls the computer operating system. The operating system and all software are typically written in C++ or in machine language. Often multiple copies of the bootstrap program reside in radiation hard Fuse Link PROM for redundancy; the main program is in Flash EPROM. All spacecraft bus operations are performed on a time schedule, programma-

## 8.1 Functions and Software Architecture

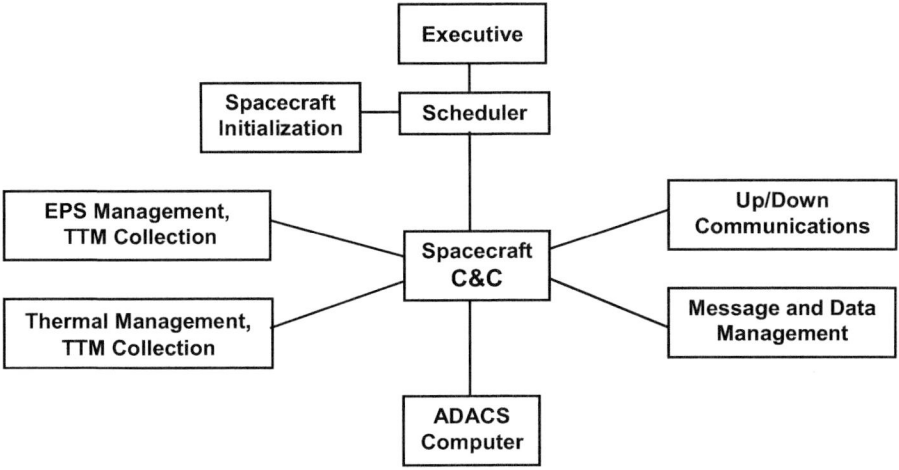

**Fig. 8.1** Software architecture and software modules

ble from the ground and employing default values on the spacecraft. Hardware counters programmed by the C&DH execute the schedule, and perform repetitive operations in response to a single command. All high speed and low level operations are relegated to the multiple, high speed communications processors that operate autonomously. They are powered up only when needed. This reduces high speed requirements of the C&DH, permitting it to operate at a relatively low duty cycle, conserving electric power.

Though not required by throughput and speed requirements, ADACS functions are performed in a separate processor that communicates with the C&DH only by two-way file transfer of telemetry and commands. C&DH and the ADACS codes are resident in both processors; however, each processor executes only its assigned functions (C&DH or ADACS). In case of hardware failure in either the C&DH or the ADACS processor, the other processor can perform the other's functions, but computer loading becomes much higher. During high communications traffic periods, operating with only one processor, highly computationally intensive functions must be eliminated. Kalman filtering of the multiple attitude sensor data is computationally intensive and is eliminated at those times, reverting to the use of the (ground designated) single sensor for attitude determination. Generally this will be the GPS attitude determination or a star tracker, if there is one on board.

The software architecture consists of the Executive and the Subroutines for accomplishing the various tasks. The Executive schedules hardware, electric power and ADAC interface initialization, Operating System initialization, State Determination, Mass Memory initialization, Default Schedule initialization, and the Idle Task. The Executive operates the Idle Task in which the Executive cycles through all of the potential tasks (subroutines) it might have to perform, to determine if there is anything it has to do. It does this at a low clock speed to conserve power. If, while cycling through potential tasks, the Executive finds one that requires execution, it switches the processor to high speed and causes the appropriate subroutine to execute the required task. It services interrupts in the same manner.

## 8.2 Performing Each Function or Module

### 8.2.1 Initialization of the CDH Processor, Hardware, and Operating System

Once the C&DH is released from the reset condition (held during ascent until the separation switch indicates that the spacecraft has been released from the launch vehicle), the C&DH performs a Flash EPROM Operating System check, and initializes the hardware and nonvolatile RAM variables. The Mass Memory is tested for unusable blocks. After all the start-up tasks have been initialized, the C&DH loads the Default Schedule. The Default Schedule consists of a C&DH State of Health (SOH) telemetry collection event, and a communications event. The C&DH performs the communications event repeatedly until communications with the ground station is successful. After the Default Schedule is loaded, the C&DH transitions to the Idle Task to cycle through the spacecraft subsystems for possible action.

The ADACS is released from reset at the same time as the C&DH and performs a similar initialization sequence. This sequence will be described later in the ADACS software section; however, the portion of the ADACS initialization that is pertinent to the C&DH initialization will be described here.

The ADACS computer will complete initialization before the C&DH because its initialization sequence is shorter. The C&DH listens for notification by the ADACS computer that its initialization is completed and is ready for commands. The C&DH synchronizes time with the ADACS and commands the ADACS to begin State of Health data collection.

The Idle Task maintains the C&DH computer in a slow clock mode to conserve power. The tasks performed include resetting the watchdog timer, validating the operating system in EDAC memory, EPS monitoring and recovery, and interrupt handling.

The interrupt handler monitors the real-time clock alarm (Event Execution Time), operating system time tick, GSE port service, communication reception and transmission service, and watches for unscheduled communications from mobile users.

### 8.2.2 Executing Scheduled Events

All scheduled events use the real-time clock alarm function to start execution. The alarm is set to the time of the next event by the Scheduler at the completion of the previous event. Prior to executing the event, the watchdog timer is reset. Since the watchdog timer duration is usually twice the longest event duration, and the watchdog timer is reset on entry to each event, all events are executed without regard to the watchdog timer duration. After the watchdog timer is reset, the event is performed. When the event is completed, the next event is removed from the schedule, and the real-time clock alarm is set for this event and the ADACS computer returns to the Idle Task.

The scheduler performs three major functions: (1) disassembles and interprets commands received during a communications event, (2) identifies the next executable event, and (3) validates the scheduler command sequence queue. Each command is validated in three ways: the satellite access password is checked to determine whether the ground station has the authority to schedule commands, the CRC of the packet containing the commands is checked to be sure it has been received correctly, and the CRC on each command is checked before placing it in the schedule.

If the command is valid, it is placed in the schedule in time-order sequence. If an error is detected, the command is discarded. The spacecraft schedule can be telemetered to the ground for verification.

After the event is executed, the scheduler readies the command to be performed next and sets the alarm for the specified event execution time. An event can reschedule itself at a later time. This allows self-scheduling repetitive events, such as telemetry collection, to only require a single command (schedule memory location). The spacecraft is capable of storing a large number of commands (counting repetitive commands as a single command) in EDAC memory. The iterative execution of commands continues until the schedule is exhausted or new commands are received. If the scheduler ever becomes empty, it loads the default schedule that contains the "Beep/Receive" communications event. Also, if a communicate event is not scheduled for the next 3 days, the scheduler places a communicate event ("Beep/Receive") in the schedule. This tells the master ground station that the satellite is out of schedule and needs servicing.

The "Beep-Receive" communications event is a repetitious spacecraft transmit event (transmitting "Routine Telemetry"), followed by activation of the spacecraft command receiver for a specified time interval. So, if communication is lost with the spacecraft, or if someone just forgot to send up a schedule, the spacecraft goes into short transmission "Beeps" every (say) 1 min, to tell the world within its line of sight that the spacecraft is here and it has no schedule to execute. This feature proved very useful in many instances when, for one reason or another, the spacecraft got "lost." Because of the intermittent nature of the transmit events, the power required to execute "Beep-Receive" is quite low.

All events are self-routing, i.e., if the event is destined for a processor other that the C&DH, the code for that event will route it to the proper destination processor.

### 8.2.3 Stored Command Execution

The satellite operates on an internal schedule, executing commands uplinked and stored in memory together with the time when commands are to be executed. Uplinked commands are processed by the spacecraft to extract pointer information with which time sequential commands can be retrieved from memory for execution regardless of the order in which these commands were sent up to the spacecraft. The time to execute the commands is determined on the ground from the spacecraft orbital parameters used by the ground station. The ground station computes the time when the spacecraft will be over the specific areas on the ground where the command is to be executed. Exceptions are commands to the ADACS to point the spacecraft to a given point on Earth and take a picture. The ADACS has all the information with which to autonomously compute the required maneuvers.

Exceptions to the time-scheduled execution of commands are satellite self-scheduled events (such as telemetry data collection) and geographically commanded repetitive events. Telemetry is collected at ground commandable intervals. This is implemented by placing a number in one of the hardware counters, and counting down until the count reaches zero. At that instant, an interrupt occurs and causes the execution of the repetitive event. The number is then reinserted into the counter so that the process may be repeated.

### 8.2.4 Housekeeping

Housekeeping functions include maintenance of on-board hardware status, telemetry data collection and formatting, electric power system management, management of the thermal control system, ADACS status maintenance and ADACS telemetry data collection.

*On-board hardware status* is maintained by a "state vector," telemetered to the ground, where each bit indicates the status of a different hardware module, and whether it is being used at this time or not. Onboard hardware status, if all components are functioning, is controlled by a default status vector that (for example) signifies that User Downlink Transmitter No. 1 is used, that the C&DH and the ADACS software are operating normally, that all batteries are on the line, that the thermal control system (heaters) are cycling with a duty cycle to maintain electronics temperature at (say) 5 ° C, and that all deployables (activated by pyrotechnic devices) have been fired.

### 8.2.5 Management of the On-Board Electric Power System

Electric Power System Management is performed in a number of different ways. Battery charging and controlling the charge regulators in accordance with the temperature and voltage of the batteries is performed by the "smart" electric power subsystem in response to its own controller. The C&DH control of the power system includes turning on or off various batteries, taking a battery off the line for conditioning, monitoring individual batteries for low voltage, taking the load off of these until they are recharged, and maintaining a battery watt-hour charge status for telemetry to the ground to indicate battery status and the level of general battery usage over different parts of the world. In addition to control of the batteries, the C&DH control of the electric power system is also used to conserve electric power. It determines the usage of individual PCB, turns them on or off depending on need, controls the clock speed to slow it down when over ocean areas or when performing only idle tasks. The spacecraft can also be commanded into Save Power mode by ground command. Finally, should power become critical, the hardware will trigger a low power condition and shed all unnecessary loads in an attempt to save the spacecraft. This computer control of the electric power system is a flight-proven feature that has resulted in achievement of high power efficiencies.

### *8.2.6 Management of the On-Board Thermal Control System*

Based on spacecraft telemetry, ground operators are provided with data about battery temperatures. The operators may decide that the battery temperature should be increased. Management of the Thermal Control System is implemented by controlling the temperature by cyclic, computer-controlled, operation of heaters. Instead of using thermostats (that would put the vital battery heater system under the control of an automatic device whose failure could jeopardize the entire mission), the computer is used to turn on and off the heaters, with time-out on the duration of execution of the heater commands. By controlling heater operation via the computer, more precise and less power consuming heater control can be achieved.

### *8.2.7 Telemetry Data Collection*

Telemetry data collection is performed by repetitive scheduling of the execution of a "status" event at ground commanded or default schedule commanded repetition intervals. This event powers up the Analog I/Os, samples each of the analog and binary telemetry points, feeds these values to a multiplexer and A/D converter for input to the C&DH computer. In the C&DH, the sets of telemetry point values are stored, and telemetry statistics are updated. These are the minimum, maximum, average and most recent values of all the telemetry point time histories. In this way, if only routine telemetry is required, the stored telemetry dump to the next ground station has always the same number of data points, regardless of the time between ground contacts, yet the data contains the important SOH statistics of every telemetered point. The default time interval between telemetry data collection instances is typically quite long, so that 90 sets of samples cover one orbit.

If there is an anomaly, or the ground station operator wishes to examine a particular subsystem in more detail, he or she can command that specific subsystem to collect and downlink telemetry from a specified time to another specified time with a sample rate that can also be commanded. In this way, the Engineering Telemetry data may contain much more granularity. Telemetry from ADACS, from the Communications Processor and from the Propulsion System are downlinked and displayed in a similar manner.

## 8.2.8 Communications Software

The communications software is divided into up/down communications and on-board message handling. While specific details of the communications protocol are usually a joint decision between the customer and the spacecraft builder, general concepts of operation are described below. Since this spacecraft has multiple and different store-forward communications functions, there are also different message types that must be serviced. Since communications involves a large number of steps, Higher Level Commands are used to facilitate implementing communications.

The commands provided to the spacecraft by the ground control station are high level commands, interpreted by the spacecraft. For example, to communicate and contact a ground station at time $T_0$ using transmitter No. 2 and send data at 56 kbps on frequency $f_0$ and time out after 10 min is a single high level command. The spacecraft implements this command by turning on a TCXO oscillator at $T_0$-X (to permit it to stabilize), connect the TCXO to transmitter number 2, turn on the B+ to number 2 at $T_0$-Y, set up the protocol control software to transmit at 56 kbps, select the modulator to be used, and select the first piece of data to be transmitted so that it would be ready for transmission when the transmitter key-on switch is turned on at $T_0$. These high level commands generally follow the concept of implementing the answers to a sequence of specifications about the action to be implemented, as a structured set of instructions to When? What? and How? This high level, menu-driven command structure has extensive flight history. Command confirmation takes only a short time. It contrasts sharply with the older methods of controlling satellites, where separate ground commands were needed to perform each switch or setup function, and where each element of the command had to be repeated by the satellite for confirmation by the ground before it was executed, thus limiting the number of (equivalent) high level commands that could be executed per unit time.

Message deliveries are attempted a predetermined number of times, and if they are undeliverable, they are recorded as undeliverable in the Communications History. Message responses not received within 2 seconds are retransmitted until the retry count is exhausted. Once the retry count expires, no more messages are sent to that user on the current pass.

The major function of the processor associated with the command receiver is to relieve the C&DH of all low level processing associated with frequency selection, receiver tuning, decrypting, removing Forward Error Correction coding and message or command interpretations. The communication processor is only powered up by C&DH command. After the processor powers up, it notifies the C&DH that it is ready for receiver frequency assignment. The C&DH sends the frequency, modulation type, data rate of the expected uplink command or message. The communication processor programs the receiver to the specified frequency, initializes the digital hardware and DMA, and waits for the receiver squelch threshold to be exceeded. The processor receives data until detection of packet termination. Only user packets received error free, i.e., have a valid CRC, are forwarded to the C&DH. During packet reception, the receive signal strength for each packet is monitored and appended to the packet that is forwarded to the C&DH. The number of good and bad packets received are included in telemetry when requested by the C&DH. The processor collects SOH information requested by the C&DH. SOH consists of information pertaining to the Executive, interprocessor communication errors, invalid command or data transfers, and other miscellaneous information.

The communication processor program memory is divided between two types of storage, Fuse Link ROM and Flash EPROM. The Fuse Link ROM contains the software necessary to initialize the processor, communicate with the C&DH, and houses critical software functions. The Flash EPROM contains two copies of the software. When the communication processor is powered up and detects an unrecoverable error condition, or is reset by the C&DH, it scans both copies of the operating system. The processor executes the most correct version. The communication processor software is written in C with low level drivers or time critical portions of code written in assembly language.

### 8.2.9 Attitude Control System Software

Functions performed by the ADACS were already described. Here, only those aspects that propagate the orbit from GPS data are described. The on-board GPS provides position (in ECEF coordinates), velocity and time information. First the coordinates are converted to J2000 coordinates, then the spacecraft accelerations are double integrated to obtain position in the future, and thus the orbit is propagated.

### 8.2.10 Uploadable Software

From time to time, it may be necessary to change the way the on-board software operates. New software must be sent to the C&DH. This is made easy by the fact that the on-board software is organized as a set of independent modules, called into service by the Idle Task sequencer. The executive addresses each module in turn. This means that a new module can be sent up to the spacecraft and located anywhere there is free memory space available. Only the pointer that causes the Idle Task to go to the next module must be changed to point to the location of the new code. The C&DH should have the capability to update the code resident in Flash EPROM. The sequence is initiated by the C&DH issuing a command to receive new software. The function that receives, stores, and programs the Flash EPROM resides in Fuse Link ROM. After all the software has been received, the C&DH commands to program the Flash EPROMs. The C&DH then resumes normal operations using the new software patch or subroutine.

## 8.2.11 Propulsion Control System Software

Commands to the propulsion system are to provide a delta V burn through the CG of the spacecraft, using multiple thrusters. Since the net force of the thrusters misses the CG by an undetermined small amount, providing a net thrust through the CG involves duty cycling the multiple thrusters and correcting the thrust error by changing the duty cycles in response to an attitude or rate sensor. Since the ADACS includes an IMU, attitude errors are readily determined. The propulsion control system computes the length of the thrust required to produce a commanded delta V. It does this from the tank pressure, using the ideal gas law. This time interval is divided into many shorter intervals to apply to the thrusters. Then, the IMU is used to provide the input to the computer to determine which thrusters are to be fired for these shorter time intervals to result in straight flight. When thrusting is initiated, the processor activates the shutoff valve to open gas flow from the fuel tank, measures the gas pressure and temperature (a normal SOH measurement) and computes the duration of the burn from the gas law. If the required burn duration is longer than 0.1s, the program computes the number of burns required so as not to upset the spacecraft due to any one burn being too long. Then, starting at the programmed time, the set of short burns are executed by opening appropriate thruster valves for specified durations of time. The time between burns is controlled by angular velocity, so that consecutive burns should result in opposing torques, resulting in minimum attitude error from the burn sequence.

## 8.3 Software Development

Software should be developed in an orderly manner with adherence to a set of principles. The software should be:

(a) Broken up into different modules, each of which performs one of the functions described earlier in this chapter.
(b) Each module should be described in full (defining the module requirements, the inputs to the module, the algorithm or logic the module will perform and the outputs of the module). This is usually called the pseudocode. The pseudocode may also define the method of testing the code and the ICD (Interface Control Document) for interfacing to the hardware.
(c) From the pseudocode, the software can be reviewed before it is written to catch conceptual mistakes. Also, the size of the code (as Source Lines of Code - SLOC) can be estimated.
(d) If not included in the pseudocode, separate ICD's should be written to define interfaces between software and hardware modules.

8.3 Software Development

(e) Based on whether this code is new or is reused from an earlier program, an estimate of the time to write the code can be made. Estimating programming effort depends on the assigned programmer's experience. For this reason, programming time estimated based on SLOC is, at worst, a WAG, and at best, it is not much better. The SLOC can also be estimated by comparing each module with similar modules written in the past.

(f) Software written should be archived regularly and should have a version number and date so it could be readily identified. Several good version control COTS software are available.

(g) Software progress reviews should be held frequently to determine progress and to identify incompatibilities or problems. If a software person tells you that he or she is almost finished, interpret this to mean that the software does not work.

(h) Changing software modules is dangerous, if not rigorously controlled, and could invalidate software written by another member of the team. For this reason, software changes should follow the Change Process (of writing down what the change is, why it is made and what other part of the software will be affected by the change). There are several Change Control Systems (Subversion, Git) that can establish and maintain the change control rigor.

(i) The completed software modules should be tested under realistic situations.

(j) While expensive, a complete software simulation of the spacecraft hardware should be built. This enables testing the individual software modules as well as the complete software system.

(k) From the pseudocode, software should be written that can perform some (even if not all) the functions of each module. This permits an early testing of the entire system, verifying that the system as conceived will work. Later improvements to the code will just make the system work better. This iteratively improving software development process is called the Spiral process (since the initial software spirals in toward the final software). The advantage of the Spiral process is that it creates code very soon after starting the development process.

# Chapter 9
# Spacecraft Structure

Selection of the most suitable spacecraft structure depends on many factors. The main ones are (1) the launch vehicle payload envelope and interface separation system dimensions, (2) the launch vehicle loads to the spacecraft structure, (3) the weights of the spacecraft bus and payload components, (4) thermal design and ability to get rid of incident and internally generated heat, (5) the way solar arrays will be mounted, deployed or rotated to provide the required Orbit Average Power, (6) instrument pointing and (7) any other special requirements, such as the possible requirement to keep propulsion, bus and payload separated from one another.

This section will go through the process of analyzing a typical spacecraft structure with common-sense explanations for each step.

## 9.1 Introduction

During the launch sequence, the structure is exposed to a complex dynamic environment. It starts with the launch vehicle at rest on the pad with a 1 G vertical static load. At the moment that the rocket engines begin to fire, before the rocket begins to rise, the structure is exposed to severe vibration loads. Some of these are structure-born, because engine vibrations pass through the launch vehicle, but a significant portion are acoustic, due to the engine sounds reflecting off the ground around the launch vehicle and impinging upon the spacecraft. There is a relationship between the height of the launch vehicle and the acoustic loads on the spacecraft at the top of the rocket. As the launch vehicle begins to rise, it accelerates slowly as a result of having to lift the high mass of the initial fuel. As the ground falls away, the reflected acoustic loads dissipate. As the velocity increases, the relatively dense air hitting the nose of the launch vehicle at lower altitudes, creates a dynamic pressure, "Q", inducing a compressive load on the upper structure. It is common practice for the engines to be throttled back until sufficient altitude is reached, where the air density is less, to minimize the effects of this pressure. During this entire time, as fuel burns

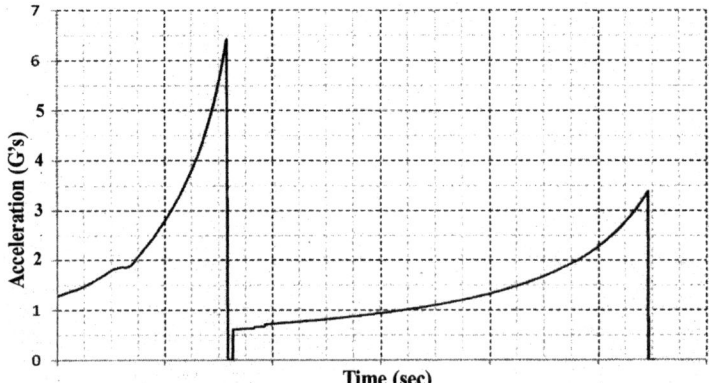

**Fig. 9.1** Axial acceleration vs. time of the Falcon 1 launch vehicle

off, the launch vehicle is getting lighter and acceleration increases for a given engine thrust. A typical axial acceleration profile for the Falcon 1 launch vehicle, given in Fig. 9.1, illustrates this sequence. It is noted that the peak axial acceleration of 6.5 g's occurs at first stage separation.

During this time, the spacecraft is attached (together with the other manifest payloads) to the launch vehicle, which has its own stiffness and mass. The complete assembly is subjected to dynamic loads over a large frequency range. The dynamic responses of the launch vehicle and the manifested components couple with each other, producing a set of loads unique to that specific launch. In order to enable the analysis during spacecraft design, estimated design loads are published by the launch vehicle supplier. These are conservative estimates of the loads the structure will actually see. Depending upon the maturity of the launch vehicle/manifest, they may be based on similar launch vehicles, actual measured values from previous launches, or coupled loads analysis (CLA) of the launch manifest. During the design cycle, additional CLA cycles may be performed, and the published loads updated. The loads used for testing and to verify the structure are based on the latest loads estimates at the time of testing.

## 9.2 Requirements Flow-Down and the Structure Design Process

Structural Requirements are flowed down from the Launch Vehicle and from Spacecraft Requirements.

Launch Vehicle requirements are defined by the launch vehicle provider. Typically, they include:

1. **Minimum Spacecraft Resonant Frequency**: (typically 50 Hz for a small satellite): This is specified to avoid coupling with launch vehicle modes. In the

## 9.2 Requirements Flow-Down and the Structure Design Process

case of the ISS, there are additional requirements to preclude coupling with the Station's robotic arm modes. Finally, there may be on-orbit deployed configuration frequency requirements for instrument control and performance.

2. **Static Loads**: (provided by the launch vehicle) are based on estimates, past experience, or Coupled Loads Analysis. These are conservatively estimated numbers and take into account the coupled effects of the entire manifest and launch vehicle. Coupled Loads Analyses include uncertainty factors. These factors are higher for the preliminary CLA at the beginning of a project, because the design and manifest are not finalized. By the final CLA, there are fewer variables, and the factors of safety used can be lower.

3. **Sine Specification**: (may or may not be required by the launch provider.) Typically, a sine spec may be a few G's in magnitude, in the 0 to 50 or 100 Hz frequency range. It is intended to include the low frequency/high displacement dynamic loads on the spacecraft.

4. **Random Loads**: These loads include the higher frequency dynamic loads on the spacecraft. As a rule, the larger the spacecraft-instrument, the less the effect these loads have. When a specific vibration spectrum is not available from the launch vehicle provider, the designer should begin with the General Environmental Verification Specification, GEVS, shown in Fig. 9.2.

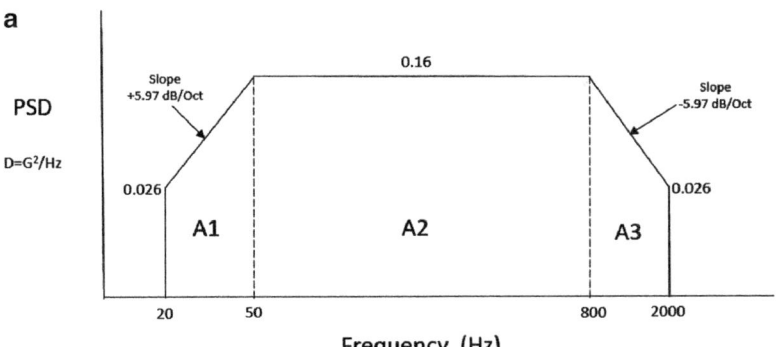

Fig. 9.2 (a) A launch vehicle random vibration spectrum (GEVS).
(b) Random vibration spectrum

5. **Acoustic Loads**: caused by acoustic pressure impingement, may be critical for lightweight items with large surface areas (solar arrays). An effective static pressure can be calculated from the acoustic sound pressure, using Eq. 9.1

$$\text{Lp} = 20^* \log(\text{p}/\text{Pref}) \text{sound pressure}(\text{dB}), \text{where}$$
$$\text{p} = \text{sound pressure}(\text{Pascal}), \text{and} \quad (9.1)$$
$$\text{Pref} = 2.0\text{E} - 5 = \text{reference sound pressure}(\text{Pascal})$$

6. **Pressure Loads**: propulsion systems, sealed and vented containers. Propulsion systems have specific multi-tiered requirements, depending on the propellant used. A hazardous propellant, such as hydrazine, has higher safety requirements than an inert propellant such as compressed nitrogen. Safety requirements also depend on the contained energy of the propellant system. Fuel lines must be analyzed for 4 times, and proof tested to 2 times maximum expected pressure. For use near manned spacecraft, they require a "leak before burst" analysis to demonstrate that, in the event of a crack, the crack will grow linearly to a length that would allow the pressure to leak out before unstable crack growth will occur. Propellant tanks are typically proof tested to 1.5 times their Maximum Design Pressure.

   Sealed containers are volumes not vented during launch; they remain at one atmosphere internal pressure on-orbit. Sealed containers must be qualified by testing to two atmospheres internal pressure.

   Vented container analysis consists of calculating how much vent area is required for a given enclosed volume. The venting pressure loads are calculated based upon the maximum pressure decay rate and the ambient pressure during launch.

   Note: Pressure load requirements may be tiered based on the amount of energy contained in the system.
7. **Factors of Safety**: (defined by the project)

## 9.3 Structure Options, Their Advantages and Disadvantages

Structural configurations can vary from a simple box containing the spacecraft electronics (several examples are shown in Fig. 9.3) to a complex set of shapes that contain or support the electronics, propulsion and various deployables, shown in Fig. 9.4.

If the spacecraft is small, almost any structural configuration will work. However, as the size of the structure and the weight of its electronics and payload grow to several hundred or more pounds, the placement of the electronics affects the strength required and the resulting resonant frequencies of the spacecraft. For instance, in Fig. 9.5a, the spacecraft structure consists of bottom and top plates, longerons and side panels.

## 9.3 Structure Options, Their Advantages and Disadvantages

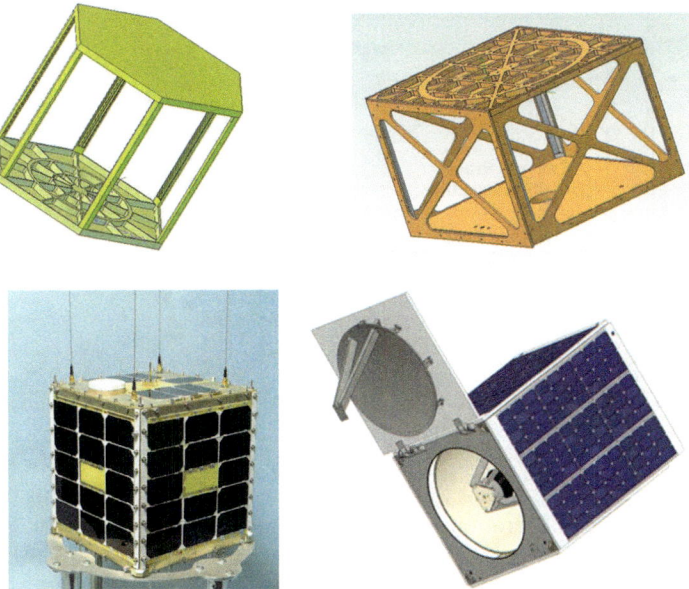

**Fig. 9.3** Simple box spacecraft structures

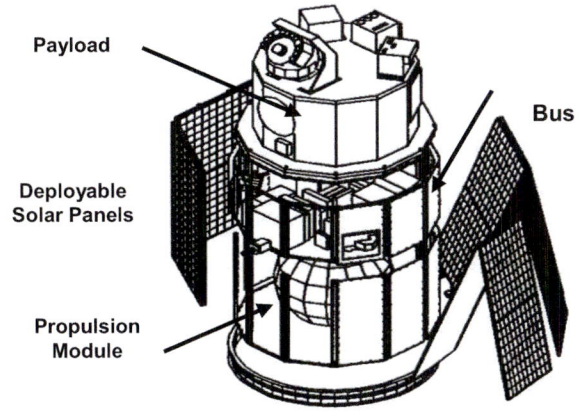

**Fig. 9.4** A more complex spacecraft structure

Under the load of the electronics mounted to the inside of the side panels, the bottom plate will bend and vibrate. This can be mitigated by using corner braces. However, as shown in Fig. 9.5b, by mounting the electronics to internal panels at or near the separation system diameter, the load path will be straight and will permit reducing the thickness, strength and weight of not only the bottom plate but also the weight of the side panels. In addition, in Fig. 9.5a solar incidence will heat the outer panels, which will create a difficult environment for the electronics, while in Fig. 9.5b the electronics is protected from direct solar heating.

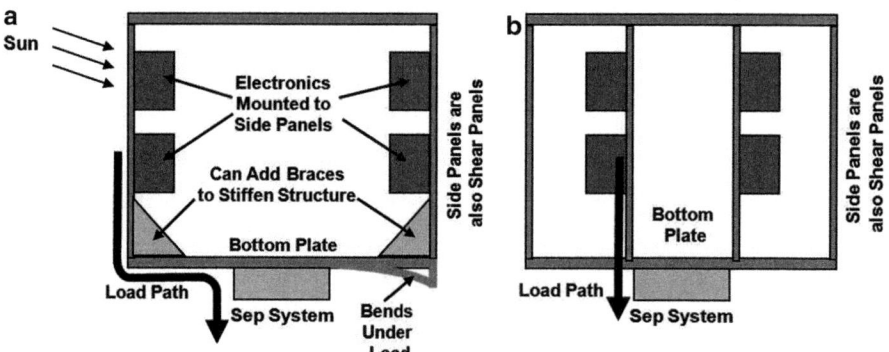

**Fig. 9.5** (**a, b**) Load paths in two different structures and two electronics mounting methods

In selecting and designing the spacecraft structure and the layout of the electronics, intuitive consideration must be given to the spacecraft stiffness, resonant frequency and thermal performance even before undertaking a detailed analysis. There are some "rules of thumb" to aid in this process:

- Avoid long, unsupported structural members
  - Include stiffeners, where possible, to avoid buckling
  - Stiffeners should be strong enough to support diagonal tension buckling of shear webs
- Milled aluminum is strong and provides a good thermal path
  - Use materials that are resistant to Stress Corrosion Cracking per MSFC-STD-3029, "Guidelines for the Selection of Metallic Materials for Stress Corrosion Cracking Resistance in Sodium Chloride Environments"
- Honeycomb panels feature high stiffness to weight ratios
  - Honeycomb face sheets may be metallic or composite
  - Carry high stress levels without buckling
  - Honeycomb core needs to be perforated to allow venting during launch
  - Handle acoustic pressure loads well
  - Require local inserts for bolt attachments
  - Composite face sheets may require an electrical barrier ply when used with solar cells
  - May require additional testing to verify fabrication workmanship
  - Be sure to accommodate thermal expansion between materials
- Use many screw fasteners
  - Fastener joints should be bearing critical (not shear critical)
  - Use high strength A-286 steel fasteners (avoid 300 series stainless bolts)
  - Cadmium plated fasteners are not permitted around manned spaceflight

9.3 Structure Options, Their Advantages and Disadvantages

- Use self-locking inserts and nuts to attach bolts
- Fasteners should be redundant to avoid single point failures

• Graphite Epoxy structures are light and strong

- Their coefficient of Thermal Expansion (CTE) can be tailored for optical applications
- Poor thermal or electrical conductors
- May require additional testing to verify the workmanship of the composite layup
- Higher manufacturing costs

• Assure that there is a good thermal path to get rid of the heat

An example of a complex structure is the double satellite manifest launched by NASA in 2009 and shown in Fig. 9.6. In this mission, the Lunar Reconnaissance

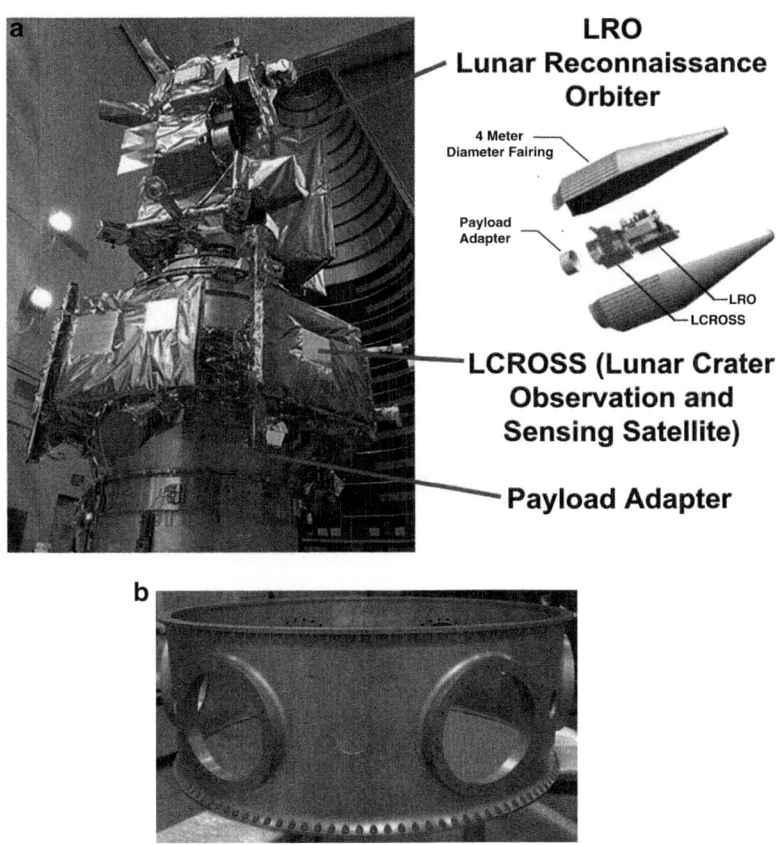

Moog Inc. Space and Defense Group

**Fig. 9.6** (**a**) The LCROSS and LRO stack of satellites. (**b**) ESPA (EELV secondary payload adapter) ring. LCROSS structure supports LCROSS modules and the LRO spacecraft. (**c**) Each bracket supports the load of only one subsystem. (**d**) Details of the LRO propulsion system. (**e**) Four external structure modules comprise the LRO structure configuration

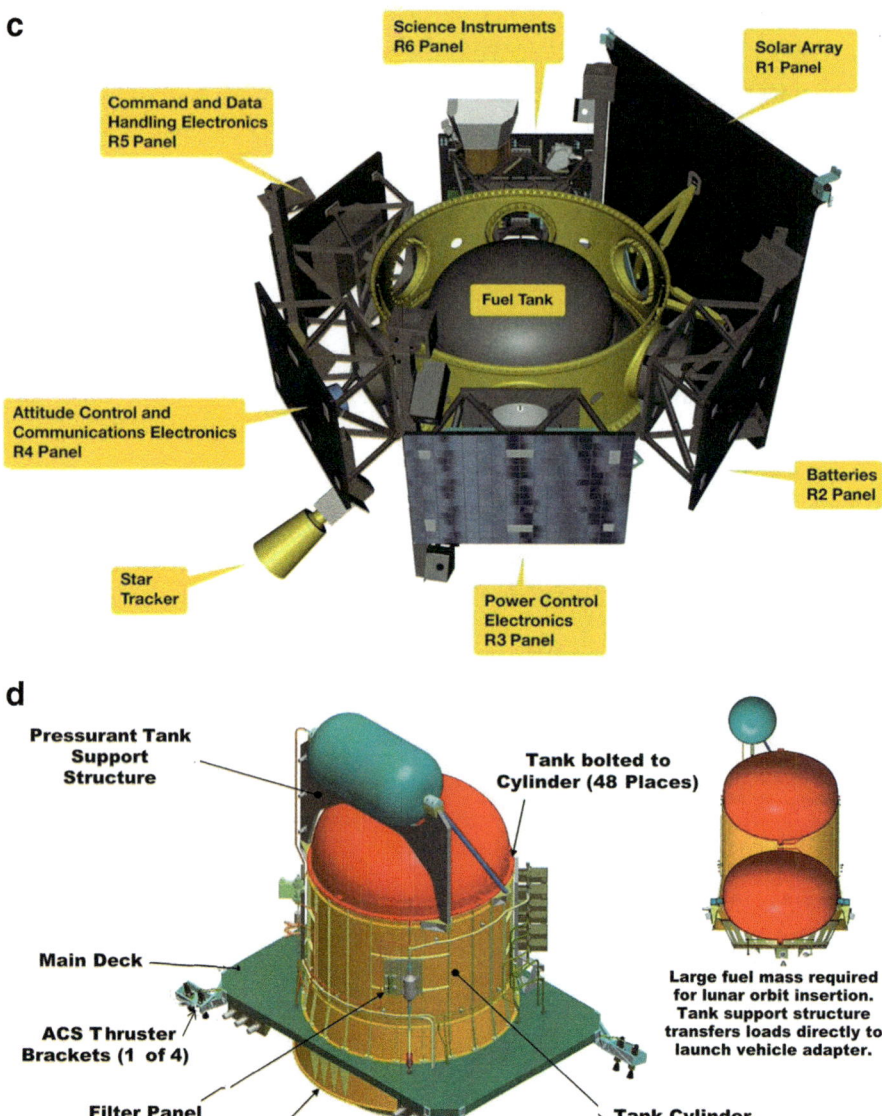

Fig. 9.6 (continued)

## 9.3 Structure Options, Their Advantages and Disadvantages

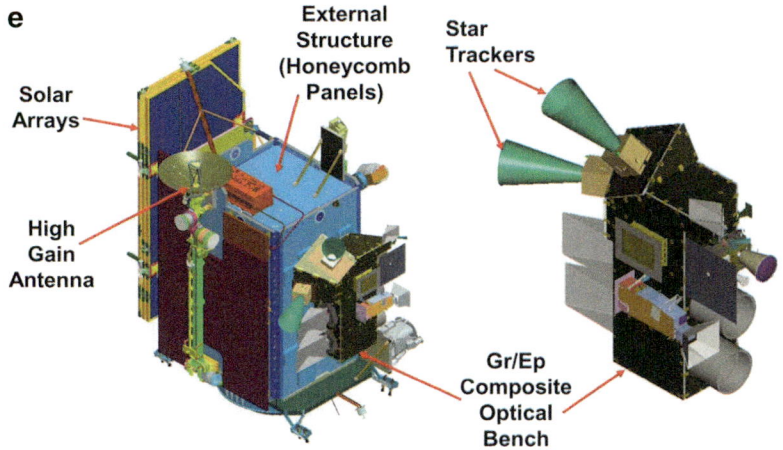

**Fig. 9.6** (continued)

Orbiter (LRO) was launched atop the Lunar Crater Observation and Sensing Satellite (LCROSS). LRO orbited the moon and mapped its surface. The LCROSS mission was to approach the moon. The upper stage then impacted the Moon and LCROSS flew through the ejecta looking for signs of water. The two vehicles were stacked for launch.

The LCROSS structure used an ESPA ring with components mounted radially around its perimeter. The upper flange of the ESPA ring supported the LRO spacecraft. Both spacecraft required large fuel tanks to supply thrust for lunar orbit insertion.

The fuel tanks were located inside in LCROSS, and multiple components were mounted on the outside of the ring. From a structures load point of view, this provided modularity to LCROSS and an efficient load path to support LRO.

The LRO was a 4-sided structure and consisted of 4 vertical honeycomb panels containing the LRO subsystems and instruments. The four vertical panels were mounted to a horizontal main deck, which interfaced with the LCROSS through the Launch Vehicle Adaptor (LVA). The large fuel loads were carried through to the separation system of LRO through the central structure (in a manner similar to the example in Fig. 9.5).

The alignment accuracy between the star trackers and the optical bench instruments was critical for mission success. The composite bench material had a very low CTE, thereby minimizing the thermal effects on pointing accuracy. The bench was attached to the spacecraft using flexures to isolate deflections of the spacecraft from the optical bench. Alignment accuracy of other components on the spacecraft, such as solar arrays and high-gain antennae, were not as critical as those of the optical instruments.

This example of a complex spacecraft illustrates that use of direct load paths is a good design practice.

## 9.4 Structure Materials and Properties

Machined aluminum, aluminum honeycomb and graphite epoxy laminates are commonly used structural materials (Fig. 9.7).

| Alloy | Density(kg/M³) | E (GPa) | γ | Ftu (MPa) | Fty (MPa) | CTE (in/in/ °C) |
|---|---|---|---|---|---|---|
| 6061-T6 aluminum | 2770 | 68.3 | 0.33 | 289 | 241 | 2.25E-05 |
| Beryllium | 1850 | 241.0 | 0.08 | 342 | 211 | 1.15E-05 |
| Invar | 8069 | 141.0 | 0.23 | 518 | 276 | 8.90E-06 |
| 6AL-4V Titanium | 4432 | 110.3 | 0.31 | 896 | 813 | 8.60E-06 |
| A-286 Steel (fasteners) | 8310 | 206.8 | 0.31 | 1100 | 827 | 1.65E-05 |

**Fig. 9.7** Typical metal material properties

Composite materials have unique structural properties that can be tailored for specific applications. While there are some very specialized composites for use in extreme environments, this discussion will focus on laminated composites. The two most common structural composites used on spacecraft are fiberglass and graphite/epoxy (Gr/Ep). Both composites are assembled from fibers and an epoxy matrix. Fiberglass has less strength than Gr/Ep, but its cost is relatively low and it is commonly used to provide thermal isolation. Gr/Ep structural composites come in two configurations: fabric and tape. The fabric versions use a graphite fabric, impregnated with epoxy. While not quite isotropic, their strengths in both in-plane axes are similar. In the tape versions, the ply fibers run in one direction and they are impregnated with epoxy. They are orthotropic and both their strength and Coefficient of Thermal Expansion (CTE) values vary dramatically with orientation. Along their fiber axis, tape composites have extremely high strength and stiffness and have very low (sometimes even negative) CTE. Along axes normal to their fiber direction, where only the epoxy matrix carries the load, the plies are relatively weak and have a higher CTE. Because of the weakness in the tape off axis direction, tape laminates usually feature multiple plies at different angles to tailor the required strength for each axis. This tailoring requires that the laminate be custom made, which is expensive.

The fact that the GrEp fibers have a negative CTE and the epoxy matrix has a positive CTE means that it is possible to design a composite laminate with near zero CTE. This property is used extensively in on-orbit optical applications where pointing must be maintained even with thermal gradients across the structure (Fig. 9.8).

9.5 Fasteners

| Property | Units | Std CF Fabric | HMCF Fabric | E Glass Fabric | Kevlar Fabric | Std CF UND | HMCF UND | M55 UND | E Glass UND | Kevlar UND |
|---|---|---|---|---|---|---|---|---|---|---|
| Young's Modulus 0° | GPa | 70 | 85 | 25 | 30 | 135 | 175 | 300 | 40 | 75 |
| Young's Modulus 90° | GPa | 70 | 85 | 25 | 30 | 10 | 8 | 12 | 8 | 6 |
| In-Plane Shear Modulus | GPa | 5 | 5 | 4 | 5 | 5 | 5 | 5 | 4 | 2 |
| Mayor Poisson's Ratio |  | 0.1 | 0.1 | 0.2 | 0.2 | 0.3 | 0.3 | 0.3 | 0.25 | 0.34 |
| Ultimate Tensile Strength 0° | MPa | 600 | 350 | 440 | 480 | 1500 | 1000 | 1600 | 1000 | 1300 |
| Ultimate Compressive Strength 0° | MPa | 570 | 150 | 425 | 190 | 1200 | 850 | 1300 | 600 | 280 |
| Ultimate Tensile Strength 90° | MPa | 600 | 350 | 440 | 480 | 50 | 40 | 50 | 30 | 30 |
| Ultimate Compressive Strength 90° | MPa | 570 | 150 | 425 | 190 | 250 | 200 | 250 | 110 | 140 |
| Ultimate In-Plane Shear Strength | MPa | 90 | 35 | 40 | 50 | 70 | 60 | 75 | 40 | 60 |
| Ultimate Tensile Strain 0° | % | 0.85 | 0.4 | 1.75 | 1.6 | 1.05 | 0.55 | - | 2.5 | 1.7 |
| Ultimate Compressive Strain 0° | % | 0.8 | 0.15 | 1.7 | 0.6 | 0.85 | 0.45 | - | 1.5 | 0.35 |
| Ultimate Tensile Strain 90° | % | 0.85 | 0.4 | 1.75 | 1.6 | 0.5 | 0.5 | - | 0.35 | 0.5 |
| Ultimate Compressive Strain 90° | % | 0.8 | 0.15 | 1.7 | 0.6 | 2.5 | 2.5 | - | 1.35 | 2.3 |
| Ultimate In-Plane Shear Strain | % | 1.8 | 0.7 | 1 | 1 | 1.4 | 1.2 | - | 1 | 3 |
| Thermal Expansion Coefficient 0° | M/M/°C | 2.1 | 1.1 | 11.5 | 7.4 | -0.3 | -0.3 | -0.3 | 6 | 4 |
| Thermal Expansion Coefficient 90° | M/M/°C | 2.1 | 1.1 | 11.5 | 7.4 | 28 | 25 | 28 | 35 | 40 |

Std CF: Standard Modulus Carbon Fiber
HMCF: High Modulus Carbon Fiber
UND: UniDirectional

**Fig. 9.8** Typical composite material properties

## 9.5 Fasteners

Typical English fastener sizes are given in Fig. 9.9.

| SIZE | | Coarse Threads (UNC) | | | Fine Threads (UNF) | | |
|---|---|---|---|---|---|---|---|
| Size | Major Diameter (in) | Threads per inch | Tensile Stress Area (in2) | Minor Diameter Area (in2) | Threads per inch | Tensile Stress Area (in2) | Minor Diameter Area (in2) |
| 0 | 0.0600 | - | - | - | 80 | 0.001 | 0.001 |
| 1 | 0.0730 | 64 | 0.002 | 0.002 | 72 | 0.002 | 0.002 |
| 2 | 0.0860 | 56 | 0.003 | 0.003 | 64 | 0.003 | 0.003 |
| 3 | 0.0990 | 48 | 0.004 | 0.004 | 56 | 0.005 | 0.004 |
| 4 | 0.1120 | 40 | 0.006 | 0.005 | 48 | 0.006 | 0.005 |
| 5 | 0.1250 | 40 | 0.007 | 0.006 | 44 | 0.008 | 0.007 |
| 6 | 0.1380 | 32 | 0.009 | 0.007 | 40 | 0.010 | 0.008 |
| 8 | 0.1640 | 32 | 0.014 | 0.011 | 36 | 0.014 | 0.012 |
| 10 | 0.1900 | 24 | 0.017 | 0.014 | 32 | 0.020 | 0.017 |
| 12 | 0.2160 | 24 | 0.024 | 0.02 | 28 | 0.025 | 0.022 |
| 1/4 | 0.2500 | 20 | 0.031 | 0.026 | 28 | 0.036 | 0.032 |
| 5/16 | 0.3125 | 18 | 0.052 | 0.045 | 24 | 0.050 | 0.052 |
| 3/8 | 0.3750 | 18 | 0.077 | 0.067 | 24 | 0.087 | 0.080 |
| 7/16 | 0.4375 | 14 | 0.106 | 0.093 | 20 | 0.118 | 0.109 |
| 1/2 | 0.5000 | 13 | 0.141 | 0.125 | 20 | 0.159 | 0.148 |
| 9/16 | 0.5625 | 12 | 0.182 | 0.162 | 18 | 0.203 | 0.189 |
| 5/8 | 0.6250 | 11 | 0.226 | 0.202 | 18 | 0.256 | 0.240 |
| 3/4 | 0.7500 | 10 | 0.334 | 0.302 | 16 | 0.373 | 0.351 |
| 7/8 | 0.8750 | 9 | 0.462 | 0.419 | 14 | 0.509 | 0.480 |
| 1.0 | 1.0000 | 8 | 0.606 | 0.551 | 12 | 0.663 | 0.625 |

**Fig. 9.9** English screw sizes

Note: Tension areas are calculated taking plastic deformation at the thread section into account, which makes them larger than the minimum thread diameter areas. The tensile areas should be used for calculating tensile stresses. If the threads are in shear, the minor diameter areas should be used when calculating shear stresses.

Note: If a fastener with an unthreaded section in shear is used, the diameter of the unthreaded section should be used for shear stress calculation.

## 9.6 Factors of Safety

In a spacecraft design, the required factors of safety are based on whether the structure is strength tested or qualified by analysis only. If the added mass incurred by using "No-test" factors is acceptable, time and cost can be saved by not running a structural test program.

Structures that include composite laminates usually require a structural test. This is because the strengths of composite laminates vary significantly, based on the quality of their manufacture. If multiple quantities of a composite laminate are produced, each unit, most likely, requires testing for the same reason.

Typical factors of safety used in the design are given in Fig. 9.10. (different values may be specified by different projects.)

| Type | Env | Remarks | Static/Static | | Random/Acoustic | |
|---|---|---|---|---|---|---|
| | | | Yield | Ultimate | Yield | Ultimate |
| Paletalic | All | Tested | 1.25 | 1.4 | 1.6 | 1.8 |
| Paletalic | All | No Test | 2 | 2.6 | - | - |
| Stability | Launch | Tested | - | 1.4 | - | 1.8 |
| Thermal Stress | On Orbit | Tested | 1.25 | 1.5 | - | - |
| Thermal Stress | On Orbit | No Test | 2 | 2.6 | - | - |
| EVA Inadvertent Kick Loads | On Orbit | No Test | 2.25 | 2 | - | - |
| Nominal EVA Operation | On Orbit | No Test | 2.25 | 2 | - | - |
| EVA Impact Load Failure Mode | On Orbit | No Test | - | 1 | - | - |
| Fail Safe | - | No Test | - | 1 | - | - |

(1) For analysis-only approach, all MS for Safety Critical Structure (excluding fasteners) are >0.15 Fastener MS must be positive, but may be less than 0.15 due to use of preload in the MS calulation.
(2) Factors of Safety for glass and structure glass bonds specified in NASA-STD-5001.
(3) Factors shown should be applied to statistically derived peak response based on RMS value. As a minimum, the peak response should be calculated as a 3-sigma value.
(4) Factors shown assume that qualification/protoflight testing is performed at acceptance level +3 dB If the difference between acceptance and qualification levels is less than 3dB, then the above factor may be applied to qualification level -3dB.
(5) Factor of Safety for pressurized systems from NASA-STD-5020. Factors of safety for pressurized systems to be compliant with AF5PCMAN 91-710 (Range Safety)
(6) Lines less than 1.5" outer diameter

**Fig. 9.10** Typical factors of safety table

Static load test on metallic structures: The main issue is whether a structural test will be performed (allowing use of lower factors of safety). A structural test may be performed using a static pull test (simplest and cheapest), but that may not test entire structure. A sine-burst vibration test (next cheapest) tests the entire structure. A centrifuge test (may require multiple test cycles at various orientations), is the most expensive.

Random/Acoustic: If using the acceptance level loads, random/acoustic factors of 1.6 yield and 1.8 ultimate must be used. If qualification loads (acceptance +3 dB) are used, factors of 1.25 yield and 1.4 ultimate should be used for margin of safety calculations.

**Stability:** The factor ensures that the structure does not buckle below the ultimate stress levels.

**Thermal Stresses:** a thermal vacuum test allows use of lower tested factors.

**EVA kick Loads and nominal EVR operations:** Typically not tested; therefore they carry higher no-test factors.

**EVR Impact:** This is considered an unplanned failure mode. Therefore; the structure may be damaged (hence no yield requirement) but must remain intact.

**Fail-Safe:** This is typically used for bolted joints. It must be shown that the joint will remain intact and carry the full load when the highest loaded fastener has failed.

## 9.7 Structural Analyses

### 9.7.1 *Structural Analysis Overview*

The purposes of the Structural Analyses are:

- To determine if the spacecraft structure is strong enough to withstand vibrational, acoustic and static loads during launch to orbit
- To determine if the deformation of the structure during ascent to orbit is small enough to meet structure and payload deformation criteria
- To determine if the resonant frequencies are high enough to meet launch vehicle requirements
- To determine the critical stress levels in the structure, and where they are located
- To determine where the structure needs to be strengthened and where it could be made lighter
- To determine the equivalent G loads induced by random and acoustic loads.
- To determine whether the stiffness (modes) and thermal distortion are acceptable for on-orbit instrument performance

The main tool for performing the above analyses is the Finite Element Model (FEM). It is a mathematical model constructed from CAD drawings of the spacecraft, its electronics and payload. It is used to provide outputs that answer the above questions. Methods of constructing the FEM will be discussed later.

After the FEM is constructed, the analysis sequence is:

1. Run the FEM model to determine the resonant modes (solution of an Eigenvalue problem). This results in the resonant mode frequencies and the amount of the spacecraft mass that participates in each mode.
2. Run the Static Loads to determine whether the structure has the required strength.
3. Conduct a Sine vibration analysis to simulate the low frequency/high displacement response of the spacecraft structure during launch. The excitation is applied by driving the model at the spacecraft-launch vehicle interface in each axis separately.
4. Run a random vibration analysis, with the excitation at the spacecraft-launch vehicle interface, to simulate the higher frequency responses during launch. In this analysis the spacecraft is exposed to all the vibrations that it will experience during launch. If different modes couple to each other to create large deflections, this analysis will produce these modes. One of the purposes of the random vibration test is to obtain accelerations over the entire frequency spectrum and then calculate from these an effective static acceleration that can be used in further analyses. This analysis is run in each axis separately.

## *9.7.2 Structural Analysis Steps in Detail*

The analysis is performed using a Finite Element Model (FEM), based on the spacecraft's CAD model. The FEM can be created automatically from the CAD model or the geometry can be entered manually using points, lines and solids. This geometry can then be used to create structural elements.

The choice of how to construct the FEM model depends on the maturity of the mechanical design. If it is the final configuration, the fastest procedure may be to create a model by auto-meshing the geometry file. However, if the design is preliminary, and changes are expected, a finite element model (FEM) should be constructed that can easily accommodate changes. Using plate and bar elements allows making changes to the FEM quickly, permitting rapid design iteration.

1. **Creation of Finite Element Model (FEM)**

A finite Element Model is a mathematical simulation of the structure. It is composed of a stiffness and a mass matrix which can be used to calculate the natural frequencies, distribution of loads, and stresses within the structure. Software from many sources is available today for creating, solving, and post processing finite element model results. Their prices and capabilities vary significantly. Some capabilities include:

1. Graphic model creation (creating/meshing FEM from CAD models)
2. Frequency analysis (eigenvalues)

## 9.7 Structural Analyses

3. Static Load stress analysis
4. Sine Load stress analysis
5. Random vibration base-drive analysis
6. Design optimization
7. Graphic post processing of results

Three steps are involved in performing a FEM analysis:

1. Create a model and use it to create a data deck file. (A data deck file is an ASCII text file which includes model information and a case control section to tell the computer what analysis to run)

2. Solve the data deck file (the stiffness and mass matrices in the deck are processed for the desired solution)

3. Take the analysis results and map them back to the model to display the results (examples: animated plot, deformed plot, stress contour plot).

The FEM can also be created by hand or from a CAD model of the design. Most programs (such as FEMAP) can create a solid element FEM from a solid CAD model. The FEM program will assign a default mesh size, based upon the geometry of the CAD model being meshed. There are a few considerations to note when using an auto-meshed FEM:

If the FEM model has a very fine mesh, it may take a long time to run analysis solutions. As personal computers have become faster and more powerful, this has become less of a concern for static runs, but dynamic solutions can still end up taking many hours. Most auto-meshing routines allow one to override the default mesh size for a component. In order to reduce the size and run times of the FEM, input a larger mesh size than the default size. Most models take only seconds to mesh. If the input mesh size is too large, the software will not be able to mesh the CAD geometry. By changing the mesh size input, one can quickly iterate to the largest mesh size that can be processed, thereby reducing the size of the FEM.

One major weakness of auto-meshed solid models is that they are difficult to edit. Every time the design changes, a new model must be created. Even though auto-meshing the structure is fast, the fastener elements and boundaries must still be recreated. If a final design is analyzed, this may be less of an issue.

A FEM can also be created by importing the CAD geometry into the FEM software and creating a structural model on top of the imported geometry. While this method takes longer than auto-meshing, it gives the analyst complete control over the size of the elements. They can be tailored to have a finer mesh in critical areas to calculate stress gradients, and a larger mesh in less critical areas, to minimize the size of the model. Also, the hand meshed model can be easily modified when making changes to the design.

An initial CAD model is shown at the left, Auto-meshed and Hand-meshed models are shown at the right in Fig. 9.11.

Regardless of the software used, the steps used in creating a model are similar, and the sequence is intuitive.

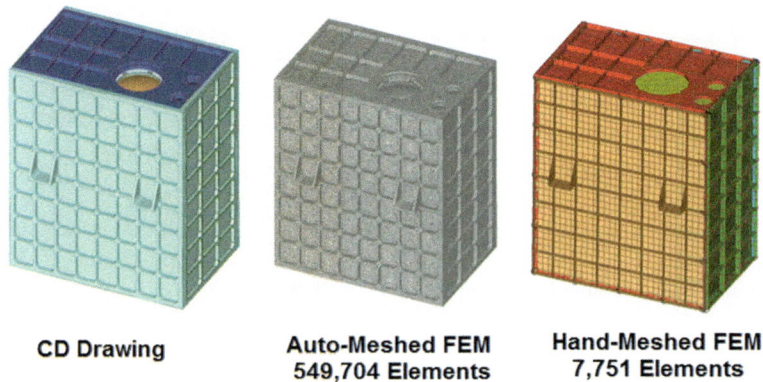

**Fig. 9.11** Model creation method comparison

## 2. Defining the Element Geometry

When creating a model by hand, as shown in Fig. 9.12, the geometry is created using points and lines that can be manipulated (extruded, revolved, rotated, copied) to create the elements.

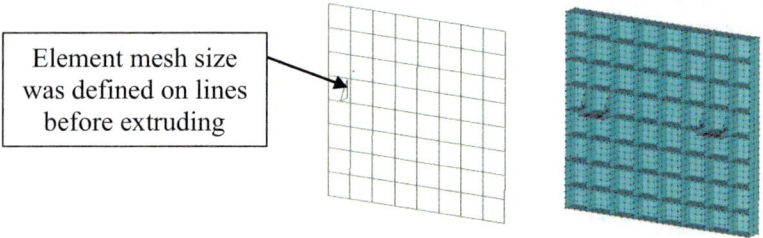

**Fig. 9.12** Manually input elements and elements created from geometry

## 3. Defining Material and Element properties

The properties of each material are entered. A separate material card is used for each material used (the software may have material databases that can be loaded into material cards). Figures 9.13 and 9.14 illustrate entering element properties and materials into the model.

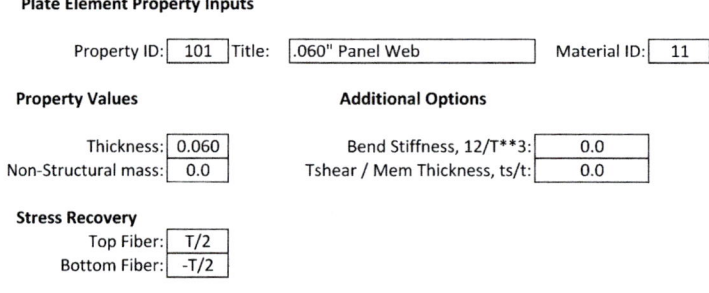

**Fig. 9.13** Example of element property input

## 9.7 Structural Analyses

**Isotropic Material Property Inputs**

Material ID: 11  Title: 6061-T6 Aluminum

**Stiffness**

Young's Modulus, E: 9900000
Shear Modulus, G: 4000000
Poisson's Ratio: 0.3

**Limit Stress**

Tension: 35000
Compression: -35000
Shear: 27000

**Thermal**

CTE: 1.25E-04
Conductivity:
Specific Heat, Cp:

Mass Density: 0.100

**Fig. 9.14** Example of material property input

### 4. Elements to Model Joints and Attachments

Three types of elements are used to attach components and may be used to calculate loads for later joint analysis. The first two are rigid elements (RBE2 & RBE3). An RBE2 rigidly attaches dependent and independent nodes. In an RBE3, the dependent node's displacement is the average of the motions of the independent nodes, without rigidly attaching the independent nodes to each other. The third element type for attaching nodes is a CBUSH element. Figure 9.15 illustrates the FEM model displays when creating CBUSH elements.

**Spring Damper Element Property Inputs**

ID: 12  Title: X-Axis bolts
Type: CBUSH  Orientation Cord. Sys: 0: Basic

**Property Values**

| DOF | Stiffness | Damping | Structural Damping |
|-----|-----------|---------|--------------------|
| 1   | 1.00E+07  | 0.0     | 0.0                |
| 2   | 1.00E+07  | 0.0     | 0.0                |
| 3   | 1.00E+07  | 0.0     | 0.0                |
| 4   | 1.00E+07  | 0.0     | 0.0                |
| 5   | 1.00E+07  | 0.0     | 0.0                |
| 6   | 1.00E+07  | 0.0     | 0.0                |

**Spring Damper Element Inputs**

ID: 106  Property: 12
Nodes: 1034  2501

Orientation: From Property

**Fig. 9.15** Example of CBUSH property and element cards

## 5. Adding Masses and Non-Structural Mass

Two methods are used for attaching a mass to a FEM. For a simple electronics box, the box can be modeled as a concentrated mass, using a rigid element to attach it to its mounting bolt locations. In this case, the RBE3 should be used. An RBE2 would rigidize all the mounting bolt locations, thereby reinforcing them by creating a false load path. Using an RBE3 attaches the box's mass, but does not reinforce the structure. This is illustrated in Fig. 9.16.

Mass can also be added to a model as Non-Structural Mass. In a solar array, the mass of the solar cells and wiring can be smeared over the array's surface by inputting NSM in the property card.

**Mass Element Property Inputs**

| | | | | | |
|---|---|---|---|---|---|
| **ID:** | 27 | **Title:** | Reaction Wheel | | |
| **Orientation Cord. Sys:** | | 0: Basic Rectangular | | | |

**Property Values**

| | | | | | |
|---|---|---|---|---|---|
| Mass, M or Mx | 4.23 | Inertia, Ixx | 472.0 | Ixy | 87.0 |
| My, (blank = Mx) | | Iyy | 472.0 | Iyz | 56.0 |
| Mz, (blank = Mx) | | Izz | 125.0 | Izx | 92.0 |

**Rigid Element Inputs**     **Type:** RBE3

**Dependent Node:** 1205

| **Degrees of Freedom:** | Tx | X | Rx | |
| --- | --- | --- | --- | --- |
| | Ty | X | Ry | |
| | Tz | X | Rz | |

| **Independent Nodes:** | 287 |
|---|---|
| (nodes to average) | 205 |
| | 321 |
| | 541 |
| | 650 |
| | 712 |
| | 901 |

**Fig. 9.16** Example of mass element and RBE3 rigid element cards

## 9.7 Structural Analyses

### 6. Defining Boundary Constraints

The typical model is constrained using Single Point Constraints (SPC's) to define which model nodes are constrained. In Fig. 9.17 the model is constrained at four locations (two on the front panel and two on the rear panel) (Figs. 9.18 and 9.19)

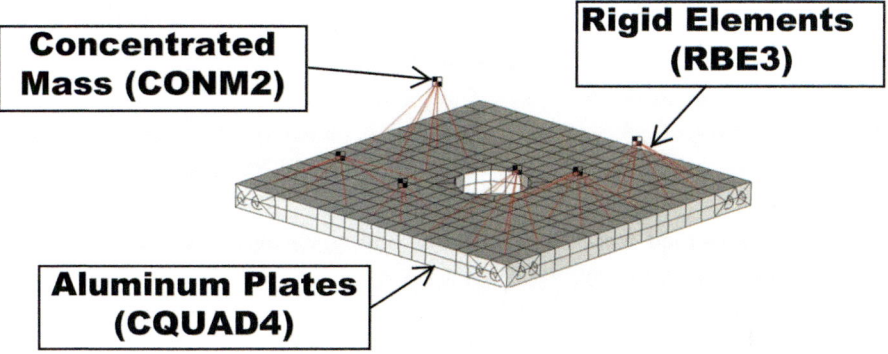

**Fig. 9.17** Machined aluminum deck with electronics components modeled using concentrated masses and RBE3 rigid element attachments

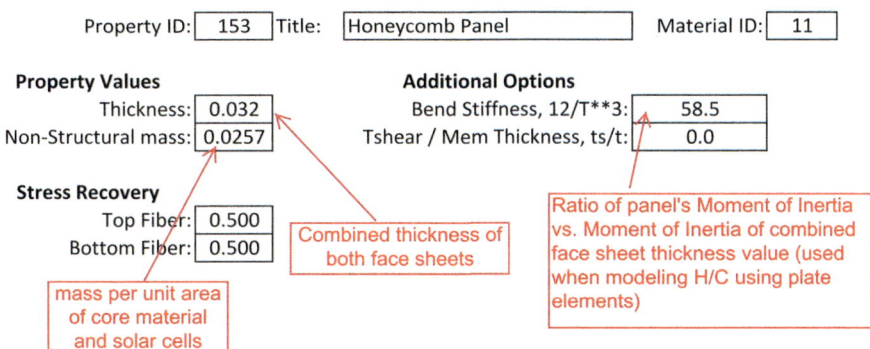

**Fig. 9.18** Plate property card used to define a honeycomb solar array panel (with aluminum face sheets)

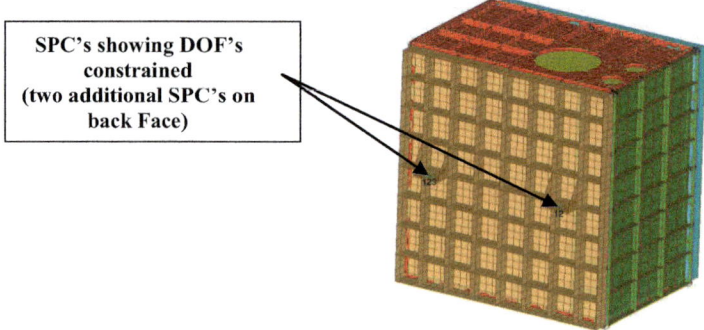

**Fig. 9.19** Example of single point constraints

While this method works well for static and eigenvalue solutions, it does not work in sine or random analyses solutions where the model must be driven from a single node. For sine and random analysis, an RBE2 rigid element can be used to attach the model's interface mounting locations to a single node. In this case, all the interface locations should be rigid in relation to each other so an RBE2 is used instead of an RBE3 (Fig. 9.20).

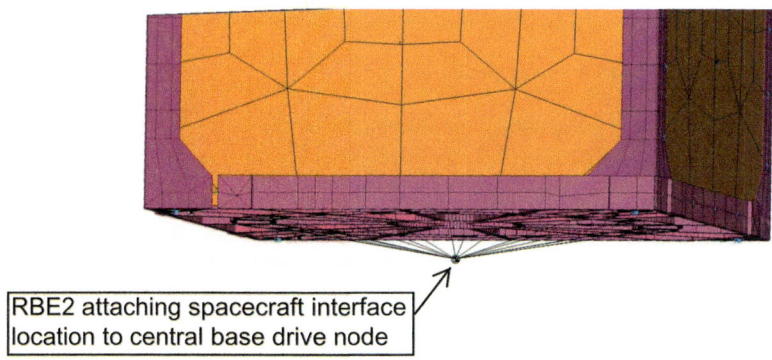

**Fig. 9.20** Attaching spacecraft mounting interface locations to a single node for analyses, requiring a base drive (sine and random)

### 7. Lowest Resonant Frequency (Eigenvalue Analysis) Requirement

Once the FEM is created, the eigenvalue (modal) solution is usually run first. The eigenvalue run calculates the natural frequency modes of the FEM. It is also useful as a model check. If the first vibrational mode of the structure is very low (less than, say, 0.001 Hz), it is probably caused by one of two factors: the model may not be properly constrained at its boundaries, or an element is not properly attached. This will become very evident when looking at the animated or deformed mode shapes. If it looks like the entire model translates wildly when animated, it is probably an issue with the constraints. If it looks like a specific component is moving

## 9.7 Structural Analyses

wildly, the component is probably not attached properly. An example of the deformed shape of the structure is shown in Fig. 9.21. The modal effective weights of each mode can also be calculated during an eigenvalue run. The modal effective weight summary lists the frequency of each mode and how much of the total model's mass is participating. It is common practice to consider any mode where more than 5% of the model's mass participates as a primary mode (Fig. 9.22).

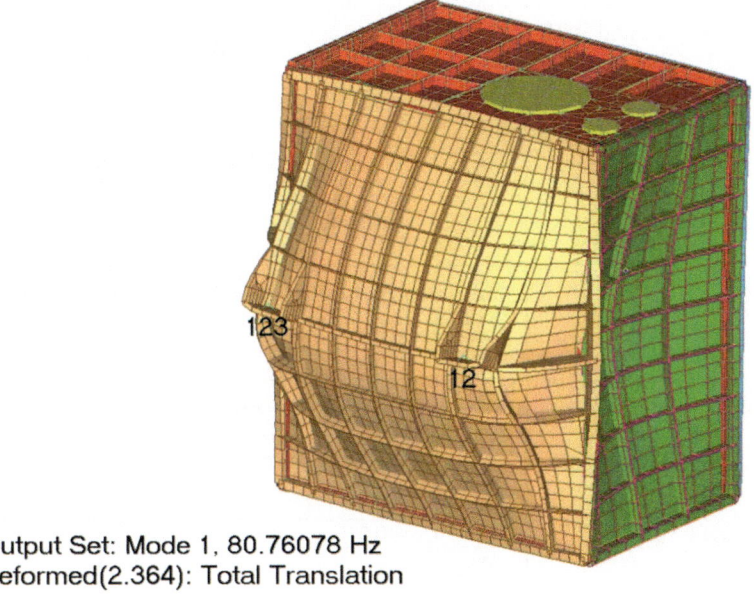

Output Set: Mode 1, 80.76078 Hz
Deformed(2.364): Total Translation

**Fig. 9.21** Vibrational mode exhibiting the lowest resonant frequency

| MODE | FREQUENCY | X | Y | Z | Rx | Ry | Rz |
|---|---|---|---|---|---|---|---|
| 1 | 80.76 | 0% | 97% | 0% | 38% | 0% | 11% |
| 2 | 158.34 | 5% | 0% | 0% | 0% | 42% | 1% |
| 3 | 197.20 | 3% | 0% | 0% | 0% | 2% | 77% |
| 4 | 210.56 | 0% | 0% | 51% | 1% | 1% | 0% |
| 5 | 235.21 | 81% | 0% | 0% | 0% | 45% | 5% |
| 6 | 284.67 | 0% | 0% | 32% | 48% | 3% | 0% |
| 7 | 327.00 | 0% | 0% | 0% | 0% | 0% | 0% |
| 8 | 371.23 | 0% | 0% | 0% | 0% | 0% | 0% |
| 9 | 399.26 | 0% | 0% | 0% | 0% | 0% | 2% |
| 10 | 402.73 | 6% | 0% | 0% | 0% | 2% | 1% |

Modes with >5% participation

**Fig. 9.22** Modal effective weight table for first ten modes

At this point, the FEM has not been correlated to test data. While it should be relatively accurate at low frequencies, it may be significantly inaccurate above 500 Hz.

## 8. Static Analysis

Once an eigenvalue solution has run successfully, running static solutions is the next step. The elements and constraints for the static run are the same as those for the eigenvalue run. Most FEM software support the creation of a variety of applied static loads. Examples of body, nodal and elemental load types are shown in Figs. 9.23, 9.24 and 9.25.

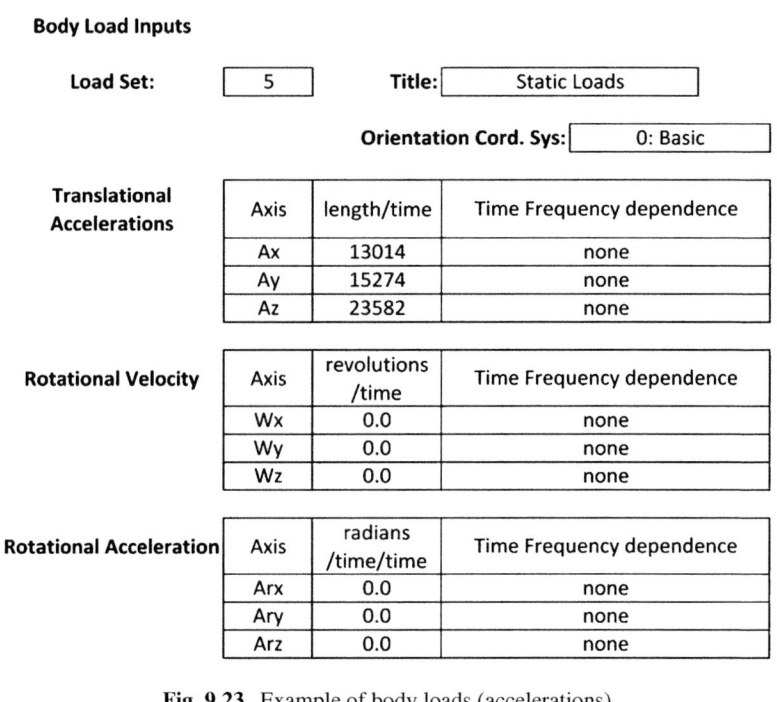

Fig. 9.23 Example of body loads (accelerations)

Fig. 9.24 Example of nodal load types

## 9.7 Structural Analyses

**Elemental Load Input**

| Load Set: | 6 |
|---|---|
| Element | 732 |

| Direction: (choose one) | |
|---|---|
| Normal to Element Face | X |
| Vector | |
| Along Curve | |
| Normal to Plane | |
| Normal to Surface | |

| Pressure: | 14.7 |
|---|---|
| Plane: | 1 |

| Title: | 14.7 psi Pressure (1 atm.) |
|---|---|

| Load Type: (choose one) | |
|---|---|
| Distributed | X |
| Pressure | |
| temperature | |

| Method: (choose one) | |
|---|---|
| Constant | X |
| Variable | |
| Data Surface | |

**Fig. 9.25** Example of elemental load types

Once the loads and constraints have been applied, a static solution can be run to calculate, loads, displacements and stresses. The stresses can then be used to calculate margins of safety (Fig. 9.26).

Output Set: NX NASTRAN Case 1
Elemental Contour: Plate Top VonMises Stress
Second Contour: Solid Von Mises Stress

**Fig. 9.26** Example of stress contour plot

## 9. Sine Analysis

Sine analysis is a base-driven dynamic load case. It may or may not be required by a launch provider. It simulates the low frequency/high displacement portion of the launch environment. While the input levels may seem low, if the frequency range includes any of the spacecraft primary modes, the response levels may be significant, depending upon the load amplification of the design. The results from a sine analysis are response load/displacements versus frequency. The response load values can be applied to the spacecraft FEM as a static load to calculate stresses and deflections.

One of the reasons for running a sine test on a spacecraft is that it simulates the deflections of the structure at the low frequency/high displacement range. There have been cases where restraint systems and latches have actually deployed during sine tests due to excessive manufacturing tolerances (Fig. 9.27).

```
INIT MASTER(S)
NASTRAN SYSTEM(319)=1
ID Sine Deck
SOL SEMFREQ
TIME 10000
CEND
  TITLE = basedrive_sine_ax
  ECHO = NONE
  SET 1 = 3,1681,5221,5230,5492,5501,5506,38773,40175,
  DISPLACEMENT(SORT1,PUNCH,PHASE) = 1
  ACCELERATION(SORT1,PUNCH,PHASE) = 1
  SPC = 20001
  DLOAD = 1
  METHOD = 1
  SDAMPING = 6
  FREQUENCY = 4
BEGIN BULK
$
PARAM,POST,-1
PARAM,OGEOM,NO
PARAM,AUTOSPC,YES
PARAM,MAXRATIO,1.+8
PARAM,GRDPNT,0
PARAM,WTMASS,.002588
EIGRL     1              10    0              MASS
$ Femap with NX Nastran Load Set 4 : basedrive_sine_ax
PARAM,RESVEC,YES
$ Femap with NX Nastran Function 6 : Sine Damping Function
TABDMP1   6    CRIT                           +
+      0.   .05     1.    .05ENDT  ◄——[0.05 Corresponds to Q=20 damping]
PARAM,HFREQ,50.
$ Femap with NX Nastran Function 3 : Loading Function
TABLED2   3    0.                             +
+      0.   1.     1.    1.ENDT
RLOAD2    102    101              3       ACCE
SPCD      101    1    1    1.
DLOAD     1    1.    1.    102
$    2    3    4    5    6    7    8    9
FREQ2     4    5.    50.    80  ◄——[5-50 Hz Frequency Range]
```

**Fig. 9.27** Example of sine solution NASTRAN data deck

## 10. Random Vibration Analysis

Random analysis is also a base-driven dynamic load case. It is used to simulate the higher frequency portion of the launch environment. Random analysis is the only analysis where the spacecraft is exposed to all its modal frequencies at the same time. Modes may couple with each other, creating very high responses. The purpose of the random vibe analysis is to calculate an effective static acceleration and to assess the modal responses over the entire launch frequency spectrum to exhibit modes with high responses (Fig. 9.28).

```
INIT MASTER(S)
NASTRAN SYSTEM(442)=-1, SYSTEM(319)=1
ID Random Deck
SOL 111
TIME 10000
CEND
  TITLE = basedrive_random_ax
  ECHO = NONE
  SET 1 = 3,1681,5221,5230
  DISPLACEMENT(SORT1,PUNCH,PHASE,RALL) = 1
  ACCELERATION(SORT1,PUNCH,PHASE,RALL) = 1
  SPC = 20001
  METHOD = 1
  DLOAD = 1
  SDAMPING = 4
  RANDOM = 200
  Freq = 40
OUTPUT(XYPLOT)
  XYPUNCH ACCE PSDF/
  3(T1),3(T2),3(T3)/
  1681(T1),1681(T2),1681(T3)/
  5221(T1),5221(T2),5221(T3)/
  5230(T1),5230(T2),5230(T3)
BEGIN BULK
$
PARAM,POST,-1
PARAM,OGEOM,NO
PARAM,AUTOSPC,YES
PARAM,MAXRATIO,1.+8
PARAM,GRDPNT,0
PARAM,WTMASS,.002588
PARAM,RESVEC,YES
RANDPS    200    1    1   1.    0.    7
$ Femap with NX Nastran Function 7 : Random ASD (20+kg)
$
TABRND1    7   LOG   LOG                         +      ──10 Grms Input Spectrum
+      19.99  .013   50.   .08   800.   .08  2000.  .013+
+      ENDT
$
EIGRL  1    20.   2000.
$ Femap with NX Nastran Function 4 : Random Damping Function
TABDMP1    4   CRIT                              +
+      0.    .05   1.    .05ENDT◄──0.05 corresponds to Q=20 damping
PARAM,HFREQ,2000.
$ Femap with NX Nastran Function 3 : Loading Function
$
TABLED1 3     LOG   LOG                          +
+      20.   1.0   2000.  1.0  ENDT
RLOAD2       102   101              3       ACCE
SPCD         101   5632    1    1.
DLOAD         1    1.      1.   102
FREQ2        40    20.    2000.  80◄──20-2,000 Hz Frequency Range
$ Femap with NX Nastran Constraint Set 20001 : Constraints
SPC       20001  5632 123456   0.
```

**Fig. 9.28** Example of random solution NASTRAN data deck

The results from the random run include a punch file and an F06 file. The punch file contains the response frequency vs. $G^2$/Hz data, and it can be graphed. The F06 file contains the $G_{RMS}$ values of the energy under the graphed punch file curve. Both values are 1-sigma levels. The accepted procedure is that the structure is analyzed using 3-sigma $G_{RMS}$ levels as a static load. The 3-sigma accelerations in each axis are applied simultaneously to calculate stresses. Figure 9.29 illustrates a 1-sigma response for a 10 $G_{RMS}$ input

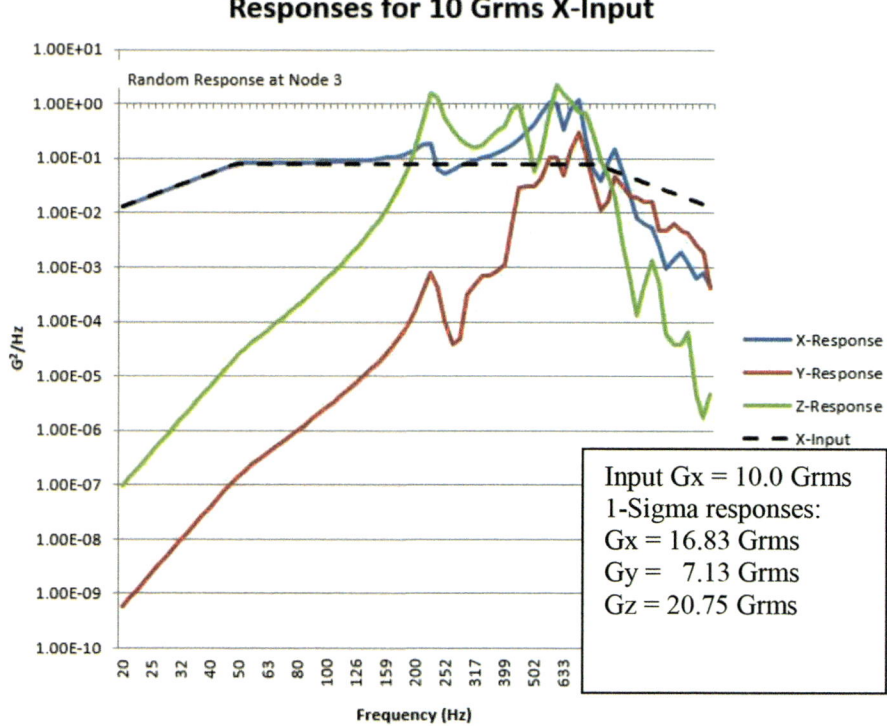

**Fig. 9.29** Example of 1-sigma response curves for 10.0 grms acceptance level X-Input

From the response shown in the table above, the component at node 3 should be analyzed, using a 3-Sigma static acceleration load of:

$$Gx = 3 \times 16.83 = 50.49 \text{ G's}$$

$$Gy = 3 \times 7.13 = 21.39 \text{ G's}$$

$$Gz = 3 \times 20.75 = 62.25 \text{ G's}$$

9.7 Structural Analyses

When choosing node locations to calculate responses, take into account the effective mass at the location. For example, a location at the center of a web will resonate at its own natural frequency at some point over the random analysis frequency range. This resonance may create very high response levels, but the active mass of the web is negligible, and the stresses induced by the resonance are insignificant.

11. **Fastener Analysis**

Fastener loads can be calculated easily by the FEM using rigid elements (RBE2, RBE3) and scaler spring (CBUSH) elements. When using rigid elements, requesting a grid point force balance (GPFB) at the constrained node will produce a listing of all the forces acting on the node. The MPCFORCE load will be the load from the RBE. For scaler spring elements, requesting the element forces will calculate the forces in the CBUSH elements.

Once the forces are calculated, fastener margins are usually calculated by hand. Fastener shear and tensile stresses can be calculated using the fastener area values shown in Fig. 9.9. From these stresses, principal stresses can be calculated and used to produce margins of safety. Margins of safety are calculated for tension (yield and ultimate) as well as shear (ultimate).

Fastener margins can also be calculated taking the bolted joint stiffness into account. When the fastener is tightened (torqued), it is normally preloaded to approximately 65% of yield stress. This preload creates compression in the bolted joint. When the joint is loaded, part of the load is carried by the fastener and part is carried by the preloaded joint components. There are spreadsheets available on the web for calculating bolted joint margins based on joint stiffness.

Although bearing stress margins are not presented as commonly as fastener margins, understanding bearing stress is important when analyzing fastener patterns. Bearing stress is calculated by dividing the fastener shear load by the area of the hole that it is installed in.

$$\text{Bearing stress} = f_{bearing} = P/(d^*t) \text{ where P is the fastener shear load,}$$

d is its diameter and t is the thickness on the thin side of the joint

When the hole at the highest loaded fastener in a fastener pattern yields in bearing, the hole elongates, but continues to carry load. This yielding redistributes the joint load among the adjacent fasteners, thereby allowing the joint to carry a higher load before joint failure occurs.

Another mechanism for redistributing fastener loads is fastener yield, when the fasteners themselves yield and redistribute the load to adjacent fasteners in the pattern. It is always desirable to use a fastener material with a large range between the yield and ultimate stress allowables. If the yield allowable is close to the ultimate allowable, the fastener will be brittle. If the brittle fastener fails, its load is suddenly passed onto the adjacent fastener. This sudden load may cause the adjacent fastener to fail, beginning a chain reaction where the bolted joint unzips, causing catastrophic failure.

Two of the most commonly used fastener materials are high strength A286 steel and 300 series corrosion resistant steel (CRES):

>A286 high strength steel: Fty = 120,000 psi and Ftu = 160,000 psi
>300 Series CRES: Fty = 30,000 psi and Ftu = 80,000 psi

A-286 high strength steel comes in heat treats where Ftu is greater than 160,000 psi. Their yield strengths are closer to their ultimate strengths; however, they are more brittle and are not commonly used for flight structures.

## 9.8 Weight Estimate

Estimating and updating the weight of a spacecraft should be done regularly and accurately during development. It is essential to not exceed the spacecraft weight agreed to with the launch vehicle supplier, and to meet the CG location requirements in the XY plane. Typically, the $CG_{XY}$ should be within 0.25″ of the separation system axial centerline. A $CG_{XY}$ location error will cause spacecraft tipoff (rotation in the XZ and YZ planes).

The procedure for accurately estimating spacecraft weight, Center of Gravity and Pitch or Roll Moments of Inertia are outlined below. Create a spreadsheet:

- List all of the spacecraft parts and components
- State the weight of each
- Apply weight margins to each part
  - 2% if the part exists, was weighed or if a vendor specification of the part exists
  - 10% if the weight of the part was computed
  - 20% if the weight is estimated
- Compute the margined weight of each part
- Determine the XYZ location of the part CG in the spacecraft coordinate system
- Determine the component moments (weight*location) in all 3 coordinates about the origin of the spacecraft coordinate system
- Determine the spacecraft CG in each of the 3 coordinates
  (Sum the component moments in each coordinate Divide the 3 sums by the spacecraft margined weight)
- Compute the Pitch/Roll Moment of Inertia of the spacecraft
  - For each component compute its inertia, $\mathbf{W/g*(Z-Z_{CG})^2}$
  - Sum the inertias in each coordinate
  - Divide the three sums by the spacecraft margined mass.

This process is illustrated for a 20″ × 20″ × 36″ spacecraft consisting of (1) a Propulsion Module, (2) an Electronics Module and (3) a Payload Module, as shown in Fig. 9.30.

## 9.8 Weight Estimate

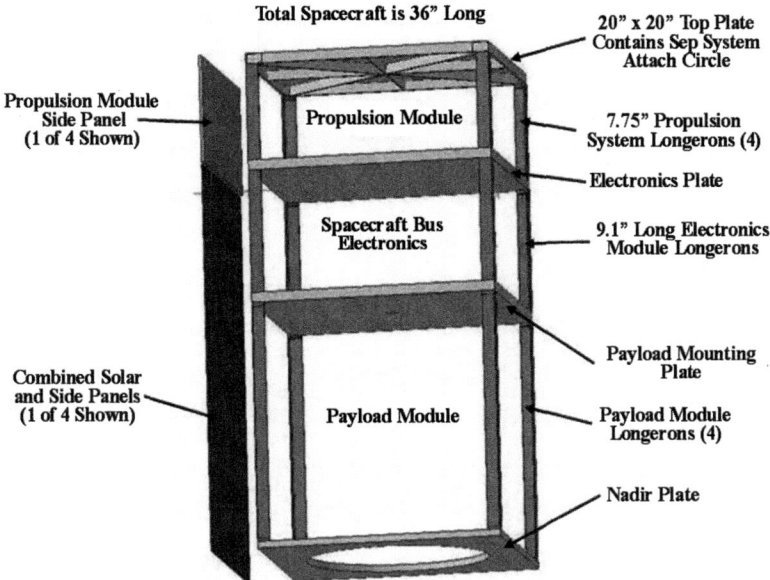

**Fig. 9.30** Spacecraft for which the weight statement is constructed

The spreadsheet implementing the process described above is shown in Fig. 9.31. The origin of the coordinate system is at the center of the Top Plate, and Z is positive going down

The spacecraft CG is 12.928" from the top. Its Moment of Inertia is 1.4299 slug-ft$^2$ (1.910 kg-m$^2$). In the XY plane, the CG is off the center by 0.389" in X and 0.379" in Y. Thus, the CG is 0.543" from the XY plane center. It does not meet the launch vehicle specifications that the CG should be within 0.25" of the center. For this reason, the locations of some of the spacecraft components should be moved in the XY plane.

The weight distribution of the spacecraft among its various subsystems is shown in Fig. 9.32.

The difference between $CG_Z$ and Center of Pressure is 5.072". This is large enough that at low altitudes (500 km) atmospheric drag will try to tip the spacecraft, and this has to be compensated by a more frequent unloading of built up momentum in the reaction wheels. At altitudes above 600 km, this effect is very small and probably negligible.

156  9 Spacecraft Structure

| Weight Statement for Example Spacecraft (20" x 36") | | | | | | | | | Z | | X | Y |
|---|---|---|---|---|---|---|---|---|---|---|---|---|
| Components and Parts | lbs | Meas | Calc | Est | lbs | X | Y | Z | Moment | m*(z-zo)² | Moment | Moment |
| Separation System (15" Lightband) | 1.150 | 1 | | | 1.173 | 0.000 | 0.000 | -0.500 | -0.0489 | 0.046 | 0 | 0 |
| Propulsion Module (8" high) | | | | | | | | | | | | |
| Propulsion Top Plate (0.76") | 6.302 | | 1 | | 6.932 | 0.000 | 0.000 | 0.375 | 0.2166 | 0.236 | 0 | 0 |
| 2 Fuel Tanks (4.58 dia, 12.3 lon | 6.000 | 1 | | | 6.120 | 0.000 | 0.000 | 3.040 | 1.5504 | 0.129 | 0 | 0 |
| Fuel for 2 Tanks | 3.300 | | 1 | | 3.630 | 0.000 | 0.000 | 3.040 | 0.9196 | 0.077 | 0 | 0 |
| Tank Saddles | 1.480 | | 1 | | 1.628 | 0.000 | 0.000 | 3.040 | 0.4124 | 0.034 | 0 | 0 |
| Fill Valve | 0.500 | | 1 | | 0.550 | 10.000 | 0.000 | 4.250 | 0.1948 | 0.009 | 5.500 | 0 |
| Pressure Reducer | 0.500 | | 1 | | 0.550 | 2.000 | 0.000 | 4.250 | 0.1948 | 0.009 | 1.100 | 0 |
| Pressure Sensor | 0.500 | | 1 | | 0.550 | 0.000 | 0.000 | 4.250 | 0.1948 | 0.009 | 0 | 0 |
| Shutoff Valve | 0.507 | | 1 | | 0.558 | -2.000 | 0.000 | 4.250 | 0.1975 | 0.009 | -1.115 | 0 |
| Thruster X1 | 0.050 | 1 | | | 0.051 | 9.500 | 9.500 | 0.000 | 0.0000 | 0.002 | 0.485 | 0.485 |
| Thruster X2 | 0.050 | 1 | | | 0.051 | 9.500 | -9.500 | 0.000 | 0.0000 | 0.002 | 0.485 | -0.485 |
| Thruster Y1 | 0.050 | 1 | | | 0.051 | -9.500 | 9.500 | 0.000 | 0.0000 | 0.002 | -0.485 | 0.485 |
| Thruster Y2 | 0.050 | 1 | | | 0.051 | -9.500 | -9.500 | 0.000 | 0.0000 | 0.002 | -0.485 | -0.485 |
| Propulsion Electronics | 0.500 | | | 1 | 0.600 | 2.000 | 0.000 | 5.000 | 0.2500 | 0.008 | 1.200 | 0 |
| Propulsion Plumbing | 2.500 | | | 1 | 3.000 | 0.000 | 0.000 | 6.000 | 1.5000 | 0.031 | 0 | 0 |
| Propulsion +X Side Panel | 0.480 | | 1 | | 0.528 | 10.000 | 0.000 | 4.000 | 0.1760 | 0.009 | 5.280 | 0 |
| Propulsion -X Side Panel | 0.480 | | 1 | | 0.528 | -10.000 | 0.000 | 4.000 | 0.1760 | 0.009 | -5.280 | 0 |
| Propulsion +Y Side Panel | 0.480 | | 1 | | 0.528 | 0.000 | 10.000 | 4.000 | 0.1760 | 0.009 | 0 | 5.280 |
| Propulsion -Y Side Panel | 0.480 | | 1 | | 0.528 | 0.000 | -10.000 | 4.000 | 0.1760 | 0.009 | 0 | -5.280 |
| Electronics Module Structure (10" high) | | | | | | | | | | | | |
| Electronics Plate (milled Al) | 3.060 | | 1 | | 3.366 | 0.000 | 0.000 | 8.250 | 2.3141 | 0.016 | 0 | 0 |
| Longeron 1 | 0.750 | | 1 | | 0.825 | 10.000 | -10.000 | 13.000 | 0.8938 | 0.000 | 8.250 | -8.250 |
| Longeron 2 | 0.750 | | 1 | | 0.825 | 10.000 | 10.000 | 13.000 | 0.8938 | 0.000 | 8.250 | 8.250 |
| Longeron 3 | 0.750 | | 1 | | 0.825 | -10.000 | -10.000 | 13.000 | 0.8938 | 0.000 | -8.250 | -8.250 |
| Longeron 4 | 0.750 | | 1 | | 0.825 | -10.000 | 10.000 | 13.000 | 0.8938 | 0.000 | -8.250 | 8.250 |
| Side Panel +X | 0.800 | | 1 | | 0.880 | 10.000 | 0.000 | 13.000 | 0.9533 | 0.000 | 8.800 | 0 |
| Side Panel +Y | 0.800 | | 1 | | 0.880 | -10.000 | 0.000 | 13.000 | 0.9533 | 0.000 | -8.800 | 0 |
| Side Panel -X | 0.800 | | 1 | | 0.880 | 0.000 | 10.000 | 13.000 | 0.9533 | 0.000 | 0 | 8.800 |
| Side Panel -Y | 0.800 | | 1 | | 0.880 | 0.000 | -10.000 | 13.000 | 0.9533 | 0.000 | 0 | -8.800 |
| Bottom (Pyld Mounting) Plate | 3.060 | | 1 | | 3.366 | 0.000 | 0.000 | 17.750 | 4.9789 | 0.017 | 0 | 0 |

Fig. 9.31 Weight statement

9.8 Weight Estimate

| Electronics Module Structure (10" high) | | | | | | | | | | |
|---|---|---|---|---|---|---|---|---|---|---|
| Electronics Plate (milled Al) | 3.060 | 1 | 3.366 | 0.000 | 0.000 | 8.250 | 2.3141 | 0.016 | 0 | 0 |
| Longeron 1 | 0.750 | 1 | 0.825 | 10.000 | -10.000 | 13.000 | 0.8938 | 0.000 | 8.250 | -8.250 |
| Longeron 2 | 0.750 | 1 | 0.825 | 10.000 | 10.000 | 13.000 | 0.8938 | 0.000 | 8.250 | 8.250 |
| Longeron 3 | 0.750 | 1 | 0.825 | -10.000 | 10.000 | 13.000 | 0.8938 | 0.000 | -8.250 | -8.250 |
| Longeron 4 | 0.750 | 1 | 0.825 | -10.000 | -10.000 | 13.000 | 0.8938 | 0.000 | -8.250 | 8.250 |
| Side Panel +X | 0.800 | 1 | 0.880 | 10.000 | 0.000 | 13.000 | 0.9533 | 0.000 | 8.800 | 0 |
| Side Panel +Y | 0.800 | 1 | 0.880 | -10.000 | 0.000 | 13.000 | 0.9533 | 0.000 | -8.800 | 0 |
| Side Panel -X | 0.800 | 1 | 0.880 | 0.000 | 10.000 | 13.000 | 0.9533 | 0.000 | 0 | 8.800 |
| Side Panel -Y | 0.800 | 1 | 0.880 | 0.000 | -10.000 | 13.000 | 0.9533 | 0.000 | 0 | -8.800 |
| Bottom (Pyld Mounting) Plate | 3.060 | 1 | 3.366 | 0.000 | 0.000 | 17.750 | 4.9789 | 0.017 | 0 | 0 |
| **Electronics** | | | | | | | | | | |
| **EPS** | | | | | | | | | | |
| EPS Assembly 2.590 | 2.790 | 1 | 3.069 | 4.000 | 4.000 | 10.000 | 2.5575 | 0.006 | 12.28 | 12.276 |
| Batteries and Housing | 1.800 | 1 | 1.980 | -4.000 | -4.000 | 10.000 | 1.6500 | 0.004 | -7.920 | -7.920 |
| Solar Panel on +X Side | 1.037 | 1 | 1.141 | 10.000 | 0.000 | 22.000 | 2.0913 | 0.020 | 11.41 | 0 |
| Solar Panel on -X Side | 1.037 | 1 | 1.141 | -10.000 | 0.000 | 22.000 | 2.0913 | 0.020 | -11.41 | 0 |
| Solar Panel on +Y Side | 1.037 | 1 | 1.141 | 0.000 | 10.000 | 22.000 | 2.0913 | 0.020 | 0 | 11.407 |
| Solar Panel on -Y Side | 1.037 | 1 | 1.141 | 0.000 | -10.000 | 22.000 | 2.0913 | 0.020 | 0 | -11.41 |
| **Digital Assembly** | | | | | | | | | | |
| C&DH | 1.440 | 1 | 1.584 | -4.000 | 4.000 | 10.000 | 1.3200 | 0.003 | -6.336 | 6.336 |
| Payload Processor/memory | 1.100 | 1 | 1.210 | 4.000 | 4.000 | 10.000 | 1.0083 | 0.002 | 4.840 | 4.840 |
| **ADACS** | | | | | | | | | | |
| 3-Axis Magnetometer | 0.645 | 1 | 0.658 | 0.000 | -5.000 | 13.000 | 0.7127 | 0.000 | 0 | -3.29 |
| Sun Senors (6) | 0.300 | 1 | 0.330 | 0.000 | 0.000 | 13.000 | 0.3575 | 0.000 | 0 | 0 |
| Star Tracker 1 & mount | 1.940 | 1 | 2.134 | 0.000 | -6.000 | 11.000 | 1.9562 | 0.002 | 0 | -12.8 |
| Star Tracker 2 & mount | 1.940 | 1 | 2.134 | 0.000 | 6.000 | 11.000 | 1.9562 | 0.002 | 0 | 12.804 |
| Reaction Wheel X & mount | 4.040 | 1 | 4.444 | 0.000 | 0.000 | 13.000 | 4.8143 | 0.000 | 0 | 44.440 |
| Reaction Wheel Y & mount | 4.040 | 1 | 4.444 | 10.000 | 0.000 | 13.000 | 4.8143 | 0.000 | 44.440 | 0 |
| Reaction Wheel Z & mount | 4.040 | 1 | 4.444 | 0.000 | 0.000 | 11.500 | 4.2588 | 0.002 | 0 | 0 |
| Z Torque Coil | 0.750 | 1 | 0.900 | 0.000 | 0.000 | 9.000 | 0.6750 | 0.003 | 0 | 0 |
| X Torque Coil | 0.750 | 1 | 0.900 | 0.000 | 10.000 | 13.000 | 0.9750 | 0.000 | 0 | 9 |
| Y Torque Coil | 0.750 | 1 | 0.900 | -10.000 | 0.000 | 13.000 | 0.9750 | 0.000 | -9 | 0 |
| ADACS computer | 0.450 | 1 | 0.495 | -4.000 | 6.000 | 12.000 | 0.4950 | 0.000 | -1.980 | 2.970 |
| GPS Receiver | 0.123 | 1 | 0.125 | -4.000 | 8.000 | 12.000 | 0.1255 | 0.000 | -0.502 | 1.004 |

**Fig. 9.31** (continued)

| Communications | | | | | | | | | | |
|---|---|---|---|---|---|---|---|---|---|---|
| RF Mounting Plate | 0.270 | 1 | 0.297 | -3.000 | -3.000 | 17.000 | 0.4208 | 0.001 | -0.891 | -0.891 |
| Payload S-Band Xmitter | 0.220 | 1 | 0.224 | -3.000 | -4.000 | 17.000 | 0.3179 | 0.001 | -0.673 | -0.898 |
| UHF CMD Receiver | 0.264 | 1 | 0.269 | -2.000 | -3.000 | 17.000 | 0.3815 | 0.001 | -0.539 | -0.808 |
| Power Spliter(s) 2 | 0.088 | 1 | 0.090 | -2.000 | -4.000 | 17.000 | 0.1272 | 0.000 | -0.180 | -0.359 |
| Limiter | 0.022 | 1 | 0.022 | -2.000 | -4.500 | 17.000 | 0.0318 | 0.000 | -0.045 | -0.101 |
| GPS Ant | 0.163 | 1 | 0.166 | 3.000 | 3.000 | 0.000 | 0.0000 | 0.006 | 0.499 | 0.499 |
| S-Band Xmit Ant | 0.170 | 1 | 0.173 | 3.000 | 3.000 | 37.000 | 0.5347 | 0.022 | 0.520 | 0.520 |
| S-Band Xmit Patch Ant | 0.160 | 1 | 0.176 | 3.000 | -3.000 | 36.000 | 0.5280 | 0.020 | 0.528 | -0.528 |
| UHF Receive Patch Ant | 0.160 | 1 | 0.176 | -3.000 | -3.000 | 36.000 | 0.5280 | 0.020 | -0.528 | -0.528 |
| Harnessing | 2.000 | 1 | 2.400 | 0.000 | 0.000 | 13.000 | 2.6000 | 0.000 | 0 | 0 |
| Payload & Mounting Structure | 24.000 | 1 | 26.400 | 0.000 | -1.000 | 23.000 | 50.6000 | 0.579 | 0 | -26.400 |
| Ballast | | | 0.000 | | | | 0.0000 | 0.000 | 0 | 0 |
| Totals | 96.242 | | 105.867 | | | | 114.0536 | 1.4299 | 41.195 | 40.163 |
| MOI (slugs-ft²) and kg-m²= | 1.4299 | 1.910 | | | | Z CG in | 12.928 | | 0.389 | 0.379 |
| Margin (lbs) = | 9.626 | | | | | Z CP in | 18.000 | | | |
| CG$_{XY}$ (in) = | 0.543 | | | | | | | | | |

Fig. 9.31 (continued)

## 9.8 Weight Estimate

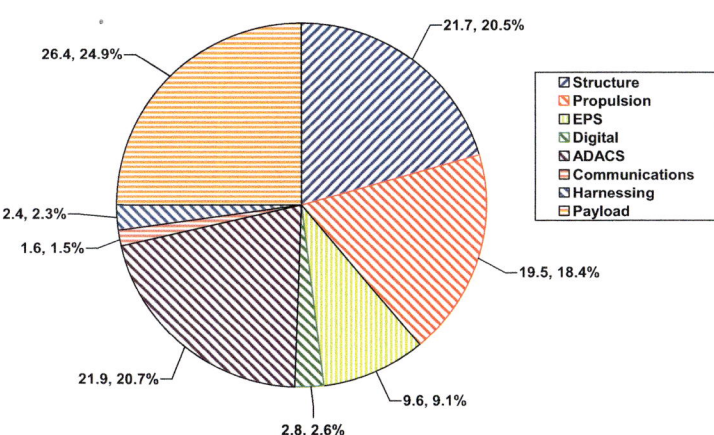

**Fig. 9.32** Weight distribution among various subsystems

# Chapter 10
# Deployment Mechanisms

Various parts of the spacecraft often have to be deployed on orbit. These may include solar panels, antennas, gravity gradient booms and various other booms and instruments of the payload. There are two main rules for designing deployables.

First, one must be sure that in the stowed state, launch vehicle vibrations should not permit separation of any part of the deployable from the spacecraft. This demands that at each point of the deployable where it touches the spacecraft, the deployable should be secured to the spacecraft with a force greater than the weight of the deployable times the launch vehicle acceleration. If the force is insufficient, during vibration testing, the deployable will "chatter," and it may be damaged.

The second rule is that during deployment, the force should be sufficient to reinitiate deployment after the deployable has been stopped. One should not rely on the inertia of the deployable to continue deployment. There should be a positive force acting on the deployable at all times.

Design practices for deployables are many. Some are listed below:

- Whenever possible, two surfaces that must slide relative to one another should use rollers to minimize friction
- Use space qualified lubricants or dry lubricants
- Avoid the use of dissimilar materials in contact with one another
- Design to preclude metal-to-metal adhesion
- Analyze the behavior of moving parts under a large temperature span
- Ensure that the deploying force or torque should have a large safety margin
- Use dampers to avoid unbounded accelerations during deployment
- If detents or latches are used, assure that slight overdeployment is used to permit latches to engage
- If assurance is required (in telemetry) that deployment has actually taken place, it could significantly complicate the mechanism

## 10.1 Deployment Devices

There are several different deployment and release devices.

### 10.1.1 Hinges

Figure 10.1 shows a typical hinge used to deploy solar panels initially held fixed to the sides of the spacecraft, and then deployed to some angle (say 30°) when on orbit. The hinge uses a coiled spring on a shaft. When deployed, there is still positive torque to keep the panel in the deployed state. The red plate is used to determine the deployment angle (a 30° deployment hinge is shown). The deployed spring residual torque must be large enough to overcome any on-orbit loads.

Alternatively, instead of using the residual torque to keep the hinge deployed, detents or latches can also be used. When using detents, the design should allow for a slightly greater than required deployment angle so that as the hinge deploys, it overrides the hole designed for the detent, thereby assuring detent penetration into the hole. One disadvantage of using detents is that the deployed structure may experience "play" due to the tolerances of holes and detents.

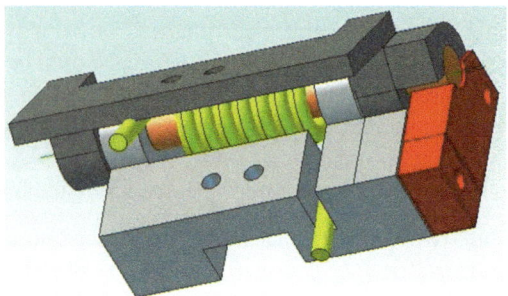

**Fig. 10.1** Spring loaded hinge to deploy from 0–30°

### 10.1.2 Deployable Booms

Deployable booms, like a gravity gradient boom or an expending structure, require light weight and stiff structural elements that stow in very small spaces and deploy to great lengths. A good example of such booms is the gravity gradient boom which must often deploy to some 50 ft or more and, when stowed, fit in just a small canister.

Developed in the early 1960's, the Stem and Bistem booms were used to create long structures stowed in small initial volumes. Figure 10.2 illustrates the Stem and Bistem mechanisms. In both, the flat spring material is curled along its longitudinal

## 10.1 Deployment Devices

axis. It straightens out when wound up on a spool. When allowed to unwind from the spool, the spring assumes it original curled state to create a round boom. The Stem boom can be quite rigid, but torsionally indeterminate. The Bistem uses two deployed booms, one inside the other.

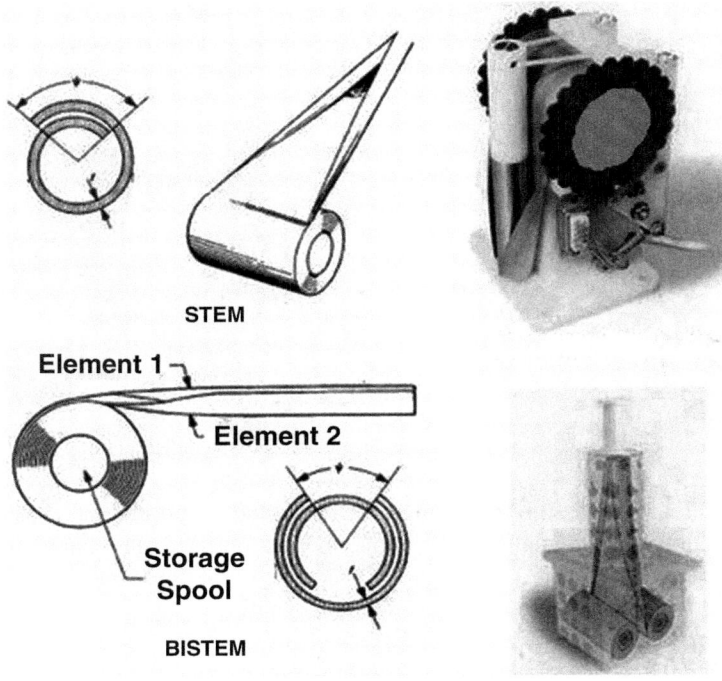

**Fig. 10.2** Stem and bistem deployable booms

Later, stiffer deployable structures were developed. Two deployable boom structures are shown. In Fig. 10.3a, two pre-bent flat springs are welded together to form a stiff tube. The tube is squashed to become a flat spring, which is then rolled onto a drum. When deploying from the drum, the squashed tube resumes its original shape. This makes for a very stiff boom. Since often the Sun will shine on one of the springs and not on the other, the boom will experience thermal bending. To mitigate this, holes are drilled into the flat springs so that sunshine incident on one side can get through the holes and impinge on the other spring also, keeping the boom straight.

In Fig. 10.3b, a "stacer" boom is shown. Here, when the flat beryllium spring is stowed, it is coiled up like a roll of postage stamps. Since the spring is coiled at (typically) 60° angle, the spring will try to deploy axially from the canister where it

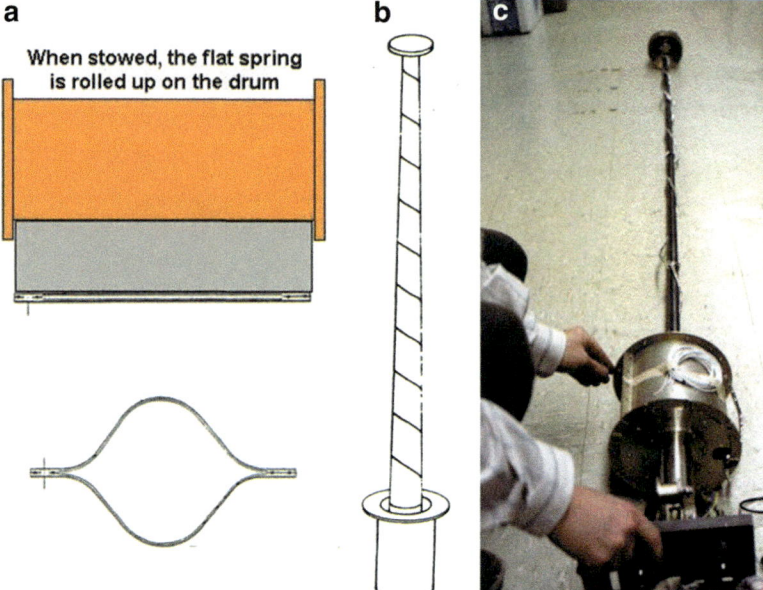

**Fig. 10.3** (a–c)Two different deployable boom configurations

is stowed. For this reason, the spring is held stowed by an internal (kevlar) line, which is cut with a line cutter when deployment is initiated. The same line with a constant force spring and a governor are used to control deployment. The internal string also prevents over-deployment and the unwinding of consecutive layers of the spring. Thermal bending is also an issue with this design, but the beryllium flat spring is a good thermal conductor, so bending is minimized. Testing the deployable boom on Earth can be difficult. In Fig. 10.3c, the tip of the boom is secured to a piece of ice, and the boom is deployed horizontally on a smooth floor. Thus, deployment in space at zero g and no friction is simulated with good results.

## 10.1.3  Large Deployable Antennas

For low power space-to-ground communications from geostationary altitude, the spacecraft antenna must be very large. To be compatible with the spacecraft, the stowed antenna configuration and the overall weight of the antenna and deployment mechanism should both be small. These seem like incompatible requirements. Nevertheless, the art of designing ever larger spacecraft antennas seems to be in good health, with novel ideas emerging frequently.

One of the early types of large deployable antennas consisted of articulated webs interconnected with a thin metal mesh, shown in Fig. 10.4.

## 10.2 Restraint Devices

**Fig. 10.4** Articulated rib with metal mesh ATS-6 antenna

Other approaches have also been developed. In one of these, the ribs are wound up in a drum and when deployed, they straighten out and drag the metal mesh with them. In another, the outer diameter is constructed from an inflatable tube, and the surface is either a mesh or a diaphragm.

The reader is referred to the references for more detailed information about large deployable antennas.

## 10.2 Restraint Devices

Explosive bolt cutters or line cutters, explosive bolts, electric burn wires, paraffin actuators, solenoid pin pullers and motorized cams are used to keep the deployables stowed on orbit.

### 10.2.1 The Explosive Bolt Cutter

The explosive bolt cutter is perhaps the most frequently used restraining device, shown in Fig. 10.5. This device uses redundant pyrotechnic initiators that ignite the charge. The charge then drives the cutter forward to cut the bolt. Bolts of 6–32″ to

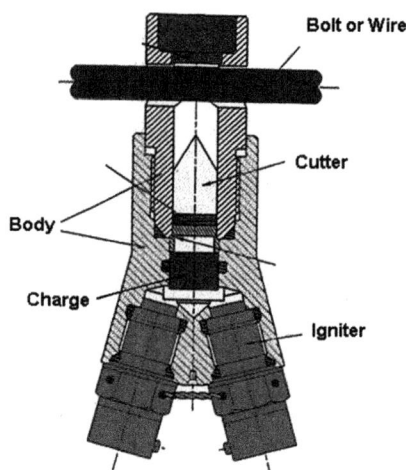

**Fig. 10.5** Explosive bolt cutter (Ensign Bickford)

0.25″ are commonly cut by such bolt cutters. While the cutters are relatively inexpensive, the reader is cautioned that a large number of cutters must be purchased to provide samples for lot qualification and testing. The pyro shock generated by these cutters must be taken into account in the design of the nearby spacecraft structure and electronics.

## 10.2.2 Electric Burn Wires

The electric burn wire is a device where a short piece of nichrome wire is wrapped around a Kevlar or Vectran line, and when current is applied to the wire, it burns the line and releases whatever is held taut by the line. Its main advantage is that it is not a pyrotechnic device. The relationships between the wire gauge, wire tensile strength and the current needed to burn the line are in Fig. 10.6.

| Gauge | lbs | Burn°C | Min Amp | Ohm (1″) | Min Volt/inch |
|---|---|---|---|---|---|
| 38 | 0.628 | 1350 | 0.846 | 3.5167 | 2.975 |
| 36 | 0.982 | 1350 | 1.182 | 2.2500 | 2.661 |
| 34 | 1.559 | 1350 | 1.673 | 1.4167 | 2.370 |
| 32 | 2.513 | 1350 | 2.387 | 0.8833 | 2.109 |
| 30 | 3.927 | 1350 | 3.345 | 0.5625 | 1.881 |
| 28 | 6.234 | 1350 | 4.731 | 0.3542 | 1.676 |
| 26 | 9.928 | 1350 | 6.706 | 0.2225 | 1.492 |
| 24 | 15.865 | 1350 | 9.533 | 0.1392 | 1.327 |
| 22 | 25.136 | 1350 | 13.425 | 0.0883 | 1.186 |

**Fig. 10.6** Nichrome (1″) burn wire cutting 65 lbs. Vectran line

## 10.2 Restraint Devices

As the required tensile strength of the nichrome wire increases, the current required to burn the line increases substantially. For this reason, the burn wire release mechanism is best suited to the release of small devices. The length of time for burning the line is several seconds, so if multiple releases are required, the exact instance of the releases cannot be ensured. For this reason, the release of the spacecraft deployable should be insensitive to the exact time of the release.

Since the nichrome wire melting point in space is less than it is at ambient pressure, care must be taken not to overheat the nichrome wire and cause it to melt.

There are two design approaches to burn wire separation systems. In one, the nichrome wire is in tension by the line. In this type, the maximum line tension is the tensile strength of the nichrome wire. In the other type, the nichrome wire is wrapped around the line. In this design, line tension is not dependent on the nichrome wire break strength. Both types are shown in Fig. 10.7.

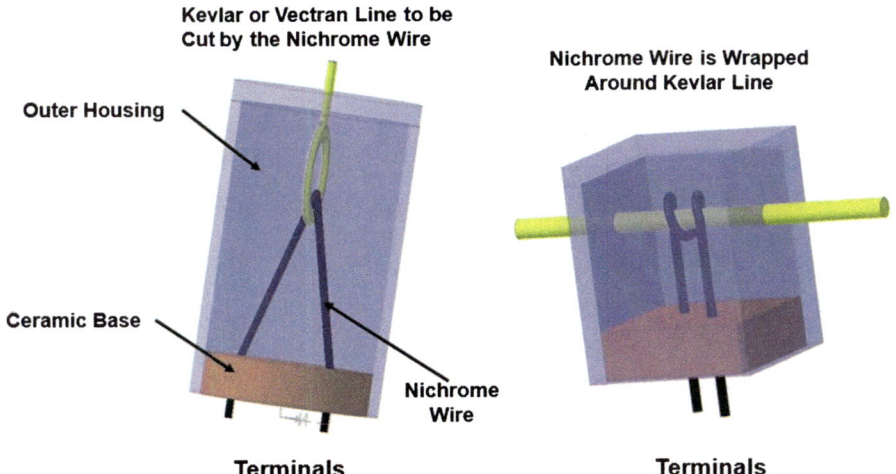

**Fig. 10.7** Two different types of burn wire systems

### 10.2.3 Solenoid Pin Pullers

A solenoid, when energized, pulls a pin that holds the restrained structure. When the pin is pulled, the restrained structure is free to move. While this pin puller appears to be simple, because of the friction between the pin and the surrounding housing, a surprising amount of force is needed to pull the pin. The solenoids tend to be large and require a lot of current.

## 10.2.4 Paraffin Pin Pushers

Paraffin is a wax that, when heated, can increase its volume up to 25% over a temperature range of 30–300 °F. This is used to push out a pin from an enclosed cylinder. Figure 10.8 shows the principle of these actuators, and the stroke and force of three different Rostra Vernatherm actuators. The advantages of the paraffin actuators are that they are small, provide a lot of force and are reusable.

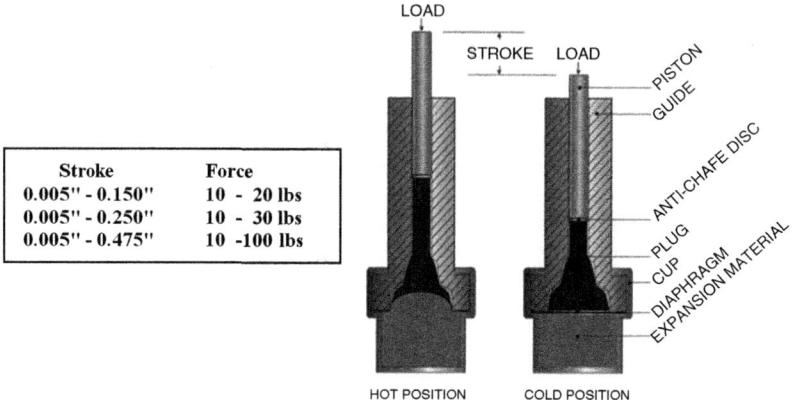

Fig. 10.8 Rostra Vernatherm paraffin actuators

## 10.2.5 Motorized Cams or Doors

To deploy larger structural members while controlling deployment speed, motorized cam releases or motorized structure deployments can be used. For example, the aperture block of a space telescope, a large structure, can be deployed by a motor-driven deployment mechanism which controls the speed of deployment. Usually, motorized deployment mechanisms can be used repeatedly for they can close as well as open the deployable structure.

## 10.2.6 Separation System

The spacecraft-launch vehicle separation system needs to push up the heavy spacecraft within the stroke of the separation system springs to a velocity of about 3 ft./sec, and it must do so without tipping the spacecraft. Figure 10.9a illustrates the Planetary Systems Inc., Lightband motorized, pyroless separation system that accomplishes the above objectives. Figure 10.9b shows the separation system released, showing the length of the pushoff springs and the part of the Lightband that stays with the spacecraft. This system uses a large number of motorized cam segments that hold the two halves of the band together.

10.2 Restraint Devices

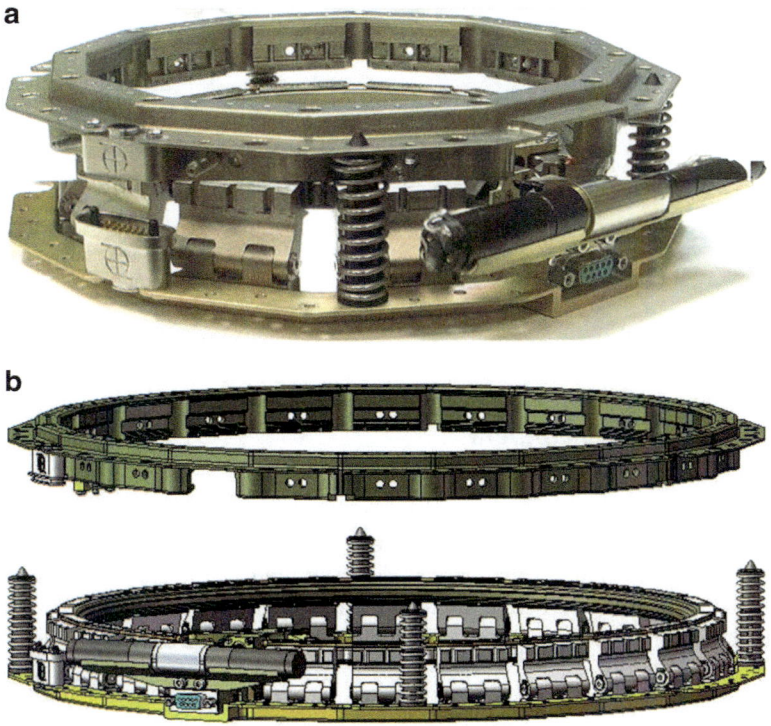

**Fig. 10.9** (**a**) Lightband motorized spacecraft-launch vehicle separation system (**b**) Motorized pyroless lightband separation system motorized cam segments visible

### 10.2.7 Dampers

To restrain the speed of deployment and to avoid the shock when the deployed structure hits the stops, dampers are used. The most often used dampers are the fluid, magnetic and inertial dampers.

### 10.2.8 Fluid Dampers

The principle of the fluid damper is shown in Fig. 10.10. When a piston is pushed in, as the attached structure is deployed, the fluid must go from one chamber to another through orifices or holes. If the holes are small, damping is significant. If the holes are larger, there is less damping. There are dampers to damp linear motion like the one shown in this figure and others, used to damp rotary motion.

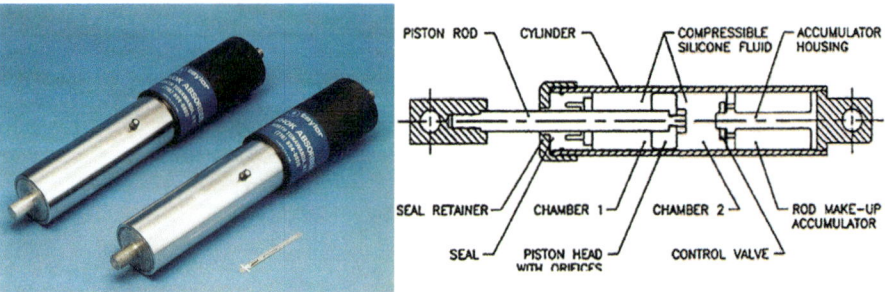

**Fig. 10.10** Fluid damper to damp linear motion (Taylor devices)

## 10.2.9 Magnetic Dampers

Magnetic dampers, now used in suspension systems of automobiles, use an oil-based magnetorheological fluid in a piston-cylinder mechanism. The fluid contains magnetic particles. When a magnetic field is applied to the fluid, its viscosity is increased to provide increased damping action. The speed with which viscosity can be changed is on the order of a millisecond. Viscosity can be controlled by the size of the magnetic field.

## 10.2.10 Constant Speed Governor Dampers

Governor mechanisms tend to rotate at constant speed. They can be attached to deployable structures to damp their deployment speed. For example, the old-fashioned rotary telephone dial mechanism was used to control the deployment speed of booms in several gravity gradient stabilized satellites using "stacer" booms. Without these constant speed governors, the booms would have torn themselves apart during deployment.

## 10.3 Choosing the Right Mechanism

Because of differences in the amount of force, the required electric power, and the size and weight of the different release mechanisms, the process of selecting the right mechanism is not simple. In Fig. 10.11, some qualitative guidelines are offered to aid the selection process.

| Mechanism | Force | Size | Weight | Power | Cost |
|---|---|---|---|---|---|
| Explosive Bolt | Large | Relatively Small | Low | Moderate Pulse | Moderate |
| Explosive Bolt-Wire Cutters | Large | Relatively Small | Low | Large Pulse | Low |
| Solenoid Pin Pullers | Moderate | Medium | Medium | Large | Low |
| Paraffin Pin Pushers | Medium | Relatively Small | Medium | Moderate | Moderate |
| Motorized Cams | Large | Medium-Large | Large | Low | High |
| Fluid Dampers | Large | Medium | Medium | High | High |
| Magnetic Dampers | Medium | Medium | Medium | High | High |
| Governors | Medium | Medium | Large | None | High |

**Fig. 10.11** Properties of retention-release mechanisms

The large electric power needed to operate some of the release mechanisms suggests that those mechanisms that need to be deployed simultaneously be staggered, if possible. In addition, since some of the devices require large currents (at low voltage), several may be connected in series rather than in parallel.

## 10.4 Testing Deployables

Because of their size and the need to simulate a zero G environment, testing large deployables can be a challenge. Mechanical Ground Support Equipment (MGSE) has to be designed to support the deployable structure and counterbalance gravity. Each structural member is suspended from cables that are counterbalanced with weights to simulate zero gravity.

More is said about testing deployables in the chapter on Integration and Test.

# Chapter 11
# Propulsion

## 11.1 The Basics

In LEO spacecraft, the main uses of a propulsion system are to raise or lower the orbit and to maintain the spacecraft on station in a constellation of spacecraft. The Rocket Equation, given below, describes the fundamental property of a propulsion system. It gives the magnitude of change in spacecraft velocity as a function of the spacecraft and expanded fuel masses.

$$\Delta V = -g^* I_{SP}^* \ln(1 + m_P / m_o) \quad (11.1)$$

where $m_o$ is the spacecraft initial mass, $m_P$ is the fuel mass, $I_{SP}$ is the Specific Impulse of the fuel used and $g$ is the acceleration of gravity. The Specific Impulse is a measure of fuel efficiency. It measures the fuel exhaust velocity (when multiplied by $g$) and its unit is in seconds. Values of $I_{SP}$ for various fuels are given in Fig. 11.1.

$$V_E (\sec) = I_{SP}^* g \quad (11.2)$$

| Fuel | $I_{SP}$ (sec) | Comments |
|---|---|---|
| Nitrogen gas under pressure | 70 | Used in small spacecraft for station keeping |
| Butane | 70 | Used in small spacecraft instead of Nitrogen (higher density) |
| Hydrazine | 160 | Used in larger spacecraft for drag makeup, station keeping |
| Bipropellant | 230 | Used in Launch Vehicles and larger spacecraft |
| Liquid LOX/LH$_2$ | 268 | Used in Launch Vehicles |
| Rocket Fuel (Castor) | 280 | Used in Launch Vehicles and as an apogee kick motor |

**Fig. 11.1** Specific impulse of different fuels

The fuel mass required to achieve a given ΔV is given by the equation below:

$$m_P = m_o \left[1 - e^{-(\Delta V * g * I_{SP})}\right] \quad (11.3)$$

By conservation of momentum, when a specific mass of fuel is expelled at a given velocity, producing an impulse of $J_{Fuel}$, then spacecraft velocity changes in the opposite direction by $\Delta V = J_{Fuel}/m_o$.

The ΔV can be used to change the orbit of the spacecraft or to move it back and forth for station keeping. Figure 11.2 shows the geometry due to a "burn" tangential to an initial circular orbit. The ΔV places the spacecraft in an elliptical orbit, with the perigee of the orbit at the point of burn and the apogee 180 degrees later. The period of the elliptical orbit is:

$$P = 0.0001658^* (a)^{1.5} \quad (11.4)$$

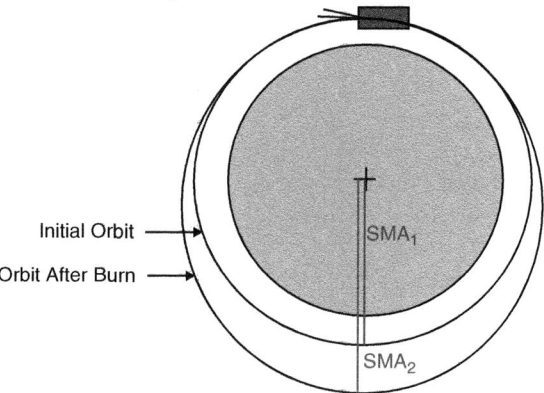

Fig. 11.2 Geometry of a "burn" tangent to a circular orbit

For small ΔV burns the Semi-Major Axis of the elliptical orbit, **a**, changes by Δa, given by the equation:

$$\Delta a = 2^* a^* \Delta V / V \quad (11.5)$$

Since the Semi-Major Axis of the ellipse is greater than that of the initial circular orbit, the period of the elliptical orbit will be greater, and it will take longer to return to the point of burn. By this time the spacecraft, had it remained in the initial circular orbit, would have advanced forward by V*ΔP. This is a perplexing result, because thrusting to move the spacecraft forward, makes it actually fall behind.

## 11.1 The Basics

| Property | Quantity | Dimension |
|---|---|---|
| Radius of Earth ($R_E$) | 6,378.137 | km |
| Spacecraft Mass | 50 | Kg |
| Initial Orbit Altitude | 600 | km |
| Initial Orbit period | 96.68900 | minutes |
| Initial Orbital Velocity | 7.557724 | km/sec |
| $\Delta V$ | 1 | m/sec |
| $\Delta a$ | 1.846624 | km |
| Period of Ellipse | 96.72739 | minutes |
| $\Delta P$ | 2.302965 | seconds |
| Distance Spacecraft in Initial Orbit would have traveled Forward | 17.40517 | km |
| Apparent Reverse Speed of Spacecraft Resulting from $\Delta V$ Burn | 3.00 | m/sec |

**Fig. 11.3** Spacecraft velocity resulting from $\Delta V$ burn

Figure 11.3 illustrates this situation with a numerical example. The result is that the spacecraft received a thrust to move it forward at 1 m/s faster, and, instead, it seems to have moved backwards with an apparent velocity of 3 m/s.

When a second burn is used at the Perigee to circularize the orbit with the original Semi-Major Axis by thrusting in the opposite direction, the spacecraft maintains its station 17.4 km behind its original position.

The basic steps in designing a propulsion system for a spacecraft are to:

1. Determine the $\Delta V$ requirements
2. Choose the fuel type, and thus the $I_{SP}$
3. Compute the fuel mass and volume required
4. Determine thrust vector accuracies/alignments required
5. Determine thrust granularity requirements
6. Configure the propulsion system (fuel tanks, valves, thrusters, heaters, electronics)

In the following, an example will be carried through to illustrate these design steps.

**Example**

It is required that a 125 lb. (56.818 kg) spacecraft of 5 square foot (0.465 square meter) cross sectional area be kept in a circular orbit at 560 km for 7 years. The spacecraft should not drop below 540 km.

Determine $\Delta V$ requirements. At 560 km altitude, mean atmospheric density is about $3 \times 10^{-13}$ kg/m$^3$. To simplify calculations, variations of atmospheric density over the mission lifetimes are ignored. The drag force acting on the spacecraft, F, is given below, where $\rho$ is atmospheric density, D is the drag coefficient (typically equals to 2.2) and V is spacecraft velocity. The period at 560 km is 95.855997 min and the velocity is 7.580 km/s.

$$\text{Drag Force} = F = 0.5^* \rho^* D^* A^* V^2 \approx 4.007 \times 10^{-6} \text{ Newton} \qquad (11.6)$$

This force decelerates the 56.818 kg spacecraft at $7.052 \times 10^{-8}$ m/s². Over 7 years the total loss of velocity is 15.57 m/s. This has to be made up by the propulsion system.

Determine Fuel Mass and Volume Requirements. Let us choose a simple cold gas propulsion system using nitrogen gas at an $I_{SP} = 70$ s. The fuel mass for a given $\Delta V$ for this example is given below:

$$m_P = m_o \left[1 - e^{-(\Delta V^* g/I_{SP})}\right] = 1.3273823 \text{ kg} \qquad (11.7)$$

The density of $N_2$ at 15 psi is 0.00125 g/cu cm. At 6000 psi it is 0.5 g/cu cm, and at a readily available tank of 4500 psi it is 0.375 g/cu cm. At 6000 psi a tank of 2.655 liters is needed and at 4500 psi tank pressure a tank of 3.540 liters is needed.

The inlet pressure of a typical 1 N thruster is 215 psi. If the tank pressure is to remain above 215 psi, only $100^*(6000-215)/6000 = 96.4\%$ of the fuel contained in the tank can be used. Therefore, the 6000 psi tank volume should be 2.754 liters. Similarly, the 4500 psi tank should be 3.718 liters.

The frequency of drag makeup burns depends on how low the spacecraft is permitted to drop. If the spacecraft minimum altitude is to be kept above 550 km, the maximum decrease in velocity should not exceed 6.216 m/s. This will occur in 2.795 years. Therefore, over the 7-year mission life drag makeup burns of 6.216 m/s have to be performed every 2.795 years (or smaller burns more frequently).

## 11.2 Propulsion Systems

### 11.2.1 Cold Gas Propulsion System

Figure 11.4 shows the configuration of a small cold gas ($N_2$) propulsion system. The two 2-liter fuel tanks are fueled through the check valve, and the tanks are pressurized to 6000 psi. When ready to thrust, the Tank Shutoff Valve is opened to let the high pressure gas flow to the four thrusters through a pressure reducer. The 1 lb. thrusters are operating at a 215 psi inlet pressure. For this reason, 96.4% of the fuel in the tanks is available for thrusting before the tank pressure drops below 215 psi. The thrusters are mounted at the four corners of a rectangular spacecraft. When fired together, they provide 4 lbs. of thrust. When only two thrusters are fired on the same side of the spacecraft, the spacecraft can be rotated. To make the thrust vector go through the principal axis of the spacecraft, the thrusters can be fired intermittently

## 11.2 Propulsion Systems

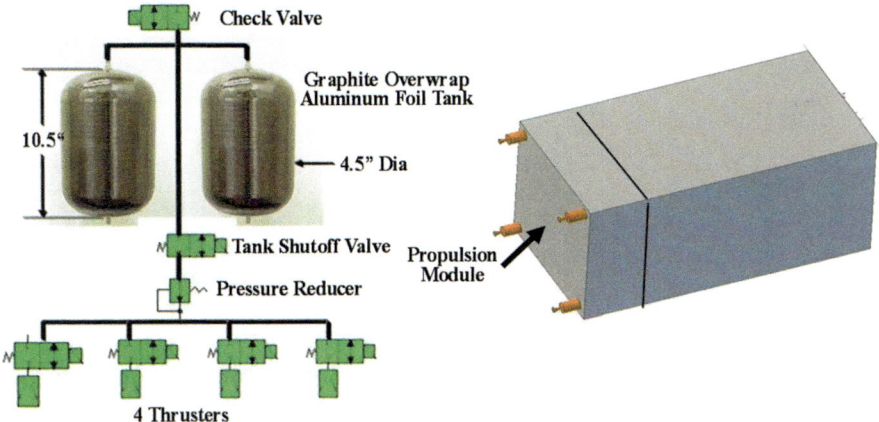

**Fig. 11.4** Cold gas propulsion system schematic

with variable duty cycles to compensate for the slight differences of thrust and differences in alignment among thrusters. Thruster valve opening and closing times are typically 3.5 ms (max) (Figs. 11.5 and 11.6).

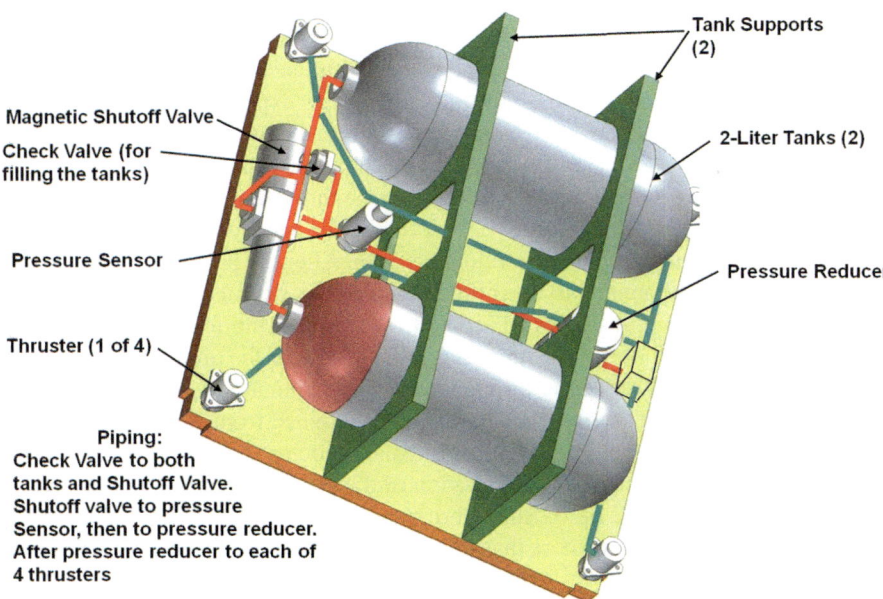

**Fig. 11.5** Layout of a cold gas propulsion system

**Fig. 11.6** Photograph of a cold gas propulsion system that can provide 4 lbs of thrust

The thrust vector should go through the spacecraft CG. Since the position of the CG is not known with great accuracy, and since the thrusters cannot be aligned perfectly, multiple thrusters intentionally straddling the expected CG position are employed. In the propulsion system above, the four thrusters are aligned to purposely miss the expected CG position by about 2 cm. The misalignment of thrusters from the CG can be compensated for by varying the thrust durations. Whenever $\Delta V$ burns are required, the spacecraft is turned (using the reaction wheels) so that its longitudinal axis is aligned with the velocity vector. After the burn is completed, the spacecraft attitude is changed to the attitude required by the mission.

The moment of inertia of the spacecraft, assuming a uniform distribution of weight in its $24'' \times 24'' \times 30''$ volume, is 3.326 slugs, and the angular acceleration imparted by a 1 lb. thruster acting through a line 2 cm from the CG is 0.01972 rad/s$^2$ (1.13 deg./s$^2$). If the other thruster, intentionally misaligned from the CG in the other direction is only misaligned by 1 cm, the angular velocity after 1 s of burn will be 0.556 deg./s. By increasing the burn duration of the other thruster to 2 s, the spacecraft angular velocity will be 0. Since thrusters typically open and close in about 3 ms, the thrust granularity is more than enough to keep the spacecraft going straight.

## *11.2.2 Hydrazine Propulsion System*

A Hydrazine propulsion system uses liquid hydrazine and expels it through heated thrusters. The hydrazine is put under (typically) 350–500 psi pressure by a gas (typically $N_2$). The gas under pressure and the hydrazine are contained in the same tank;

but the two are separated by a flexible bladder. In addition to fill and drain valves, a pressure transducer monitors fuel pressure, and a filter with a 15–25 micron rating is used to ensure that no contaminant should get into the thruster valves. Hydrazine density is 0.9 g/cm$^3$, while $N_2$ at 6000 psi has a density of 0.5 g/cm$^3$. Because of the $I_{SP}$ and density advantage of hydrazine, for a given volume of fuel, a hydrazine system is capable of ≈4 times greater $\Delta V$ than a cold gas system (Fig. 11.7)

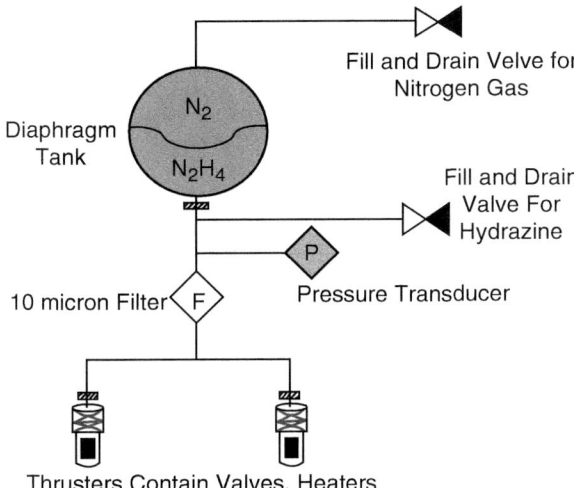

**Fig. 11.7** A typical hydrazine propulsion system

## 11.2.3 *Other Propulsion Systems*

The main type of spacecraft propulsion systems are the rocket motor, cold gas and the hydrazine systems. However, other propulsion systems such as electric, nuclear, solar pressure and others are beginning to be used. These will not be discussed here.

## 11.3 Propulsion System Hardware

Several manufacturers make components for spacecraft propulsion systems. Here, representative samples are presented to provide a feel for the size, weight, performance and power consumption of typical components used in small LEO spacecraft. Kick motors, cold gas and hydrazine tanks and thrusters are described next.

*Kick Motors* circularize the orbits of an elliptical transfer orbit into which the launch vehicle injects a spacecraft. Typical kick motors are the STAR series of motors. Characteristics of these are in Fig. 11.8.

| STAR Motor | 5C | 15G | 24 | 37 | 48 |
|---|---|---|---|---|---|
| Total Impulse (lbs-sec) | 1,252 | 50,210 | 126,000 | 634,760 | 1,303,700 |
| Effective Specific Impulse | 268.1 | 281.8 | 282.9 | 290 | 292 |
| Max Thrust (lbs) | 455 | 2,800 | 4,420 | 15,250 | 17,490 |
| Burn Time (sec) | 2.8 | 33.3 | 29.6 | 49.0 | 84.1 |
| Weight (lbs) | 9.86 | 206.6 | 481 | 2,390 | 4,780 |
| Diameter (in) | 4.77 | 15.04 | 24.5 | 35.2 | 49 |
| Length (in) | 13.43 | 31.57 | 40.5 | 66.2 | 81.7 |

**Fig. 11.8** STAR (Thiokol-Orbital-ATK) solid space motors (By permission from Orbital-ATK)

*Fuel Tanks for cold gas* propulsion systems are usually high psi-rated tanks of sizes ranging from about 1 liter to several liters. To reduce their weight, these tanks are often made of thin titanium foil with graphite epoxy over-wrap. Typical tanks made are listed in Fig. 11.9.

| Orbital-ATK Model | Volume (liters) | Weight (lbs) | Rated psi | Dia (in) |
|---|---|---|---|---|
| 80295-1 | 1.6 | 3.20 | 8,000 | 5.81 |
| 80326-1 | 3.9 | 3.38 | 3,600 | 7.66 |
| 80345-1 | 6.6 | 7.40 | 4,500 | 9.44 |
| 80202-1 | 14.5 | 15.8 | 4,500 | 12.45 |

**Fig. 11.9** SCI Orbital-ATK cold gas high pressure tanks (By permission from ATK)

For *Hydrazine* propulsion systems, the fuel tanks are divided into two parts. One contains the hydrazine fuel, the other the pressurant gas. The two parts of the tank are separated by a bladder. Typical Diaphragm tanks for hydrazine propulsion systems are shown in Fig. 11.10.

| Orbital-ATK Model | Total Volume liters | Diameter (mm) | Propellant Volume (liters) | Weight lbs | Operating Pressure (psi) |
|---|---|---|---|---|---|
| 80222-1 | 6.8 | 239 | 4.8 | 2.85 | 400 |
| 80216-1 | 17.7 | 327 | 12.5 | 6.0 | 396 |
| 80271-3 | 37.4 | 419 | 24.9 | 11.4 | 300 |
| 80308-1 | 49.1 | 419 x 508 | 37.6 kg Hydrazine | 12.4 | 320 |

**Fig. 11.10** Orbital-ATK small hydrazine fuel tanks with bladder (By permission from Orbital-ATK)

*Cold Gas Thrusters* Several companies manufacture cold gas thrusters for spacecraft propulsion systems. Figure 11.11 illustrates a typical small thruster and its technical characteristics.

Note that the power consumption of this thruster is 30 watts. While the thrusters are usually turned on for only a short period of time, making the energy required to

## 11.4 Propulsion Maneuvers

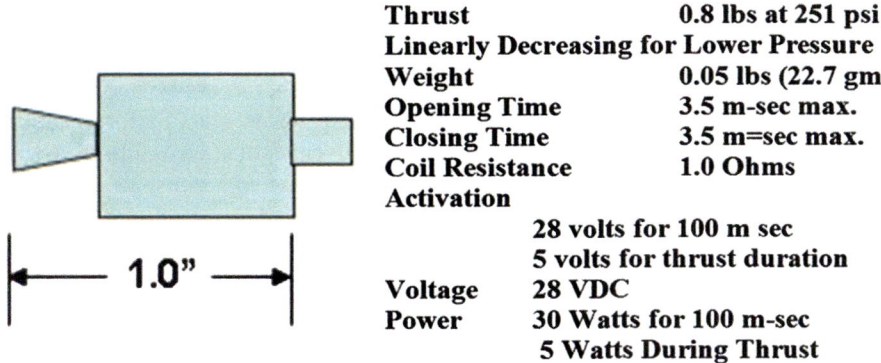

Fig. 11.11 A small (0.8 lbs) cold gas thruster (Courtesy of Moog Inc., Space and Defense Group)

operate them relatively small, the fact that typically four thrusters may be fired simultaneously, requires that the spacecraft should provide 120 watts at 28 volts to operate the propulsion system.

*Thrusters for Hydrazine Propulsion* Typical small thrusters for hydrazine propulsion system are shown in Fig. 11.12.

| Property | Moog 0.17 lb Thruster 051-271 | Moog 0.2 lb Thruster 051-346D | Moog 9.0 lb Thruster 51-288 |
|---|---|---|---|
| Operating Pressure (psi) | 386 | 400 | 500 |
| Max Response Time (msec) | 1 | 10 | 15 |
| Operating Voltage (volts) | 24-32 | 24-37 | 22-32 |
| Power Consumption (watts) | 10.4 | 8.69 | 26.5 |
| Weight (gm) | 30 | 218 | 230 |
| Operating Temperature (°C) | 4.4-149 | 4.4-149 | 4.4-149 |

Fig. 11.12 Typical small hydrazine thruster characteristics (Courtesy of Moog Inc., Space and Defense Group)

## 11.4 Propulsion Maneuvers

### 11.4.1 Maneuvers for Spacecraft in a Constellation, Maintaining and Getting to Station

#### 11.4.1.1 Station Keeping

Let us start with an example. Suppose we have a constellation of 24 Polar-Orbiting spacecraft, distributed in three planes, 120° apart. Each plane has eight equally spaced spacecraft. Their true anomalies are 45° apart. Each spacecraft is at a nominal 659.2 km altitude. The objective is to keep each spacecraft within ±10% of the inter-satellite distance at all times. To reduce ground station work load, we specify that station-keeping maneuvers on each spacecraft should not be needed more often than once in 72 days. In this way, the ground station must perform station keeping every 3 days on one of the 24 spacecraft. This situation is illustrated in Fig. 11.13. The red and the green arcs are the footprints of adjacent spacecraft, and the footprints overlap.

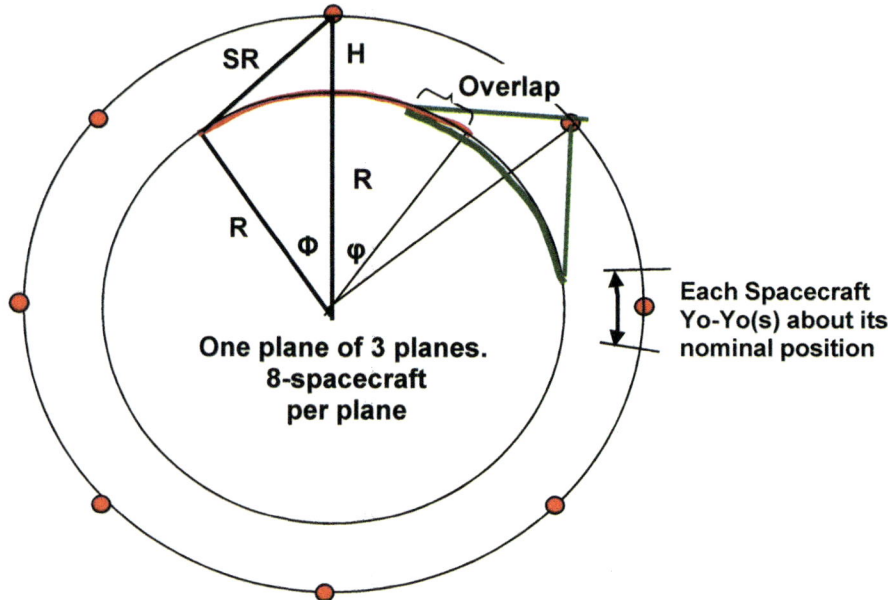

| Altitude H | 659.2 | km |
|---|---|---|
| Earth Radius | 6,378.137 | km |
| Slant Range = $SR_H$ at horizon | 2,973.799 | km |
| Subtended Angle, φ | 25.0 | deg |
| SC Footprint angle (2*φ) | 50.0 | deg |
| Number of spacecraft | 8 | |
| Total Satellite Footprints | 400.0 | deg |
| Satellite Footprint Overlap, OL | 5.0 | deg |

**Fig. 11.13** Eight spacecraft per plane to maintain station by each moving back an forth

Slant range to the horizon is 2973.799 km. The Earth Subtended Angle by the spacecraft horizon-to-horizon footprint is:

$$2^*\varphi = 2^* \operatorname{acos}\left[ R^2 / (R+H)^2 \right] = 50.0°. \qquad (11.8)$$

## 11.4 Propulsion Maneuvers

The ground footprint is 50/360*2*π*R = 5566 km. If each spacecraft position is to be maintained to ±10% of the nominal (44,216.892 km/8 = 5527 km) inter-satellite distance, then each is permitted to deviate from its nominal position by ±552.7 km. Therefore, the average speed variance from the nominal orbital speed, must be less than:

$$2*552.7 / (72*24*60*60) = 0.177 \text{ m/sec}. \quad (11.9)$$

We already saw that to move a spacecraft in its orbit, a forward thrust of $\Delta V$ results in an elliptical orbit which puts the spacecraft behind its original burn position by a specific distance after each orbit. To determine the $\Delta V$ required to achieve a spacecraft apparent speed of 0.177 m/s, we proceed as follows. Details of the calculations are shown in Fig. 11.14.

| | | |
|---|---|---|
| **Nominal Orbit Period, $P_o$ =** | 97.9220209 | min |
| **Orbital Velocity, $V_o$ =** | 7.52586804 | Km/sec |
| **Required Apparent SC Speed, S =** | 0.177 | m/sec |
| **Distance Moved Per Period, D =** | 1,039.932 | meter |
| **Delta Period, $\Delta P$** | 0.138181 | sec |
| **Period of Elliptical Orbit, $P_E$** | 97.92432392 | min |
| **The Semi-Major Axis of the Ellipse, $a_E$ =** | 7,037.447 | km |
| **$\Delta$ Semi-Major Axis = $\Delta a$ =** | 0.110 | km |
| **$\Delta V$ burn required = $\Delta a * V/(2a)$** | 0.059 | m/sec |

**Fig. 11.14** Calculating $\Delta V$ required for station keeping

The orbit period at the nominal 659.2 km altitude is 97.9220209 min and the orbital velocity (circumference of orbit/Period) is 7.525804 km/s. Since at the apparent speed of 0.177 m/s the spacecraft should fall behind by 1039.9319 meters per orbit, the $\Delta P$ should be 1.38181 s, giving an elliptical orbit period of 97.9243239 min. This corresponds to a Semi-Major Axis of 7037.447 km. The $\Delta$ Semi-Major Axis, $\Delta a = 0.110$ km; and the $\Delta V$ required to achieve this is 0.059 m/s.

The reverse burn required after 72 days is 0.118 m/s. Half is required to re-circularize the elliptical orbit at the altitude of the Perigee, and the other half is required to make the spacecraft move in the opposite direction. So, the station-keeping maneuver requires 0.598 m/s $\Delta V$ per year or 4.187 m/s over a 7-year mission. This is a very modest amount of fuel.

### 11.4.1.2 Getting on Station

Suppose we require that the spacecraft that must travel the furthest to get on station must travel 22,264 km (4 inter-satellite distances); and suppose we require that this spacecraft get on station in 10 days. The average required spacecraft speed should be 25.7685 m/s. The spacecraft must be put into an elliptical orbit so that it would fall back 151.398 m/orbit. The ΔP should be 20.117044 s and the period of the ellipse should be 98.25730497 min. This results in the Semi-Major Axis of the ellipse to be 7053.392 km. The Δa is 16.055 km, and the ΔV required to produce this is 8.565 m/s.

It is seen that getting this spacecraft on station quickly takes twice as much fuel as station keeping for 7 years. Considerable fuel could be saved by relaxing the time interval allowed to get this spacecraft on station. The innocuous requirement to establish the constellation in 10 days drives the propulsion system requirements.

Figure 11.15 illustrates the details of the station-keeping maneuver. The left-hand figure shows the spacecraft nadir pointing and performing its mission. It travels from left to right. Note that the center of pressure is at the center of the spacecraft; but the CG is below it. For this reason, at low orbit altitudes there is an atmospheric torque that tends to rotate the spacecraft counterclockwise. Below about 550 km altitude this can be significant, and to counter it may take fuel or the use of the reaction wheels. When it comes time to initiate a burn to the right, the spacecraft is turned 90° so that the thrusters are tangential to the orbit. Note that both thrusters have thrust vectors that purposely straddle the CG. If any of the thrusters are slightly misaligned, varying the duty cycles of the different thrusters, the net thrust vector can be kept parallel to the spacecraft velocity vector.

The spacecraft in the second position is ready to thrust to move the spacecraft to the right and into an elliptical orbit. Note that in this attitude the CG and the CP are both on the line of the velocity vector, eliminating atmospheric drag torque. Also, the spacecraft cross-section is smaller, reducing atmospheric drag and prolonging mission life. In fact, if the mission permits it, the spacecraft could be flown horizontally most of the time, and it could be put in a nadir-pointing position only when the mission must be performed.

In the third position, the spacecraft has been turned around 180° and is ready to burn to stop (and reverse) spacecraft station-keeping motion. At the end of the reverse burn, the spacecraft is again nadir pointed to permit it to resume its mission.

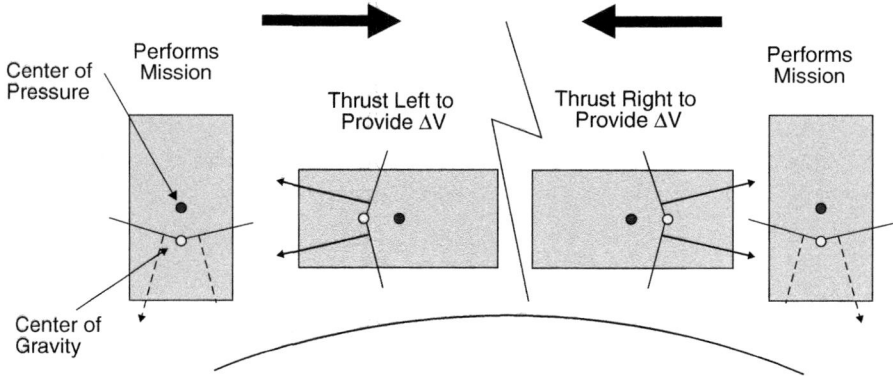

**Fig. 11.15** Details of station-keeping burns

11.5 Other Propulsion Requirements

### 11.4.1.3 Thrust Duration

Let us again assume that the spacecraft weighs 125 lbs. Using all four thrusters producing 4 lbs. of thrust, the required station keeping $\Delta V$ of 0.059 m/s can be achieved in 17.5 ms. The duration of the burn to put the furthest spacecraft on station is 2.533 s. Typical cold gas thrusters can be opened and closed in 3 ms. Thus, there might be a problem of executing a 17.5 ms burn precisely. The thrust schedule has to be planned and the thrust strategy has to be tested to implement the required thrust.

If the size of each thruster were reduced to (say) 0.2 lbs., then burn times for station keeping and for getting on station would increase to 85 ms and 12.665 s, respectively.

### 11.4.1.4 Hohmann Transfer Orbit Maneuver

A frequent application of a spacecraft propulsion system is to change the altitude of the orbit. The minimum energy transfer is a Hohmann transfer, where a $\Delta V_P$ burn puts the spacecraft in an elliptical orbit where the point of burn becomes the Perigee of the elliptical orbit. A second $\Delta V_A$ burn is applied at the Apogee of the elliptical orbit to circularize the orbit with the Apogee altitude. In the example below, it is required to raise the altitude of a 600 km circular orbit to a 700 km circular orbit. Figure 11.16 illustrates the geometry and the calculations to obtain the $\Delta V_P$ and $\Delta V_A$.

From initial and final altitudes the elliptical transfer orbit semi-major axis, a, is computed. This is (7028.137 km). The change in semi-major axis from the initial circular orbit is $\Delta a$ (50 km). The first burn takes place at a point that becomes the Perigee of the elliptical transfer orbit; and the size of this burn is $\Delta V_P = 26.884$ m/s. The second burn, $\Delta V_A$, takes place at the Apogee and is approximately the same size. Thus the total $\Delta V$ required to raise the orbit from 600 km to 700 km is 53.768 m/s. It takes 48.865 min to perform the orbit transfer.

## 11.5 Other Propulsion Requirements

There are other, less often used, propulsion requirements. These include the *Spin Up and Spin Down* of spin-stabilized spacecraft. Generally, a cold gas system has adequate $\Delta V$ to implement spin ups.

There is an increasing need to *deorbit spacecraft* at end-of-life to reduce space debris. The energy required to do this maneuver usually requires the use of a rocket motor, and the $\Delta V$ required is substantial.

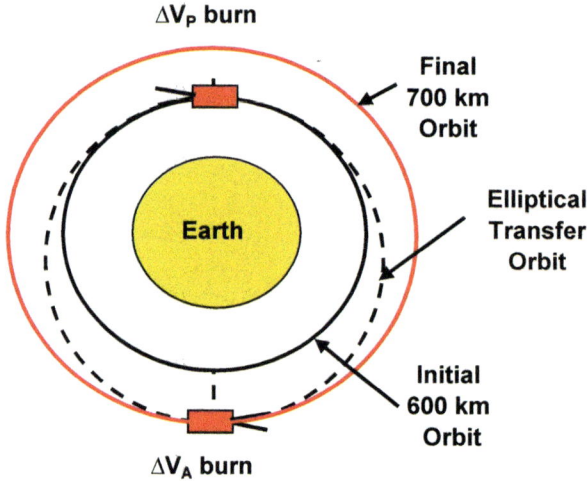

| Property | Equation | Value | Units |
|---|---|---|---|
| Earth Radius | R | 6,378.137 | km |
| Initial Orbit Altitude | $H_P$ | 600.000 | km |
| Final Altitude | $H_A$ | 700.000 | km |
| Initial Orbit Period | $P_1=0.00016587*(R+H_P)^{1.5}$ | 96.689003 | min |
| Initial Orbit Velocity | $V_1=2*\pi*(R+H_P)/P_1/60$ | 7.55772399 | km/sec |
| SemiMajor Axis | $SM=R+(H_P+H_A)/2$ | 7,028.137 | km |
| $\Delta a$ | $\Delta a=(H_A-H_P)/2$ | 50 | km |
| $\Delta V_P$ | $\Delta a=2a\,\Delta V_P/V_1;\ \Delta V_P=\Delta a*V/2a$ | 26.884 | m/sec |
| Time to Apogee | $P_E/2$ | 48.865 | min |
| Time to Apogee | $P_E/2/60$ | 0.814 | hours |
| Period of Higher Orbit | $P_2=0.00016587*(R+H_A)^{1.5}$ | 98.77483 | min |
| Final Velocity Required | $V_F=2*\pi*(R+H_A)/P_2/60$ | 7.5041463 | km/sec |
| Period of Higher Orbit | $P_2=0.00016587*(R+H_A)^{1.5}$ | 98.77483 | min |
| Velocity of Higher Orbit | $V_F=2*\pi*(R+H_A)/P_2/60$ | 7.5041463 | km/sec |
| $\Delta V_A$ | $\Delta V_A \approx \Delta V_P$ | 26.884 | m/sec |
| Total $\Delta V$ | $\Delta V \approx 2*V_P$ | 53.768 | m/sec |

**Fig. 11.16** Calculation of $\Delta V$ required for Hohmann transfer orbit

*Collision Avoidance* is an emerging requirement. No clear avoidance strategy has yet been devised. However, a strategy of raising or lowering the orbit a little bit is a possible, low energy strategy, if the protected spacecraft can be alerted in time to perform such a maneuver.

# Chapter 12
# Thermal Design

The altitude, inclination, epoch and angle of the orbit plane to the Sun (the Beta angle) determine the thermal environment of a spacecraft. In the direction of the Sun, the heat incident on a spacecraft can be huge, while on the side opposite the Sun, the spacecraft faces cold space. The objective of the thermal design is to bring these temperatures into a range where the spacecraft components can safely operate, and to ensure that internally generated heat from the spacecraft components is conducted or radiated out to maintain all spacecraft temperatures within the operating temperature ranges of its components. Typical component temperature operating ranges are given in Fig. 12.1. The goal is to design for component temperatures in the 10°–20 °C range.

| Component | Typical Operating Temperature Range °C |
|---|---|
| Electronics | -20 to +40 |
| Special Electronics (like Reaction Wheels) | 0 to 35 |
| Solar Panels | -100 to +100 |
| Hydrazine | 10 to 70 |

**Fig. 12.1** Typical temperature ranges of spacecraft components

In steady state, the heat absorbed from the Sun, Albedo and Earth IR, plus the heat generated in the spacecraft equals the heat radiated out for some spacecraft temperature. The thermal design process is to vary surface finishes and radiators to achieve this balance at the desired spacecraft temperature. The design process consists of the following steps:

1. Flow down the thermal requirements from component thermal specifications (to determine the acceptable spacecraft and component temperature ranges)
2. Assess the solar flux, Albedo and Earth IR radiation incident on the spacecraft in its orbit, and compute the orbit average heat on each of the spacecraft outside



G. Sebestyen et al., *Low Earth Orbit Satellite Design*, Space Technology Library 36, https://doi.org/10.1007/978-3-319-68315-7_12

surfaces. The environment depends on the Beta angle. Assess the range of Beta over the mission life of the spacecraft.
3. Compute the heat absorbed by the spacecraft surfaces
4. Determine the heat generated by the spacecraft electronics
5. The sum of the heat from steps 3 and 4 is the heat that has to be rejected for steady state. Iterate radiator and surface finishes to reject that much heat.
6. Construct a thermal model, including locations of its heat generating components
7. Compute spacecraft temperatures as a function of time
8. Apply techniques to bring these temperatures within component specifications. These techniques include:

    (a) Applying paint or other surface finishes to reduce heat absorption or increase heat radiation
    (b) Apply heaters to increase temperatures of cold components (such as batteries)
    (c) Use metallic or other heat conductors to take heat from hot components to points from where heat can be radiated out to space

9. When the spacecraft is completed, conduct thermal vacuum tests to determine the actual spacecraft and component temperatures as a function of time and in the steady state
10. Correlate the thermal model to bring it in conformity with actual measurements obtained during test
11. Generate adjusted model predictions of the spacecraft thermal behavior in all conditions relevant to mission life in orbit

## 12.1 The Thermal Environment

At normal incidence, average Sun radiated power is 1367 Watt/m² (0.882 Watt/square inch). Depending on the absorptivity, $\alpha$, of the material on which it is incident, and on the emissivity, $\varepsilon$, of the surface that can radiate the heat into space, the temperature of the surface is given by Eq. 12.1.

$$T(°K) = \left(S^* (\alpha/\varepsilon)^* (A_i/A_r)/\sigma\right)^{0.25} \quad \text{where} \tag{12.1}$$

$S$ = the Sun radiance = nominally 1367 Watt/m²
$A_i$ = Area of surface of Sun incidence normal to Sun
$A_r$ = Area of surfaces radiating to space
$\alpha$ and $\varepsilon$ absorptivity and emissivity of the surfaces
$\sigma = 5.67 \times 10^{-8}$ W/m²/K⁴

## 12.1 The Thermal Environment

To appreciate how large the Sun incident radiation is, consider an example of a 1 m² plate of absorptivity and emissivity both equal to 1.0. Applying Eq. 12.1, the temperature of the surface becomes T = 393° K or +120 ° C. That is pretty hot.

The thermal environments in typical LEO orbits are discussed below.

From a thermal design point of view, the space environment seen by a spacecraft can be defined by the sum of the orbit average solar flux, the Albedo and Earth IR incident on each side of the spacecraft in its orbit. Of these, solar flux is the major contributor. The amount of sunlight illuminating each side of a spacecraft can be computed for any orbit altitude and Beta with the aid of equations (and spreadsheet) developed in Sect. 3.6. For example, the Orbit Average solar fluxes incident on each of the sides of a 20″ × 20″ × 36″ nadir-pointing spacecraft for Beta angles ranging from 0° to 90° and 70° inclination are given in Fig. 12.2.

| Incl deg | Altitude km | Beta deg | +X OA watts | -X OA watts | +Y OA watts | -Y OA watts | +Z OA watts | -Z OA watts | Total OA watts |
|---|---|---|---|---|---|---|---|---|---|
| 70 | 600 | 0 | 141.3 | 141.3 | 0 | 0 | 111.7 | 9.6 | 403.9 |
| 70 | 600 | 30 | 128.3 | 128.3 | 0 | 208.1 | 97.2 | 11.8 | 573.6 |
| 70 | 600 | 45 | 112.5 | 112.4 | 0 | 331.8 | 79.6 | 15.0 | 631.3 |
| 70 | 600 | 60 | 91.4 | 91.4 | 0 | 440.7 | 56.3 | 23.7 | 703.5 |
| 70 | 600 | 75 | 52.3 | 52.3 | 0 | 613.4 | 29.0 | 29.0 | 776.0 |
| 70 | 600 | 90 | 0 | 0 | 0 | 635.0 | 0 | 0 | 635.0 |

**Fig. 12.2** Orbit average incident solar flux (watts) on each side of a nadir-pointing spacecraft (20″ × 20″ × 36″)

Values in this figure are the projection of the Sun on a unit area of each side as a function of time (from which the orbit average area could be computed). Then, these projections are multiplied by the areas of the sides and by the solar flux of 0.882 Watt/square inch.

This figure shows that at Beta = 75° is the highest total incidence of 776 Watts, while at Beta = 90° is the largest thermal gradient. The −Y side will get very hot, while the +Y side will face cold space. The above table takes into account the eclipse, of up to 35.36 min duration, when none of the sides gets any Sun.

The instantaneous solar flux incident on each side of the spacecraft is shown (for Beta = 45°) in Fig. 12.3. In this figure, each curve represents the fraction of the area of a side normal to the Sun at any instant of time. The blue curve is the total flux incident on the spacecraft at any point in time.

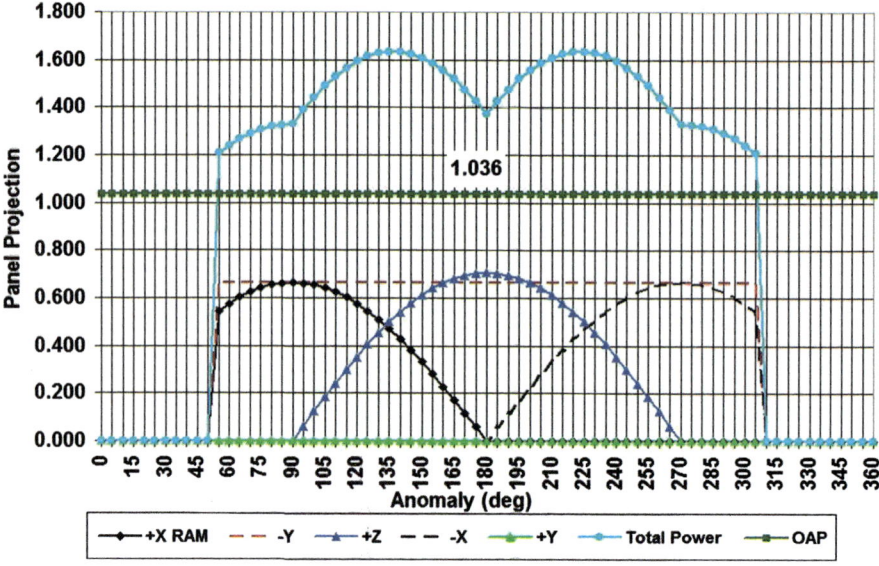

**Fig. 12.3** Solar flux incident on each side of a spacecraft as the fraction of each panel normal to the sun vs. time. Panel areas are scaled by their relative size

The incident solar flux depends on Beta, which varies over a range of ±90°. During positive Betas, the Sun illuminates the +Y side. During negative Betas the Sun illuminates the −Y side. Beta must be computed for the mission life of the spacecraft. Figure 12.4 shows Beta as a function of time for an entire year. It is seen

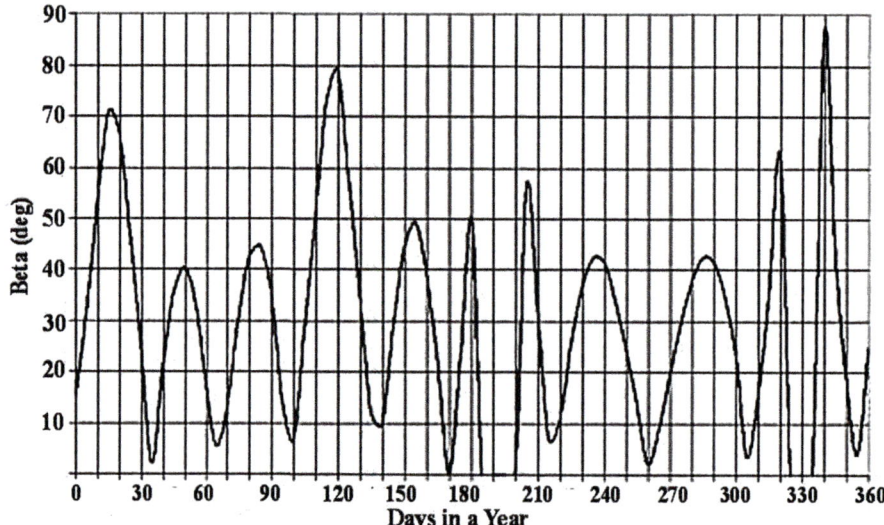

**Fig. 12.4** Beta for one year per month of the year

12.2 Heat Absorption

that for this mission Beta varies mostly from 0° to 80° with most of the time spent between 10° and 60°.

To obtain the total heat incident on the sides of the spacecraft, the contributions of Albedo and Earth IR radiation must be computed and added. Earth IR radiation and Albedo vary with latitude, time of year and orbit inclination. At 600 km, Earth IR incident is 0.119 Watts/in$^2$ on the nadir surface and 0.035 Watts/in$^2$ on surfaces normal to nadir. Figure 12.5 shows these orbit average Albedo powers for the same cases as given for Sun incidence in Fig. 12.2.

Note that Albedo fluxes are much smaller than solar fluxes, but they are not negligible. These three fluxes are average fluxes and do not consider natural variations due to cloud cover, nor the 7% seasonal variation because the Earth orbit around the Sun is elliptical.

Earth IR flux incident on this spacecraft is 149 Watts.

| Incl deg | Altitude km | Beta | +X OA watts | -X OA watts | +Y OA watts | -Y OA watts | +Z OA watts | -Z OA watts | Total OA watts |
|---|---|---|---|---|---|---|---|---|---|
| 70 | 600 | 0 | 18.5 | 18.5 | 18.5 | 18.5 | 0 | 34.5 | 108.4 |
| 70 | 600 | 30 | 16.1 | 16.1 | 14.5 | 17.9 | 0 | 30.1 | 94.6 |
| 70 | 600 | 45 | 13.3 | 13.3 | 10.9 | 15.2 | 0 | 24.5 | 77.2 |
| 70 | 600 | 60 | 9.3 | 9.3 | 6.7 | 12.4 | 0 | 17.5 | 55.2 |
| 70 | 600 | 75 | 5.0 | 5.0 | 2.5 | 9.1 | 0 | 9.7 | 31.4 |
| 70 | 600 | 90 | 2.4 | 2.4 | 0.7 | 6.9 | 0 | 3.2 | 15.5 |

**Fig. 12.5** Albedo orbit average incident flux (watts) on each side of a 20" × 20" × 36" nadir-pointing spacecraft

## 12.2 Heat Absorption

Figure 12.2 showed the solar flux incident on each side of the 20" × 20" × 36" example spacecraft. The heat absorbed depends on the surface finishes of the sides. For example, if 75% of each side of the spacecraft is covered with solar cells (absorptivity = 0.82) and the rest of the sides are painted white (absorptivity = 0.19), then the net orbit average heat absorbed is only 0.75*0.82 + 0.19*0.25 = 0.6625 times the incident orbit average solar flux. So, the flux incident on each side of the spacecraft in the inclinations and Beta angles listed in Fig. 12.2 must be multiplied by 0.6625 to calculate the heat absorbed by each side. The electric power generated by the solar cells must be subtracted from the calculation of the absorbed heat.

To determine the Earth IR absorbed, multiply together the flux, panel area and surface emittance.

If there are deployable solar panels that block incident fluxes or obscure the view from space, the absorbed heat calculation is more complicated. A thermal model of the spacecraft is needed.

## 12.3 Heat Rejection

For thermal balance, heat absorbed plus heat generated internally must equal heat rejected. The ability to reject heat from a surface of area, A square inches, of emissivity = $\varepsilon$ from a temperature T(°K) is given by Eq. 12.2. This is the emitted heat in a direction normal to the surface. Heat radiated hemispherically (which should be used in calculations of heat rejected by the spacecraft) is typically 5% less.

$$\mathbf{Q(watts)} = \varepsilon * \sigma * \mathbf{A} * \mathbf{T^4} \text{ where } \sigma = 3.66 \times 10^{-11} \text{ Watts/in}^2 \quad (12.2)$$

If the intent of the thermal design is to keep spacecraft temperatures in the range of +5 °C to +25 °C, then each of the spacecraft X and Y sides can reject about 140 watts of heat. Since the sum of the Sun, Albedo and Earth IR absorbed heat ranges from 538 watts to 779 watts (from Figs. 12.2 and 12.3) and hemispherical emissivity = 0.8, the temperature (from Eq. 12.2) will be 16 °C. The same exterior surfaces will be −7 °C to reject 538 watts. The temperature difference between the Hot and the Cold cases is 23 °C.

Where solar cells do not populate the spacecraft exterior, various surface finishes can be used to modify the spacecraft temperature range. Low absorptive finishes can be employed to reduce the absorbed solar and Albedo heat. MLI can be used to nearly eliminate absorbed solar heat; however, MLI reduces the rejection area. By appropriately choosing surface finishes, a thermal balance can be achieved in a temperature range acceptable to the spacecraft components.

The foregoing illustrated the principals involved. An accurate thermal treatment of a spacecraft can be accomplished only through use of a good thermal model.

## 12.4 Heat Generated by the Spacecraft Electronics

The electric power consumption for each spacecraft operating mode and for each component was shown for a specific spacecraft in Chap. 4, Electric Power Subsystem Design. It is repeated here as Fig. 12.6.

From this figure it is seen that the Orbit Average Power is about 20 watts, and that the instantaneous power varies from about 17.9 watts to 20.6 watts. The peak power (excluding the 10 watt fraction of the transmitter power radiated as RF power) is 26 watts. The reaction wheels consume up to about 27 watts while the wheels are accelerating. Since they do so for only very short periods of time, the excess reaction wheel power consumption over the average power can be ignored.

Solar cells convert solar power to electrical power. This reduces the power absorbed.

## 12.5 Tools Available for Altering Spacecraft Thermal Performance

| Spacecraft OAP Requirements | Watts at Voltages | | | Idle Mode Watts | | | | Imaging Mode Watts | | | | Communications Mode Watts | | | |
|---|---|---|---|---|---|---|---|---|---|---|---|---|---|---|---|
| | 5v | 12v | 28v | % | 5v | 12v | 28v | % | 5v | 12v | 28v | % | 5v | 12v | 28v |
| C&DH | 1.5 | | | 100 | 1.5 | | | 100 | 1.5 | | | 100 | 1.5 | | |
| EPS Processor | 0.2 | | | 100 | 0.2 | | | 100 | 0.2 | | | 100 | 0.2 | | |
| Imaging Payload | | | | | | | | | | | | | | | |
| Camera | | 4.0 | | | | | | 10 | | 0.4 | | | | | |
| Image Processor | 3.0 | | | | | | | 15 | 0.5 | | | | | | |
| ADACS | | | | | | | | | | | | | | | |
| Pitch Reaction Wheel | | | 3.5 | 100 | | | 3.5 | 100 | | | 3.5 | 100 | | | 3.5 |
| Roll Reaction Wheel | | | 3.5 | 100 | | | 3.5 | 100 | | | 3.5 | 100 | | | 3.5 |
| Yaw Reaction Wheel | | | 3.5 | 100 | | | 3.5 | 100 | | | 3.5 | 100 | | | 3.5 |
| Star Tracker #1 | 0.5 | | | 100 | 0.5 | | | 100 | 0.5 | | | 100 | 0.5 | | |
| Star Tracker $2 | 0.5 | | | 100 | 0.5 | | | 100 | 0.5 | | | 100 | 0.5 | | |
| Course Sun Sensors | 0.4 | | | 100 | 0.4 | | | 100 | 0.4 | | | 100 | 0.5 | | |
| 3 Torque Rods | | 0.8 | | 100 | | 0..8 | | 100 | | 0.8 | | 100 | | 0.8 | |
| Communication | | | | | | | | | | | | | | | |
| TTM & Image Xmitter | | 30.0 | | | | | | | | | | | 5.5 | 1.6 | |
| CMD Receiver | | 1.5 | | 100 | | 1.5 | | 100 | | | | 100 | | 1.5 | |
| Peak and Ave Power | 6.1 | 36.3 | 10.5 | | 3.1 | 2.3 | 10.5 | | 3.6 | 1.2 | 10.5 | | 3.2 | 3.9 | 10.5 |
| DC/DC Converter Eff. % | 87 | 85 | 85 | | 87 | 85 | 85 | | 87 | 85 | 85 | | 87 | 85 | 85 |
| OAP from Each Source | 7.0 | 42.7 | 12.4 | | 3.6 | 2.7 | 12.4 | | 4.1 | 1.4 | 12.4 | | 3.7 | 4.6 | 12.4 |
| OAP in Each Op Mode | | | | | | 18.7 | | | | 17.9 | | | | 20.7 | |
| Design for OAP of | | 20.7 | | | | | | | | | | | | | |

Fig. 12.6 Power consumption in different operating modes

## 12.5 Tools Available for Altering Spacecraft Thermal Performance

Absorption and radiation of heat can be altered by:

(a) Surface Finishes (white paint will reduce, black paint will increase the temperature)
(b) Metalic conduction from hot parts of the spacecraft to cold parts so it could radiate heat into space
(c) Heat pipes that can increase thermal conduction from one point to another
(d) Heaters (to increase the temperature of components that are too cold - like batteries)
(e) Louvers that can open or close surfaces and can change their absorptivity or emissivity
(f) MLI (Multi-Layer Insulation) to protect a wrapped volume by reducing absorbed radiation
(g) Thermoelectric Coolers (to spot cool electronics parts)
(h) Placing heat generating components where it is easy to get rid of heat

### 12.5.1 The Impact of Surface Finishes

To illustrate how surface finishes can reduce the temperature of surfaces exposed to direct sunlight, the table of absorptivity and emissivity of different finishes is shown in Fig. 12.7.

## Absorptivity, Emissivity, Thermal Conductivity

| Material | α | ε | Kw(/m K) |
|---|---|---|---|
| Aluminum | 0.16 | 0.03 | 205 |
| Aluminum Irridited | 0.14 | 0.11 | 205 |
| White Paint (GSFC) | 0.19 | 0.92 | N/A |
| Black Paint | 0.96 | 0.86 | N/A |
| Copper Foil | 0.32 | 0.02 | 401 |
| Stainless Steel | 0.42 | 0.11 | 16 |
| Mylar Tape (Aluminum Backed) | 0.19 | 0.03 | N/A |
| Delrin | 0.96 | 0.85 | 0.42 |

**Fig. 12.7** Absorptivity and emissivity of different surfaces and the thermal conductivities and specific gravities of selected metals

As an example of the effectiveness of paint in controlling the temperature of spacecraft outer surfaces, consider the previous example of a 1 m² plate; but in this example the plate is painted with Corning White Paint DC-007. The equation for temperature of the plate is now given in Eq. 12.3 where S is solar flux in Watts/m² and $\sigma = 5.67*10^{-8}$ Watts/m².

$$T(°K) = \left(S*(\alpha/\varepsilon)*(A_i/A_r)/\sigma\right)^{0.25}$$
$$= \left(1367^*(0.19/0.84)^* 1/(5.67^*10^{-8})\right)^{0.25} = 271.7° K \text{ or} -1.3°C \quad (12.3)$$

This is a significant change. By reducing the fraction of the area of the plate painted white, the temperature of the plate can be brought into the region where electronics can safely operate.

### 12.5.2 Thermal Conduction

In the thermal design, it is important to not only design for heat balance at a desired temperature, but also to move heat from a hot part of the spacecraft to a cool part, or to an area from where heat can be radiated out to space. The heat transport process has radiated and conductive paths. Considering conduction only, the fundamental equation for determining the amount of heat a metallic conductor can move from one place to another is given in Eq. 12.4.

$$q = k^* (A/L)^* dT, \text{where} \quad (12.4)$$

k = conductivity in KW change per °K (or °C) over length m
A = the cross section of the conducting structure in m²
dT = difference in temperatures between the two ends of the conductor
L = the length of the conductive path in m, and
q = is the heat conducted in watts

## 12.5 Tools Available for Altering Spacecraft Thermal Performance

The penalty on spacecraft weight that heat conduction imposes can be substantial. To illustrate this, consider the 20″ × 20″ × 36″ spacecraft example already discussed. If the goal is to move 20% of the absorbed solar heat from a sunny side to the opposite shaded side, consider using the three existing aluminum equipment decks of 20″ × 20″ each. Aluminum conductance is 3.6 W/in C°. To compute how thick these plates have to be and to ensure (for purposes of this example) that thermal gradients in these decks should be <10 °C, Eq. 12.5 must be satisfied. Equation 12.5 simply rearranges Eq. 12.4.

$$\text{Thickness} = Q^*L / (k^* \text{width}^* dT)$$
$$\text{Thickness} = (0.2 \times 363 \text{W})^* 20\text{in} / (3.6\,\text{W}/\text{in}\,C^* 20\,\text{in}^* 10C) = 2\,\text{inches}$$
(12.5)

The combined weight of these three aluminum decks is excessive. The spacecraft weight budget prohibits such a solution.

There are a number of ways to overcome this problem, including:

- Not covering the entire Sun-facing panel with solar cells
- Employing other exterior finishes (as discussed in Sect. 12.5.1)
- Using deployable arrays so that the exterior of the spacecraft could be covered with a low absorptance finish
- Isolating the body mounted array panels from the structure
- Adding heat pipes into the three interior decks (see Sect. 12.5.4)

### 12.5.3 Conducting Heat across Screwed Plates or Bolt Boundaries

Thermal conductivity through a metal structure is reduced at each joint between members of the structure. A bolted joint, for instance, creates thermal resistance. Thermal conductivity through bolted plates depends on the pressure with which the plates are bolted together. The conductance of bolted joints as a function of the bolt torquedown for various bold diameters is given in the Aerospace Thermal Control Handbook Volume 1.

### 12.5.4 Heat Pipes

Heat pipes are hermetically sealed extrusions, or pipes partially filled with a liquid used for spreading heat over relatively long distances with small gradients. In spacecraft applications, the aluminum pipes have internal grooves and use ammonia as the working fluid. Capillary forces wick the liquid along grooves from the cold end to the heat source. At the heat source, the liquid evaporates and the vapor flows down the pipe center and condenses on the colder areas of the pipe. Contrary to normal conductors (e.g. through metal plates), gradients do not increase for the distance the heat travels down the pipe. Due to the small capillary wicking forces, these pipes need to be horizontal (within 0.05″) to be tested in 1G.

Heat pipes have a transport limit. A typical ½″ OD heat pipe has a 5000 Watt/inch transport limit. One pipe can carry 250 W 20 inches. There are two gradients - one at the evaporator (where the heat enters the pipe), the other at the condenser (where the heat leaves the pipe). A ½″ OD evaporator conductance is typically 4 W/in-C° (where the heat travels into the pipe). Pipe condenser conductance is 8 W/in-C°. These conductances do not include gradients at the heat pipe flanges or interface gradients.

In the example in Sect. 12.5.2, 72 W had to be carried across 3 decks with less than 10 °C gradient. A single pipe (almost 80″ long) could be attached near the outboard edge of each deck. Each pipe would need to carry 24 watts 40 inches, approximately 1000 watt-inches, well within the transport capability. Allowing for pipe radii, the contact area along the edge would be 16 inches. The pipe gradients would be:

$$\begin{aligned} dT_{evap} &= Q/\left(h_{evap}{}^{*}L_{evap}\right) = 24W/\left(4W/in-C^{*}16\,in\right) = 0.4°C \\ dT_{cond} &= Q/\left(h_{cond}{}^{*}L_{cond}\right) = 24W/\left(8W/in-C^{*}16\,in\right) = 0.2°C \end{aligned} \qquad (12.6)$$

This is a very low gradient. The weight of 80″ of heat pipe (with 1″ × 0.040″ flanges) is 1.2 lbs. As seen, heat pipes are an effective means to mitigate gradients in the equipment decks.

## 12.5.5 Louvers

Louvers are like adjustable blinds that permit partially opening a surface of the spacecraft to the outside (either on the space side to control emissivity or on the sunny side to control absorptivity). Louvers were popular in years past, but now they have lost favor to alternate thermal control hardware options.

## 12.5.6 Heaters

Heaters are used to spot heat specific components, such as batteries, that are not allowed to drop below their respective minimum temperatures. Since heaters consume electric power, a rare commodity in a spacecraft, they should only be used where absolutely necessary. Heaters operate in a bang-bang manner; they are either ON or they are OFF. The duty cycle determines the average temperature of the heaters.

## 12.6 Constructing a Thermal Model of the Spacecraft

In the preceding examples, the spacecraft was idealized to simplify analyses. For more accurate analyses a thermal model is needed. The thermal model keeps track of complex heat flows, and it can predict temperatures, both steady state and transient.

The model uses a series of nodes that represent components, external surfaces, internal structure, items that dissipate heat, and items whose temperature is of special interest.

The nodes are interconnected with conductive (linear) and radiative (fourth power) couplings, similar to an electrical circuit comprised of resistors. The model iterates to predict heat flow between interconnected nodes, and outputs the temperatures of each node. *Thermal Desktop* is the best known software for constructing spacecraft thermal models. This graphically based program includes:

- Geometry generation (or importing common geometric models)
- Conductive coupling calculations
- Internal radiation couplings (radKs)
- External radiation couplings (radKs) to space
- External absorbed environmental flux calculations
- Temperature and heat flow calculations (using SINDA)
- It then presents the SINDA Model results in graphical, geometric or tabular format

When constructing a model, resist the urge to make it more detailed or more granular than it really has to be, lest the model become the problem rather than part of the solution. In Thermal Desktop, subdividing a surface takes but a few mouse clicks. Simplifying a subdivided surface requires deleting the article, rebuilding it, reconnecting and verifying it. Start simple to get a feel for thermal drivers. Scrutinize the results and crosscheck them against hand or EXCEL calculations. Every model has limitations and flaws. Efficient modeling requires understanding the model limitations to adjust the flaws that introduce large errors.

The thermal model is used to provide temperature predictions for various mission phases, Beta angles and operating modes. These results are incorporated into a summary of predictions that can be compared to the component temperature requirements. The thermal model is also used to predict spacecraft temperatures during thermal testing.

## 12.7 A Point Design Example

This section provides an example of how requirements must be modified, and hardware manipulated into a cohesive design. For purposes of this design, it is assumed that heater power is unavailable.

Figure 12.1 indicated that the operating temperatures for electronics is allowed to vary from −20 °C to 40 °C. Some of the Electronics may have to operate over a narrower range. So, the design should be for a − 15 °C to 40 °C allowable temperature variation.

The cause of temperature uncertainties and their likely magnitudes are discussed next, leaving the discussion of the range of orbit Betas for last.

(a) Since models are imperfect, in a thermal design, typically a 5 °C margin is used in both the hot and cold cases. So, the design must strive to keep the temperature between −10 °C and +35 °C.

(b) An additional 5 °C should be budgeted to account for variations in surface finish absorptance and emittance, and for finish degradation toward End-of-Life.

(c) Earlier in this chapter, incident fluxes were computed. However, these fluxes vary over time. Specifically, solar incident flux varies 7% annually. Earth IR and Albedo also vary considerably even at a given Beta angle. Collectively, these variations can cause up to 10 °C temperature variation for a given Beta.

(d) Eclipses are expected to cause at least 5° to 10 °C variation over a single orbit. This transient effect is not addressed by the orbit average fluxes calculated for the various Betas.

(e) Budget an additional 5 °C to account for transients due to component power consumption variations, such as turning the transmitter ON.

Grouping all of these factors together, the temperature uncertainty takes up about 35 °C of the available operating temperature range goal.

Gradients between the warmest and coldest components are on the order of 10 °C. However, satellite gradients can be reduced by adding heat pipes.

Subjecting the 20″ × 20″ × 36″ spacecraft (covered with solar cells) over a large range of Betas caused a 23 °C spacecraft average temperature change. Combining the 23 °C variation due to variations in Beta, and other factors already discussed, would exceed the allowable temperature range.

Reducing the effects of solar variations over the range of Betas is key to successful thermal design. The goal is to modify the finishes so that the Beta = 75° hot case and Beta = 0° cold case result in nearly the same spacecraft temperature.

Assume that the finish of 30% of the external surface can be modified, as shown in Fig. 12.8.

The spacecraft flies with +Z facing anti-nadir. Velocity is in the +X direction. Visualize several Betas, starting with Beta = 0°. The Sun is in the plane of the Beta = 0° orbit, and the eclipse is the longest for any orbit altitude. During each Beta = 0° orbit, the Sun illuminates the +X, then the +Z, followed by the −X, and lastly, the −Z side of the spacecraft. Since Beta = 0° is the cold case, the choice of the finish should allow collecting as much energy as possible. To keep the spacecraft from becoming too cold, the +X, +Z, −X and −Z sides should remain black, or covered with highly absorptive solar cells. The two Y sides receive no incident sun.

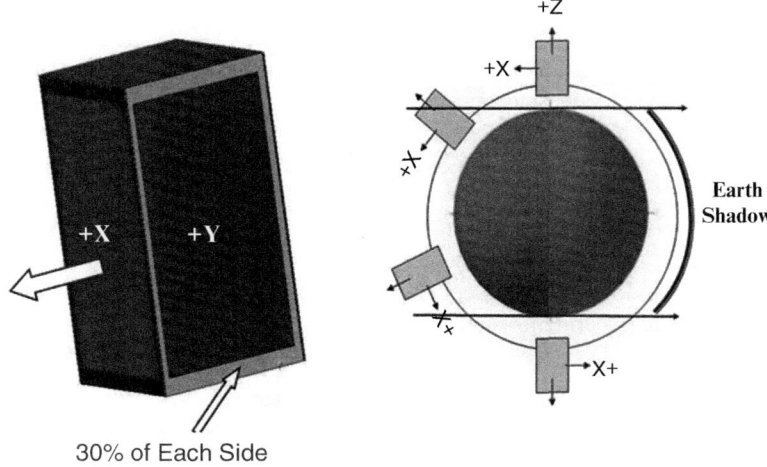

**Fig. 12.8** Nadir-pointing spacecraft flying in +X direction 30% of its surfaces available for application of surface finishes

During the Beta = +75° hot case, when there is no eclipse, the Sun illuminates the +Y spacecraft face. During the Beta = −75° hot case, the Sun illuminates the −Y face. The spacecraft receives solar illumination constantly. For this reason either a low solar absorptance finish or MLI should be applied to the +Y and −Y sides to reduce the environmental heat absorbed during the hot case.

MLI is preferred since:

(a) MLI is more effective blocking the incident Sun in the hot case
(b) MLI reduces the heat escaping from the spacecraft during the Beta = 0° cold case

A spreadsheet calculation indicates that, with the remaining 70% of the surfaces having absorptivites of 0.82, the temperature variation of 23 °C (between Beta 0° and 75°) is reduced to about half.

## 12.8 Thermal and Thermal Vacuum Testing

A thermal vacuum test series includes several test phases: Thermal Balance, Thermal Cycling and testing for Outgasing. Thermal Cycling and Outgasing are addressed in Sect. 14.5. Thermal Cycling is a spacecraft workmanship test. Outgasing prevents volatile contaminants from condensing on sensitive surfaces. The Thermal Balance test simulates the hot and cold flight environments to obtain spacecraft flight temperatures. Temperatures are recorded for later analyses to determine differences between model predictions and test results.

Thermal Balance is usually tested at two or three balance points (conditions). The balance points vary the spacecraft's radiative boundary conditions to simulate the worst hot and cold (orbit average) orbital environmental conditions. The spacecraft operates in a steady state mode (without cycling loads or heaters) until the spacecraft attains steady state temperatures. At each balance point, it takes about 12–24 h to reach stability (defined as a state where temperature variation is less than 0.1 °C/h).

The test is conducted in a vacuum chamber with an LN2 shroud. The spacecraft is surrounded by heated (and sometimes cooled) high emittance plates (painted black or black anodized). A six-sided spacecraft may have at least six heater plates surrounding the spacecraft to represent the heat incident on each of the sides. If there is more than one finish per spacecraft side, multiple plates at different temperatures may be placed opposite those spacecraft sides. The temperatures of the plates are set to present the spacecraft with an IR environment simulating the combined solar, Albedo and Earth IR fluxes. The combined simulated environment is referred to as the Equivalent Sink Temperature. Since the actual heater plate emittance is 0.85–0.90, plate temperatures are adjusted to reproduce the Effective Sink Temperature. If it is necessary to reproduce equivalent sink temperatures below −50C°, the plate needs to be cooled (usually by GN2). Spacecraft power consumption has to be monitored during the Thermal Balance test.

The resulting steady state spacecraft temperatures are compared with the model predicted temperatures in a process called *model correlation*.

## 12.9 Model Correlation to Conform to Thermal Test Data

Before the thermal model can be used to provide final flight temperature predictions, differences between actual test results, obtained during TVAC testing, and model predictions must be brought into agreement with one another. For this purpose, the model must be altered to agree with test results. The methods to alter the model require considerable experience.

As an example, however, let us say that the temperature gradients in TVAC exceed those that the model predicted. In this case, the model thermal interface conductances may be adjusted until the temperatures and gradients agree.

## 12.10 Final Flight Temperature Predictions

After the model has been adjusted to make predictions that best match the Thermal Balance Test results, final temperature predictions are made for various important points in the spacecraft and for various cases in the mission life of the spacecraft. Illustrative examples are given in Figs. 12.9 and 12.10. In Fig. 12.9, the rapid rise in temperature of the 10 Watt transmitter is seen when it is turned on.

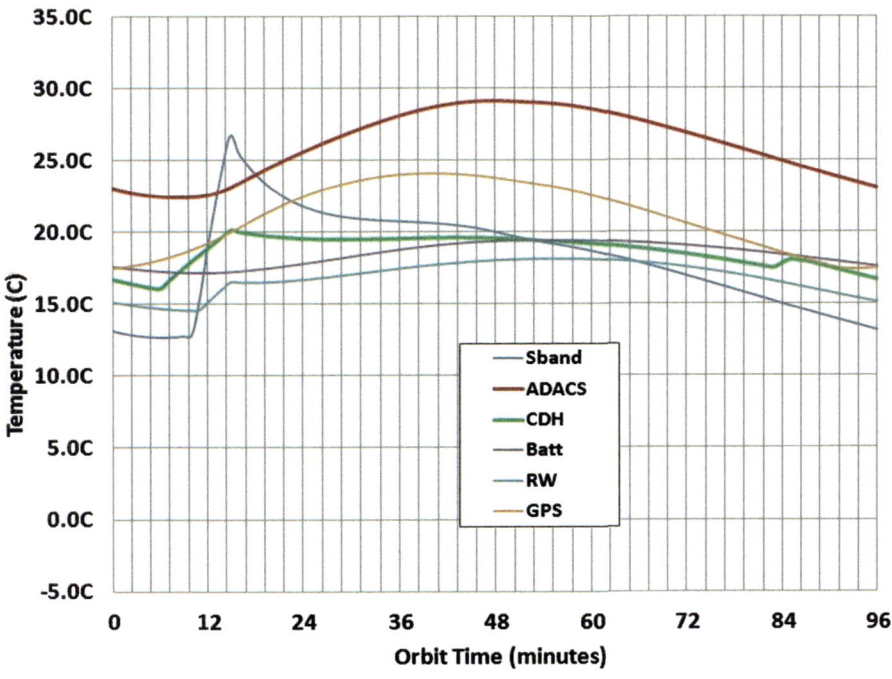

**Fig. 12.9** Temperature ranges are acceptable. The impact of the 10 Watt transmitter turning ON is evident

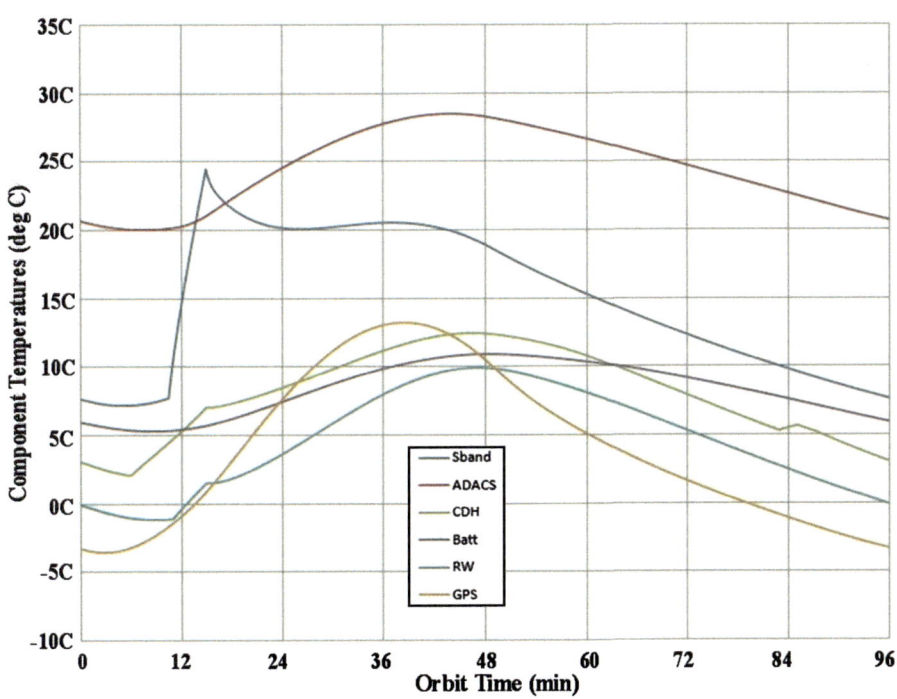

**Fig. 12.10** Hot case. ADACS temperature approaches the limit of the acceptable temperature range

# Chapter 13
# Radiation Hardening, Reliability and Redundancy

There are at least three factors that affect spacecraft orbit life. These are (1) the amount of radiation protection provided for the orbit altitude of the spacecraft, (2) the reliability of the components and of the spacecraft system and (3) the redundancy built into the spacecraft.

## 13.1 Radiation Hardening

There are two kinds of radiation effects that can cause the spacecraft to fail: (1) total radiation dose and (2) single event upset (SEU).

### 13.1.1 Total Dose

Total Radiation increases sharply with orbit altitude and linearly with time. Protection against total radiation can be achieved with shielding. The table in Fig. 13.1 shows the radiation environment in an aluminum shielded space as a function of altitude, mission life in years and the thickness of the aluminum shield. Several cases are plotted in Fig. 13.2a–c.

Electronic component vulnerabilities to radiation dose vary greatly from component to component. Most commercial CMOS parts range from 1 to 10 kRAD, while RAD hardened space qualified parts vulnerability is well over the 100 kRAD range. NASA maintains a parts list that provides data for the reliability as well as radiation hardness of components. RAD hard parts, of course, are very expensive, so a designer must consider radiation hardness mitigation by not only parts selection and shielding, but also by introducing redundancy in the electronics architecture. More will be said about this later in this chapter.

| Shielding Thickness Mil AL | Total Dose (kRad) 700 km for | | | Total Dose (kRad) 1000 km for | | | Total Dose (kRad) 2000 km for | | |
|---|---|---|---|---|---|---|---|---|---|
| | 1yr | 5yrs | 12yrs | 1 yr | 3 yrs | 5yrs | 1 yr | 5yrs | 12 yrs |
| 50 | 7.16 | 3S.80 | 85.92 | 14.00 | 42.00 | 70.00 | 164.68 | 82340 | 1976.16 |
| 60 | 6.51 | 27.66 | 66.12 | 11.00 | 33.00 | 85.00 | 125.63 | 628.15 | 1507.86 |
| 70 | 4.33 | 21.66 | 61.96 | 8.80 | 2640 | 44£0 | 97.86 | 489.30 | 1174.32 |
| 80 | 3.58 | 17,90 | 42.96 | 7.00 | 21.00 | 35.00 | 80.19 | 400.95 | 962.28 |
| 90 | 2.96 | 14.80 | 35.52 | 5.80 | 1740 | 29.00 | 65.71 | 328.55 | 788.52 |
| 100 | 247 | 12.36 | 29.64 | 5.00 | 16.00 | 25.00 | 54.34 | 271.70 | 652.08 |
| 110 | 2.08 | 10.40 | 24.96 | 4.40 | 13.20 | 22.00 | 45.55 | 227.75 | 546.60 |
| 120 | 1.83 | 9.1 S | 21.96 | 3.90 | 11.70 | 19.50 | 39.89 | 199.45 | 478.68 |
| 130 | 1.58 | 7.90 | 18.96 | 3.50 | 10.50 | 17.50 | 34.29 | 171.45 | 411.48 |
| 140 | 1.38 | 6.90 | 16.66 | 3.10 | 9.30 | 15.50 | 29.81 | 149.05 | 357.72 |
| 180 | 1.20 | 6.00 | 14.40 | 2.80 | 840 | 14,00 | 25.80 | 12100 | 309.19 |
| 160 | 1.08 | 5.40 | 12.96 | 2.50 | 7.50 | 12.50 | 23.11 | 115.55 | 277.32 |
| 170 | 0.96 | 4.80 | 11.52 | 2.20 | 6.60 | 11.00 | 20.45 | 102.25 | 245.40 |
| 180 | 0.88 | 4.40 | 10.56 | 2.00 | 6.00 | 10.00 | 18.66 | 93.30 | 223.92 |
| 190 | 0.79 | 3.96 | 9.46 | 1.84 | 5.52 | 9.20 | 16.67 | 83.38 | 200.04 |
| 200 | 0.73 | 3.65 | 8.76 | 1.70 | 5.10 | 8.50 | 14.60 | 73.00 | 175.20 |
| 210 | 0.67 | 3.35 | 8.04 | 1.60 | 4.80 | 8.00 | 13.33 | 66.65 | 159.96 |
| 220 | 0.61 | 3.05 | 7.32 | 1.54 | 4.62 | 7.70 | 12.08 | 60.40 | 144.96 |
| 230 | 0.68 | 2.90 | 6.96 | 1.50 | 4.50 | 7.50 | 11.43 | 57.15 | 137.16 |
| 240 | 0.57 | 2.85 | 6.84 | 1.44 | 4.32 | 7.20 | 11.17 | 55.85 | 134.04 |
| 260 | 0.55 | 2.75 | 6.60 | 1.40 | 4.20 | 7.00 | 10.73 | 53.65 | 128.76 |

**Fig. 13.1** kRad radiation levels in aluminum shield vs. number of years at various altitudes

Figure 13.2a shows that, if we use components that can operate to 10 kRAD total dose, a 5-year mission life can be achieved at up to 700 km altitude with a shielding thickness of 110 mils. Parts for this mission are relatively low cost. If the orbit is at 1000 km altitude, 180 mils of shielding would be needed for a 5-year mission. This would increase the card cage weight from about 0.81 lb. to about 1.33 lb.

*Single Event Upset* is caused by a single, high energy particle striking a critical point of a memory or an integrated circuit. The result can be either (1) the creation of a soft error, where a bit may change state, causing an error, but not the destruction of the circuit, or (2) a "latchup" may occur, destroying the circuit permanently. The only certain way to mitigate against an SEU latchup is to select components that are not prone to this kind of a latchup.

## 13.1 Radiation Hardening

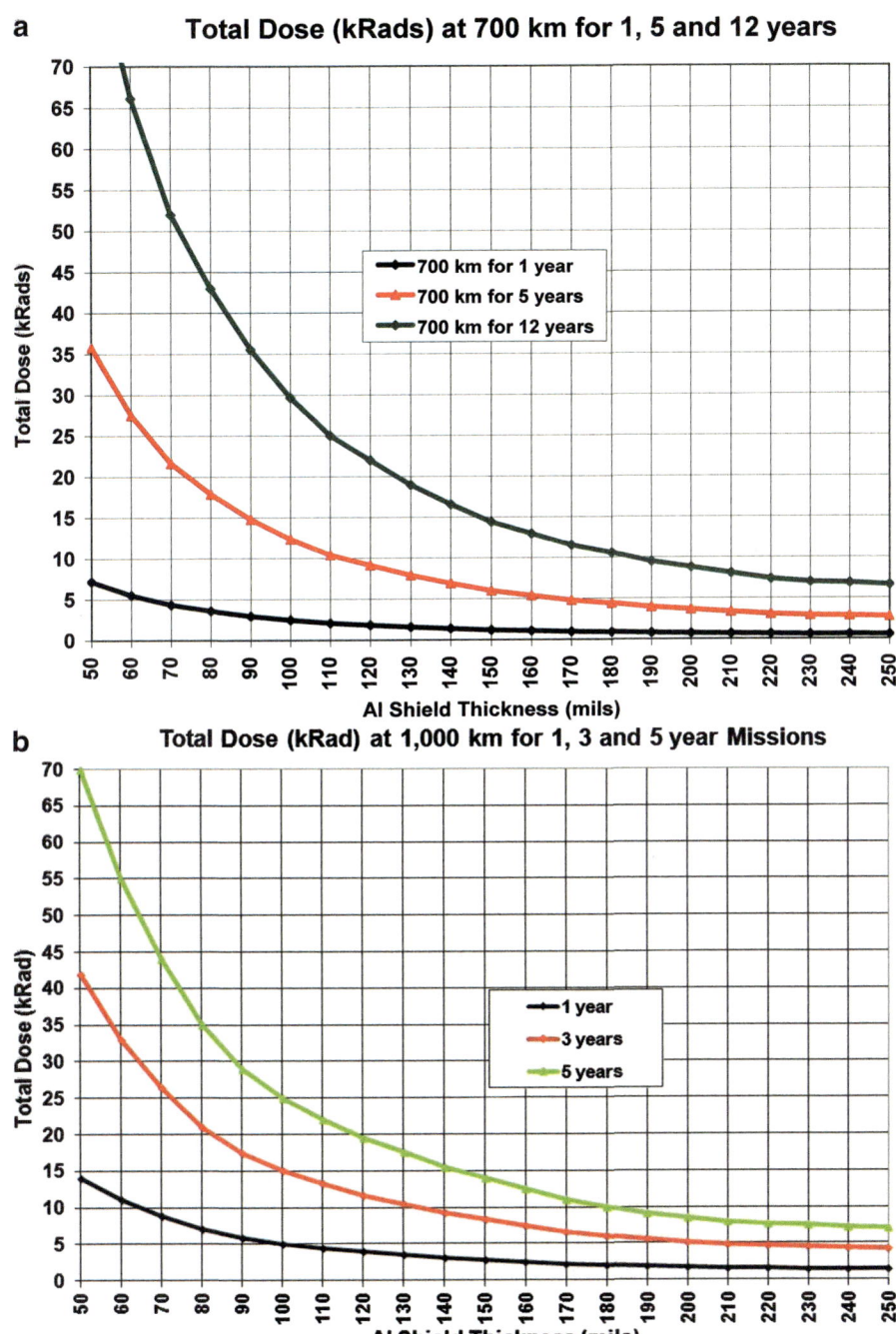

**Fig. 13.2** (**a**) kRad dose vs. various aluminum shield thicknesses and mission life (**b**) kRad dose vs. mission life and altitude for various aluminum shield thicknesses (**c**) Shield thickness vs. total kRad dose for various altitudes and mission durations

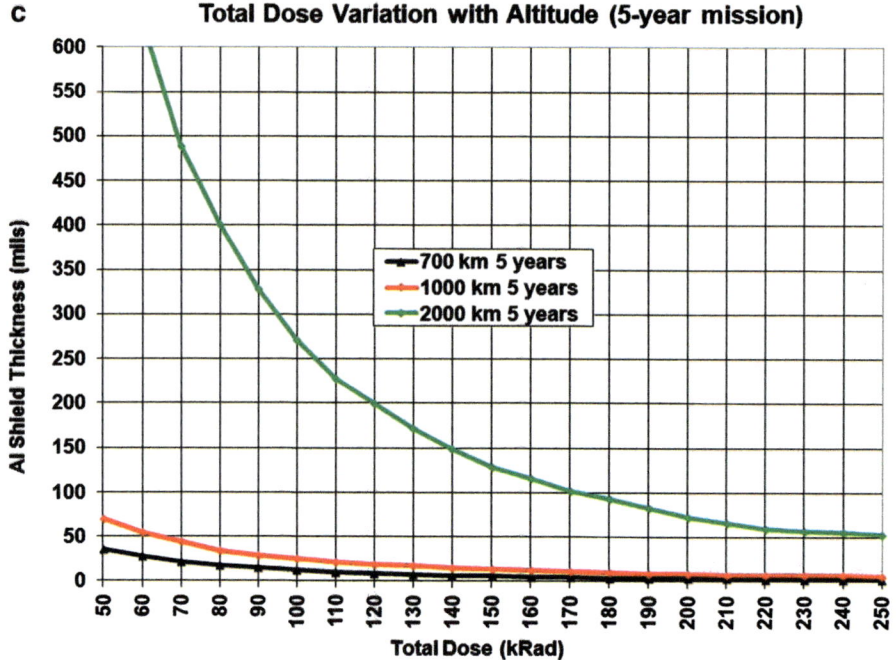

**Fig. 13.2** (continued)

## 13.2 Reliability

The designer must ensure that the spacecraft could achieve the desired mission life. In order to do so, a reliability analysis must be undertaken. A system level reliability analysis starts with a system reliability block diagram. This is a series arrangement of all the blocks in a block diagram that must each work for a spacecraft function to operate successfully. For example, a reliability diagram of the RF transmit portion of a spacecraft employing a single-string design is shown in Fig. 13.3. The diagram reflects the connections of the boxes or functions that must each work. Redundant elements, like the batteries, are shown connected in parallel, indicating that if one fails, the system may still function.

Reliability is computed from the component failure rate, $\lambda$ failures per hour. $R = e^{-\lambda T}$, where T is mission life in hours. Mean Time Between Failures (MTBF or Mean Time To Failure (MTTF) is $1/\lambda$. So, if a part has an MTBF of 500,000 H, over a 5-year (43,800 H) mission, reliability is $R = e^{-43,800/500,000} = 0.91613$. The Probability of Mission Failure over the 5 year period is $P_F = 1-R$ or 0.0837 or 8.37%. For 10 year life and a $P_F$ of 10%, we need a system reliability, $R > 0.89984$. This is a system MTTF of 830,000 h.

## 13.2 Reliability

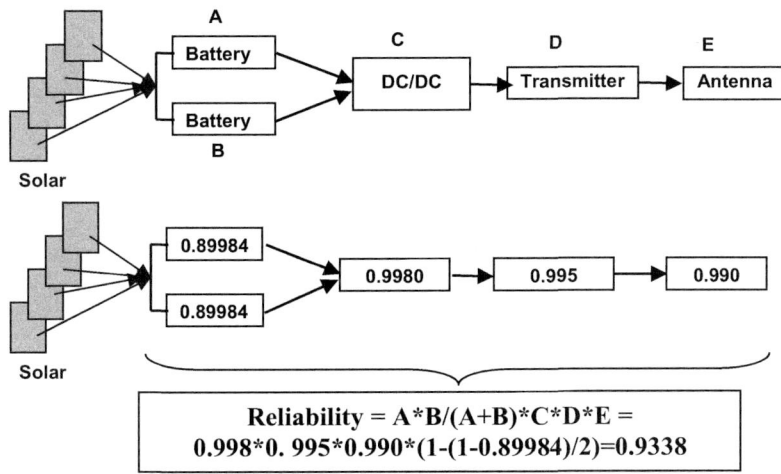

**Fig. 13.3** Reliability diagram

If a component, like a transmitter, is not used continuously, the duty cycle of utilization should be taken into account in the MTTF requirement. For example if a transmitter is used for only 10 min per 96 min orbit, a 5-year mission life only requires an MTTF of 43,500 h (for a probability of failure of 10%), and the reliability requirement can be reduced to 0.90043.

Probability that the transmitter will fail is: P(failure of transmitter)=
=1−0.995 = 0.005 = 0.5%
The probability that one battery will fail is (1−0.89984) = 0.10016 = 10.016%
The probability that both batteries will fail is 0.05 = 5%
The reliability of the 2 parallel batteries is 1−0.05 = 0.94992.
Note how Battery Reliability has increased through use of two batteries.
System reliability (from battery input to antenna output) is $R = 0.9338$, and the probability of system failure is 0.0662 or 6.62%.

The reliability numbers for components are combined to obtain the reliability of the system by:

- Multiplying the Reliabilities of Series Elements together
- If two components are in parallel, let $P_1 = 1-R_1$ and $P_2 = R_2$ and compute $R = (1-P_1*P_2/(P_1 + P_2))$

Sometimes, failure of a component does not end a mission, it only degrades performance. For example, if there are two redundant transmitters, loss of one does not reduce mission capability. If one of several solar arrays quits working, the mission is only degraded, not ended. Reliability calculations should take into account mission degradation, not only mission failure.

## 13.3 Redundancy

It is a common practice to use redundant components where the reliability of a component is less than desired. It is for this reason that, usually, redundant transmitters, batteries and other components are used in the spacecraft. While use of redundant components increases the cost, size, complexity and weight of a spacecraft, it also significantly increases its probable mission life.

Redundancy should be used judiciously and only where absolutely necessary, because redundancy increases system complexity and introduces new components whose reliability must also be taken into consideration. Some of the applications for redundancy are:

- Where a component is susceptible to failure by SEU
- Where the component reliability is too low to meet mission requirements
- To eliminate single points of failure in critical paths
- When it is inexpensive to use a redundant component

So far, only hardware reliability and redundancy were discussed. However, use of redundancy measures also applies to spacecraft software. Some of the software redundancy measures are to:

- Put multiple copies of the spacecraft operating software in EPROM memory
- Use uploadable software
- Use majority logic (perform the calculations in 3 processors simultaneously, and select the result if at least 2 of the processors gave the same result). This requires hardware redundancy as well

# Chapter 14
# Integration and Test

Integration is the process of testing the components of the spacecraft, assembling them, putting the completed spacecraft through functional testing, then performing thermal, vibration and thermal vacuum tests. After that, the spacecraft is ready to be shipped to the launch site, where it is functionally retested to ensure that no damage was done to the spacecraft. Then, the spacecraft is integrated with the launch vehicle.

This process is accompanied by writing test plans for each of these tests.

Integration and Test typically takes about 3 months on a new spacecraft, and not much less on one that was previously built. A typical schedule is shown in Fig. 14.1.

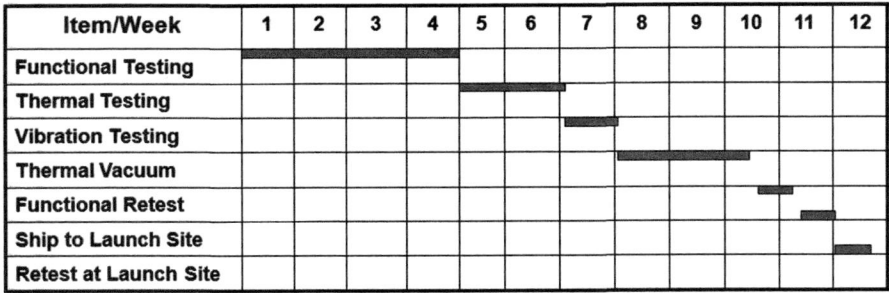

**Fig. 14.1** Typical integration and test schedule

## 14.1 Component Level Testing

Each component (computer, radio, ADACS, etc.) is tested to ensure it meets functional specifications and the thermal and vibration environment it will see on orbit. These tests are either performed by the vendor who supplies the component or by the spacecraft team. For mature components that have been used in previous spacecraft, the vibration and thermal or thermal vacuum tests can be omitted, since these tests will later be performed on a finished spacecraft level.

Deployables must be tested to ensure that they will deploy in a zero G environment in space. Zero G environment can be simulated in the laboratory by suspending each part of the deployable structure with pulleys, cables and counterweights. Deployment should be interrupted and restarted to ensure that the deployment does not rely on the inertia of a moving part. Figure 14.2 illustrates a large, deployable solar array being deployed in the laboratory while simulating zero G.

**Fig. 14.2** Simulating zero G while deploying a large solar array
(By permission from Orbital-ATK)

There should be a sufficient supply of pyro devices, if they are used, to permit repeated deployment of pyro-initiated deployables.

The spacecraft structure undergoes static load and vibration testing. These are usually performed before the spacecraft components are assembled and integrated with the structure. For vibration testing the structure, usually mass mockups of the electronics are used.

### 14.1.1 The "Flat-Sat"

It is important to start the integration and test process as soon as possible, even before most of the mechanical hardware is available. For this purpose, a "Flat-Sat" should be built. This is a flat surface or table onto which the spacecraft components (or duplicates, if available) are laid out and harnessed with a functional duplicate of the spacecraft harness. The result is the spacecraft electronics integrated on a flat surface. This is called a "Flat-Sat."

The Flat-Sat is populated with the electronics subsystems when they are ready, resulting in an early functional test of each subsystem, and, eventually, of the entire spacecraft. Availability of a Flat-Sat can speed up the integration process.

It is at this level that conceptual and technical problems can be discovered and fixed. Aside from hardware problems, most problems discovered are software related. Not that the software does not work as specified. The problem most often is that the software, as *specified*, is not as *intended*, and the software must be revised. To discover the problems and to "debug" the system a Flat-Sat level test is most useful.

## 14.2 Spacecraft Level Tests

*The Objectives of Functional Tests* are, basically, to get the assembled spacecraft working as the system it is intended to be. So, functional testing involves verifying that all components of the spacecraft can be turned on and off via ground command through the Electronics Ground Support Equipment (EGSE) speaking with the C&DH. This is followed by exercising the telemetry collection capabilities, then the ability to transmit telemetry to the EGSE (bypassing the RF link). Then, all the uploadable commands are exercised, and the ability of the spacecraft to perform those commands is tested. Similarly, correct performance of commands to the payload and outputs from the payload are verified. Finally, the entire system is operated through the RF links.

RF link performance is tested at an antenna range. Antenna patterns are measured when the antennas are mounted on the spacecraft. In this way, the effect of the spacecraft on the antenna patterns is checked.

RF signals from (and to) the spacecraft must be attenuated to simulate the minimum detectable signal levels. Often, it may be difficult to near-impossible in a laboratory, or even in an open range environment, to achieve enough attenuation to properly simulate the minimum detectable signal levels. So, one must do the best one can.

Testing the ADACS is particularly difficult. It is best done if an ADACS Dynamic Simulator is available. This hardware (and software), through the EGSE cable, takes the outputs of the spacecraft actuators (reaction wheel speeds, torque rod activities and the spacecraft instantaneous position - calculated by the on-board ADACS computer). The Dynamic Simulator, shown in Fig. 14.3a, connected to the spacecraft ADACS computer, computes the vector magnetic and solar environments corresponding to the present position of the spacecraft, simulates disturbance torques and solves the equations of motion to predict what the spacecraft attitude will be and feeds corresponding magnetic and solar vector information back to the ADACS. In this way, a "hardware-in-the-loop" simulation of the spacecraft flight is obtained. The spacecraft instantaneous attitude is also pictured on the Dynamic Simulator display, shown in Fig. 14.3b.

## 14.3 Environmental Testing

### 14.3.1 Vibration Tests

Vibration tests are typically used to qualify the spacecraft for the launch environment. These tests usually include sine, sine-burst, and random test cycles. The first decision to be made is the choice of the facility where the test will be performed. The two primary considerations in choosing a test facility are the capability of the vibration table and the number of instrumentation channels available. The vibration table must be able to drive the spacecraft and all the test fixtures to the desired vibration spectra. In the case of the sine-burst test, it should be able to drive the spacecraft to the desired acceleration at a frequency lower than 1/3rd the lowest spacecraft frequency. The second consideration is the number of instrumentation channels the facility can support. Each axis of measurement requires one channel. For example, a triax, which measures accelerations in three axes, requires three channels. The number of channels required is up to the discretion of the structural engineer. In theory, a test can be performed without instrumenting the spacecraft at all; as long as the spacecraft withstands the test environment without failing, the test is a success. It would be taking a significant risk, especially during random vibration testing, but it would be acceptable.

An interface fitting must be fabricated to attach the spacecraft to the vibration table. The test facility will provide the bolt pattern for bolting to the table. There may be two different bolt patterns, one for the lateral test table and the other for the vertical test table. The interface fitting should have a very high resonant frequency

## 14.3 Environmental Testing

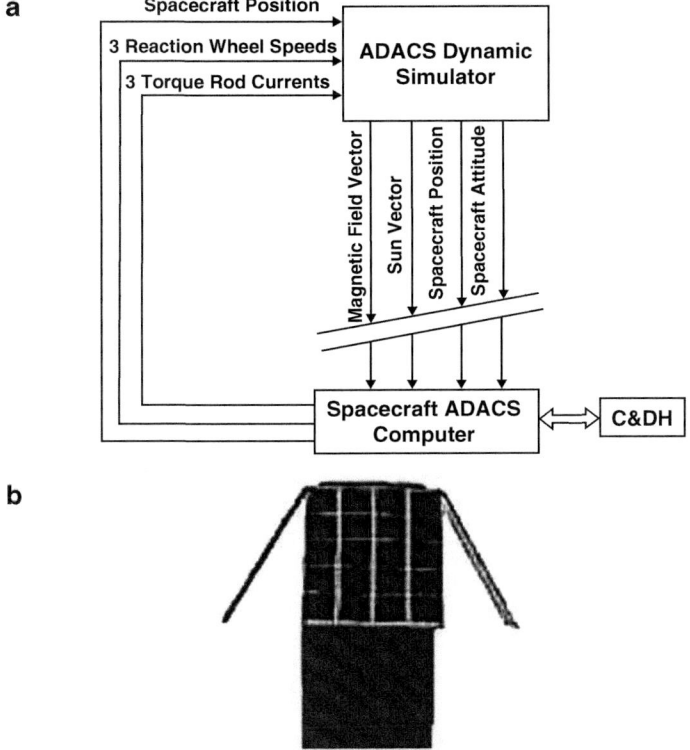

The Dynamic Simulator Display Contains:
- Commanded and Actual Reaction Wheel Speeds (3) vs. Time
- Commanded and Actual Torquer Activities vs. Time
- Earth Magnetic Field Components in Inertial Coordinates vs. Time
- Magnetic Field Components in Spacecraft Coordinates
- Reaction Wheel Torque Commands (3) vs. Time
- Sun Vector Components vs. Time
- Spacecraft Velocity vs. Time
- Spacecraft Position vs. Time
- Spacecraft Attitude vs. Time
- Spacecraft Attitude Picture vs. Time (in Ram and Side View vs. Time)

**Fig. 14.3** (**a**) Hardware in the loop simulator (**b**) Dynamic simulator "flies" the spacecraft in the "hardware-in-the loop" laboratory test setup

(>1000 Hz) when constrained at the test table bolt pattern with the spacecraft modeled as a concentrated mass, at its center of gravity, and attached to the interface fitting, using an RBE3. The fitting should be as light as possible, as the combined mass of the fitting with the spacecraft mass has to be driven during test. If possible, the interface fitting should not bolt to the vibration table beneath the spacecraft, as this would require unbolting the spacecraft from the fitting before reorienting the assembly on the table between test axes.

During tests, the spacecraft is in the launch configuration, thermal blankets are typically not installed, and the spacecraft is unpowered. A lightweight bag that allows the attachment of accelerometers is placed over the spacecraft. Lifting hardware, which is not part of the flight configuration, is removed. Solar arrays are installed and latched in the stowed configuration.

Testing is a relatively expensive activity and is usually charged by the time spent in the facility. During testing, the structural/test engineer is responsible for the hardware. Typically, the test facility people will answer questions about what the capability of their facility is, but they will not provide any guidance about the test levels or procedures. Any deviations to the written test plan will require a written note on the plan along with a signature. Because of this environment, it is imperative that a full pretest analysis be performed before arriving at the test facility. This pretest analysis should include FEM runs for all the test cases along with printed results of the responses at each accelerometer location. From the pre-test, it should become obvious which test cycle is critical (produces the lowest margin of safety). If the test is run in multiple axes, there will be one that is most critical. Beginning in the non-critical axes can be used to gain knowledge about the structure before applying the critical test loads. The order of testing in each axis can also be used to verify the instrumentation before testing at the higher levels. The first and last test cycles in each axis will be signature cycles (either sine or random). Sine signature cycles normally run from 5–2000 Hz. and random signatures run from 20–2000 Hz. Either signature method is acceptable. The *before* and *after* response plots at each accelerometer are compared against each other to verify that the hardware has not failed during the test.

The low level sine sweep is at 0.1 G over the range of 5–2000 Hz (4 oct/min) in all 3 axes.

**Typical low level sine signature vibration test levels**

| Test axis | Frequency (Hz) | Amplitude G's (peak) | Sweep rate (octaves/minute) |
|---|---|---|---|
| XYZ - axis | 5–2000 | 0.10 | 4.0 |

The test sequence is, in each axis, the sine, sine-burst, and then random vibration. The sine and sine-burst tests are easily compared to the pretest analysis predictions. They allow the engineer to verify that the instrumentation is working and producing the expected responses. During the random test, all the spacecraft modes over the entire frequency range are excited. Up until this time, all of the analysis was run using a model, uncorrelated to test data. The test results provide the first indication of how accurate the model is. Some of the modes may couple, creating higher

## 14.3 Environmental Testing

responses than predicted. While the sine and sine-burst tests will typically produce expected responses, it is not uncommon for the random responses to vary significantly from the pretest predictions.

In order to avoid accidentally overstressing the spacecraft during testing, the input levels are increased incrementally using multiple test cycles. During these cycles, the load levels are usually defined as dB levels. The table below shows the correlation between dB level and full test load level.

| dB | % Test load |
|---|---|
| −18 | 12.5% |
| −12 | 25.0% |
| −6 | 50.0% |
| −3 | 75.0% |
| 0 | 100.0% |

For example, when reviewing the −12 dB cycle results for the sine or sine burst tests, the response values should be approximately 25% of the expected full level responses.

Typical sine vibration levels are shown in the table below:

**Sine vibration test levels**

| Test | Axis | Frequency (Hz) | Level (peak) | Sweep rate octaves/min |
|---|---|---|---|---|
| Proto-flight | Thrust | 5–20 | 7.4 g | 4 |
|  | Lateral | 5–20 | 4.0 g | 4 |

Typical Sine Burst test levels are shown below:

**Sine burst test levels**

| Test axis | Test description | Test requirement (g) |
|---|---|---|
| XY | Sine burst at TBD Hz (5 cycles full-level) | 5.00 |
| Z | Sine burst at TBD Hz (5 cycles full-level) | 9.25 |

Note: The sine-burst test level is typically 1.25 times the design load level. This is accommodated during analysis by the 1.25 yield factor included in the stress margins of safety calculations.

Random testing is considered a workmanship test, not a structural test. This is important to consider when running a random test. During launch, when the spacecraft is attached to the launch vehicle, the launch vehicle is a flexible constraint. If one of the spacecraft modes takes off, its interface loads to the launch vehicle increase, but the spacecraft flexibility may damp out some of these loads. Static design loads are based on measured loads (from previous launches) or from coupled loads analysis (CLA). These take this damping into account. Therefore, from the standpoint of strength, the spacecraft is only required to sustain 1.25 times the design loads. The sine-burst test is the strength test. On the other hand, during vibration testing the spacecraft is mounted to a vibration table that provides no damping. So, if one of the spacecraft modes takes off during testing, the response may be much larger than that which would be seen during flight. Because of this phenomenon, during random testing it is permitted to limit the input levels. Limiting is accomplished in one of two ways:

1. **Force Limiting:** Force gages are installed at each of the bolts attaching the interface plate to the vibration table, and the loads measured are used to limit the overall spacecraft net CG accelerations to the static design loads. This is a relatively complicated/expensive procedure and is mostly used on large spacecraft.
2. **Notching:** The random input spectrum is manually changed to notch the input levels at discrete frequencies where responses are high. This method is inexpensive and is used frequently.

The guidelines for notching the input over a discrete frequency range:

1. Maximum limiting over a discrete frequency range is −12 dB
2. The overall $G_{RMS}$ input value cannot be reduced by more than 10% from the unnotched value

Notching input levels is an iterative operation. For a given accelerometer location, the −18 dB random load test cycle $G_{RMS}$ response results are reviewed. These $G_{RMS}$ acceleration responses are 1/8th of the full level loads. So, by multiplying the −18 dB responses by eight is an estimate of full level 1-sigma loads. Then multiply the 1-sigma loads by 3 to get the 3-sigma loads. If the structure cannot withstand the 3-sigma loads at an accelerometer location, the need for notching is indicated. Return to the −18 dB results and look at the plotted response levels; notch the input spectrum over the frequency range where the responses are significantly above the input level. When the new input spectrum is used in the table software, it will provide the overall input spectrum $G_{RMS}$ so it could be verified to be within 10% of the nominal spectrum. Rerun the −18 dB random test load cycle (using the notched input) and determine if the new 3-sigma loads are acceptable. If they are, continue testing by increasing input levels. If the response is excessive, widening the notch may be required.

While the random test is not considered a strength test, it is a very good workmanship test. It is especially good at causing fasteners to back out if they are not adequately torqued.

In the absence of specific launch vehicle vibe spectra, GEVS vibration levels should be used.

## 14.3 Environmental Testing

Random Vibration test levels for all 3 axes should be 14 (Grms). The levels are shown below (Fig. 14.4).

**Random vibration test levels**

| Frequency (Hz) | Protoflight (G²/Hz) |
|---|---|
| 20 | 0.026 |
| 20–50 | +6.0 dB/Oct |
| 50–800 | 0.16 |
| 800–2000 | −6.0 dB/Oct |
| 2000 | 0.026 |
| Overall (Grms) | 14.0 |
| Duration (min) | 1 |

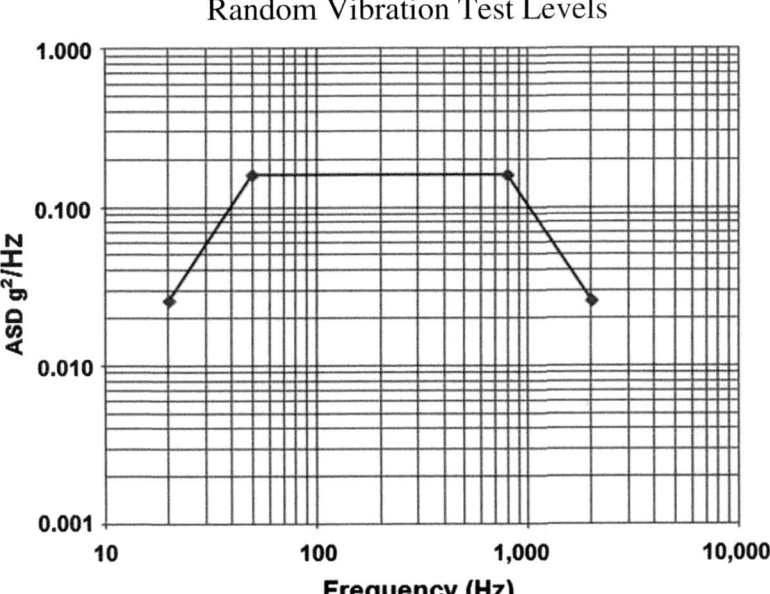

Fig. 14.4 Random protoflight test levels

In summary, the spacecraft should undergo a full suite of vibe testing in three axes, including sine, sine-burst, and proto-flight random testing. Low level sine or random surveys should be run before and after each test to verify that the modes have not changed significantly. Limited performance tests may be run after tests are complete for each axis and prior to re-configuring for the next axis.

**Typical test tolerances are shown below**

| Sine vibration | |
|---|---|
| Sinusoidal amplitude | ±10% |
| Vibration frequency | ± 2% |
| Random vibration | |
| Acceleration spectral density | +/− 3 dB |
| Overall RMS | +/− 10% |
| Duration | +10, −0% |
| Sine burst | |
| Sinusoidal amplitude | ±5% |

## 14.3.2 Thermal Test

The purposes of the thermal tests are (1) to uncover latent defects by thermally stressing the spacecraft, (2) to demonstrate successful spacecraft operation over a wide temperature range and (3) to gather temperature data to assess part, board and assembly gradients. The test success criteria are:

| Objective | Success criteria |
|---|---|
| Uncover latent defects | Demonstrate failure free three (3) cycles between HOT and COLD temperatures |
| Demonstrate spacecraft operation and establish performance | Demonstrate compliant operation over the range of temperatures and during Temperature transitions |
| Assess part, board and assembly gradients | Gather steady state temperature data |

The spacecraft in the thermal chamber is connected to the ground station (or EGSE) and the Dynamic Simulator. The ground station supplies electronic stimuli needed for functional testing.

First a thermal balance test is conducted. The thermal chamber should be able to control temperature over at least −40C and +55C at a rate of at least 3 °C/min. Chamber temperature gradients during the test should not exceed 4 °C/s. A data acquisition system should monitor and store chamber and spacecraft test point temperatures. Spacecraft telemetry should be collected by the ground station. Environmental stability is reached when the chamber is within 2 °C of desired setpoint.

*Steady State Temperature Test* The spacecraft should operate in the chamber for a minimum of 90 min at each of six steady state temperatures. These are +20 °C, +35 °C, 0°, −10 °C, +45 °C and −25 °C. The spacecraft should operate in a repeating scripted scenario of 30 min duration; and the scenario should be repeated

14.3 Environmental Testing

three times during each 90 min test. Telemetry is read and archived during each 90 min test.

*Dynamic Thermal Test* The temperature chamber is cycled three times between +35° and −25 °C while the spacecraft operates with the scripted scenario. Transition between the hot and cold temperature limits should not take less than 15 min (4° per min), dwelling 45 min at the hot or the cold temperature limits. An entire cycle takes 2 h. Three failure-free cycles should be performed. The Dynamic Test takes 6 h (Fig. 14.5).

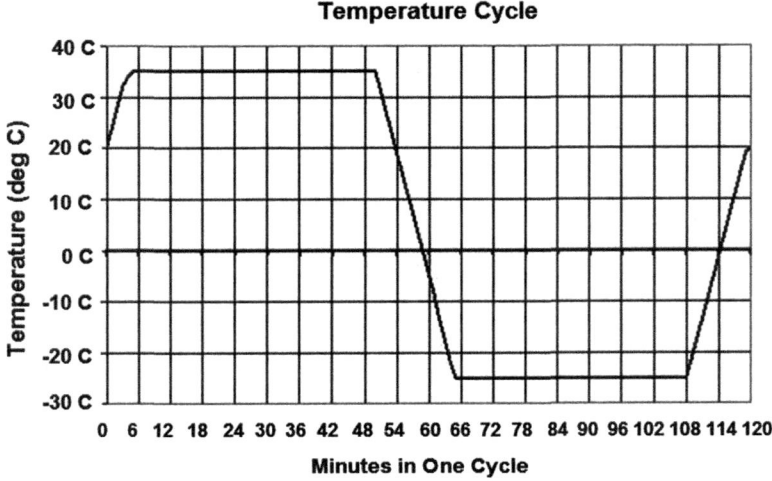

**Fig. 14.5** Dynamic thermal test temperature profile

## 14.3.3 Bakeout

The objective of this (optional) test is to outgas spacecraft components to preclude the risk of condensable volatile contamination. Prior to bakeout, the spacecraft is subjected to thermal cycling at ambient pressure. Chamber cleanliness certification is a prerequisite to integrating the spacecraft into the TVAC chamber. Immediately prior to integration, the chamber with all internal GSE should be certified to the contamination criteria of 300 Hz/h rate @60 °C on a −20 °C Quartz Crystal Microbalance (QCM).

| Objective | Success criteria |
|---|---|
| Complete outgas at 40 °C | TQCM at −20C measurement of contamination levels should drop below a specified limit |

The spacecraft and its TVAC GSE are installed in the thermal vacuum bake-out chamber. The chamber is maintained at less than $1 \times 10^{-5}$ torr and at +40 °C during test. The Data Acquisition System records temperatures in real time and continuously monitors and compares against alarms of spacecraft temperatures. The spacecraft has achieved bakeout when the TQCM rate of change is no more than 300HzTBR per hour and for at least 100 hours at TQCM rate past the knee of the curve.

### 14.3.4 Thermal Vacuum Test

The objectives of the thermal vacuum TVAC test are to:

1. Validate the thermal design by subjecting the SC to thermal test environments that conservatively simulate the flight hot and cold environments
2. Achieve hot and cold steady state temperatures and gather steady state and transient data in order to correlate the thermal models
3. Operate the spacecraft and payload in a manner similar to flight operations
4. Operate the spacecraft at temperatures in excess of those expected on orbit

The compliance matrix illustrates how the test objectives are to be achieved during test:

| Objective | Compliance criteria |
|---|---|
| Validate the thermal design | Subject the SC to test environments simulating the flight hot and cold environments and verify that all temperatures are within limits with margin |
| Achieve hot and cold steady state temperatures | Temperatures stable |
| Gather steady state data in order to correlate the thermal models. | TB stability criteria must be achieved so that data collected is useful |
| Operate similarly to flight operations | Demonstrate acceptable temperature increases during peak orbital load profile |
| Operate in environments excess of expected flight temperature | Achieve temperatures in excess of those expected in flight |
| Operation | Demonstrate failure-free operation while hot, cold and during temperature transitions |

The test is conducted with an LN2 shroud. The spacecraft is attached to the TVAC test fixture, which is conductively isolated from the chamber. The spacecraft main radiator is coupled to the chamber cold wall, and the side radiators are radiatively coupled to zone heaters which, in turn, are coupled to the shroud. The steps in the TVAC test are the following:

**Pumpdown** – during pumpdown, spacecraft operation is consistent with launch to demonstrate insensitivity to corona during ascent. The TVAC chamber is purged

## 14.3 Environmental Testing

with dry LN2 to prevent condensation; the spacecraft is in the launch configuration, the TVAC chamber is evacuated and spacecraft temperatures are monitored to make sure that temperatures do not become excessive. Once chamber pressure reaches $5 \times 10^{-5}$ torr, pumpdown is completed.

**Hot Balance** – Hot Balance demonstrates spacecraft thermal performance in the warmest expected environment. Once the hot balance begins, the spacecraft should operate with no power changes.

**Cold Balance** – The Cold Balance test demonstrates spacecraft thermal performance in the coldest expected environment.

**Cold Survival Balance** – This test proves that the spacecraft can stay above the minimum survival temperatures when the spacecraft is in the low power mode.

**Hot Cycle** – This segment of the TVAC test demonstrates proper spacecraft operation and temperature response while cycling between high dissipation and low power operation.

# Chapter 15
# Launch Vehicles and Payload Interfaces

There are two essential ingredients one must have in interfacing with launch vehicles: money and patience. The selection and manifesting process is long, schedules are uncertain, and a launch failure on an earlier flight could introduce additional years of delay.

The launch vehicle selection process depends on whether the satellite is a large spacecraft that would be the primary payload of the launch vehicle, or if it is small and would be a secondary payload. If the latter, the main criteria for selecting a launch vehicle are:

- Is there a launch already scheduled that has secondary payload volume and weight available in the time frame when your spacecraft would be completed?
- Will the (primary payload) launch go to your required inclination and altitude?
- Where is the launch site; and is it possible, including political and regulatory factors, to arrange launching from there?
- Are the launch and insurance costs within budget?

If a suitable compromise among the above factors can be reached, the main tasks in the launch vehicle selection process are completed.

## 15.1  Present Launch Vehicles

There are 10 USA, 2 Chinese, 2 EU, 4 families of Russian, and 5 other (Indian, Japanese, Israeli) launch vehicles that fly to Low Earth Orbit (LEO) active today; several others are in development. A list of these launch vehicles, their lift capabilities to LEO, maximum accelerations and their launch sites are given in the Appendix. The data was extracted from the websites of each launch vehicle. In Fig. 15.1, only representative launch vehicles to LEO are listed. The acceleration environment listed is a function of the spacecraft weight. The quantities given here are only approximate. Launch vehicle capabilities, listed on the internet sites of the launch vehicle manufacturer, change frequently, so the reader needs to consult directly with an authorized representative of the company.

| Launch Vehicle | ≈ klbs to LEO | Launch Site(s) | Max Axial and Lateral g's | Fairing Dia (in) | Comments |
|---|---|---|---|---|---|
| Minotaur I | 1.2 | VAFB/CCAFS | 9.0/±5.0 | 50 | 61" fairing available, Wallops/Kodiak possible |
| Minotaur IV | 3.5 | VAFB/CCAFS | 9.0/±3.5 | 92 | |
| Falcon 9 | 22.0 | VAFB/CCAFS | 6.0/±2.0 | 181 | 28.5°-51.6° CCAFS, 66°-145° from VAFB |
| Dnepr | 3.7 | Baikonur/Yasne | 7.5/±0.8 | 118 | |
| Atlas V | 8.3-19.0 | VAFB/CCAFS | 5.0/±2.0 | 147 | There are several versions of Atlas |
| Antares | 13.0 | Wallops/Kodiak | 6.5/±1.5 | 126 | |
| Delta II | 6.0 | VAFB/CCAFS | 7.5/±0.8 | 120 | |

**Fig. 15.1** A few of the ≈23 active launch vehicles today

A typical launch profile to a very low altitude (≈270 km) is shown in Fig. 15.2. After first stage burnout, the first and second stages separate and the fairing is cast off at an altitude of about 190 km. By this time, the rocket is out of most of the atmosphere, but there is enough atmosphere left that the spacecraft must consider

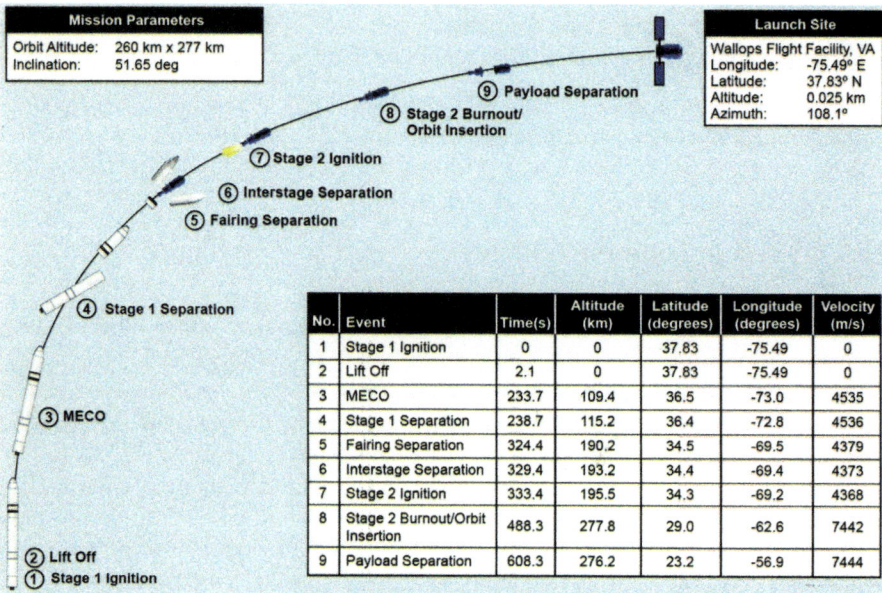

**Fig. 15.2** Typical Antares launch profile from Wallops Flight Facility WFF (Permission from Orbital-ATK)

atmospheric heating. Atmospheric pressure at fairing separation is typically down to ≈0.2 psi. The second stage lifts and inserts the spacecraft into the required orbit, and the spacecraft is separated from the launch vehicle. During ascent, telemetry from the rocket (and often from the spacecraft) is sent to the ground. If there are multiple payloads, the launch vehicle releases them in sequence. Launch vehicles that have liquid propulsion systems (like SpaceX Falcon 9) can restart the last stage engine and can deploy different secondary payloads into different orbits.

The various presently active launch vehicles, including those not listed above, are described in the Appendix. The information was extracted from each launch vehicle payload user manuals. As stated earlier, this information is approximate and subject to change.

Depending on the desired orbit inclination, there may be launch site limitations that prohibit launching into some orbit inclinations. For example, from the Eastern Test Range, Cape Canaveral Air Force Station (CCAFS), spacecraft can only be launched within an azimuth of $-39°$ and $+57°$; so Polar launches are not possible from CCAFS. The reason for the $+57°$ maximum inclination limitation is because at greater inclinations spent boosters would be dropped onto the populated US East Coast. Similarly, from VAFB, launch azimuth is limited between $-70°$ and $-104°$, so spacecraft can be launched into the popular Polar orbit from VAFB.

## 15.2 Launch Vehicle Secondary Payload Interfaces

Since the cost of a launch service is principally paid for by the primary payload, secondary payloads riding on the same launch pay much less by any measure. For this reason, it is important to review how to interface with the various launch vehicles as a secondary payload.

Several launch vehicles offer secondary payload accommodations as long as they do not interfere with the primary payload. Since the secondary payload market is a fast growing segment of the spacecraft industry, more launch vehicles will offer secondary payload launch services in the future. The payload (spacecraft) interface requirements depend on the launch vehicle. However, there is a degree of standardization emerging through development of secondary payload adapters like the ESPA ring, Secondary Payload Adapter (SPA) and the Evolved Expendable Launch Vehicle (EELV). For the Atlas V launch vehicle, these secondary payload options are shown in Fig. 15.3. Ariane uses an ESPA secondary payload ring, as shown in Fig. 15.4a. Other launch vehicles use their own approaches to secondary payload accommodation. Figure 15.4b shows multiple payloads on Dnepr.

An excellent survey of secondary adapters is contained in a report of the 17th Annual Small Payload Rideshare Symposium held at the John Hopkins University Applied Physics Laboratory. The report is up-to-date as of June 2015. Much of the information in this subsection was extracted from that report.

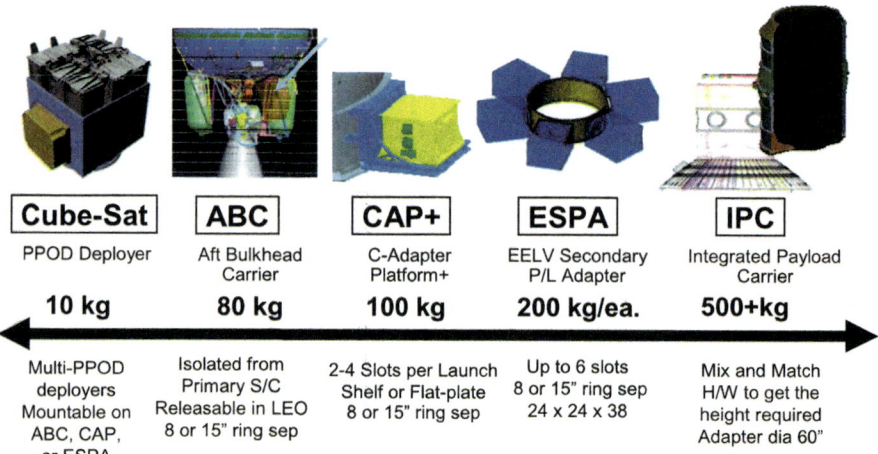

**Fig. 15.3** Secondary payload adapters for Atlas V (Atlas V launch services user's guide)

**Fig. 15.4** (a) Multiple secondary payloads on Ariane. (b) Multiple payloads on Dnepr

15.2 Launch Vehicle Secondary Payload Interfaces

**Fig. 15.4** (continued)

The ESPA ring, shown in Fig. 15.5, is becoming a leading dispenser for multiple secondary payloads. For payloads that require reduced vibration levels, Moog also builds the SoftRide vibration isolation system that can reduce the launch vehicle vibration environment as seen by the spacecraft by as much as a factor of 10. Current standard interfaces are listed in Fig. 15.6.

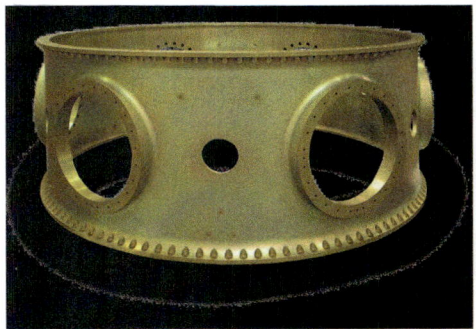 

**Fig. 15.5** ESPA secondary payload adapter (Courtesy of Moog Inc., Space and Defense Group)

Most Separation Systems with which the spacecraft is attached to the launch vehicle or to the secondary payload adapters are explosive Marman Bands or the non-explosive motorized clamp bands. The latter are the Lightband family (Planetary Systems, Corp.) motorized clamps. Both are shown in Fig. 15.7a, b.

| Diameter (in) | No Fasteners | Launch Vehicles |
|---|---|---|
| 62.01 | 120 ¼" | Atlas V, Delta IV, Falcon 9, Minotaur IV, V, VI |
| 38.81 | 60 ¼" | Minotaur I, Athena, Taurus, Pegasus |
| 24.00 | 36 ¼" | ESPA Grande, CubeStack |
| 15.00 | 24 ¼" | ESPA, Atlas V, Athena * |
| 8.00 | 12 ¼" | Small Launch ESPA, Athena |

**Fig. 15.6** Secondary payload interfaces (APL report on June 15, 2015)

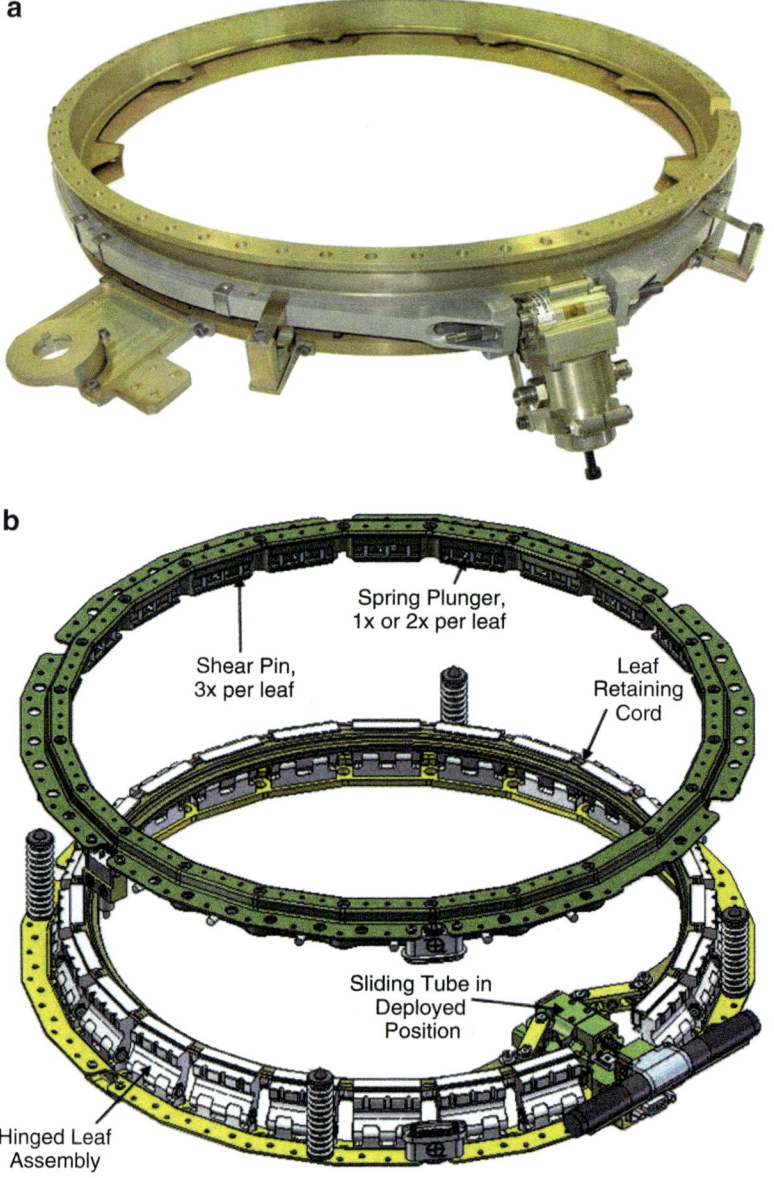

**Fig. 15.7** (**a**) A RUAG sep system. (**b**) A motorized lightband (Planetary Systems Corp)

## 15.3 Secondary Payload Environment

For the very small CubeSat (U, 2 U, 3 U, 4 U, 6 U) spacecraft, the launch vehicle interface is through the Cubesat launcher(s) shown in Fig. 15.8. The original launcher was for a single U (10 cm cube) spacecraft. More recently, launchers for larger, multiple U-size spacecraft have been developed. The interfaces between the CubeSat spacecraft and the launcher are precise and rigid and are controlled by an Interface Control Document.

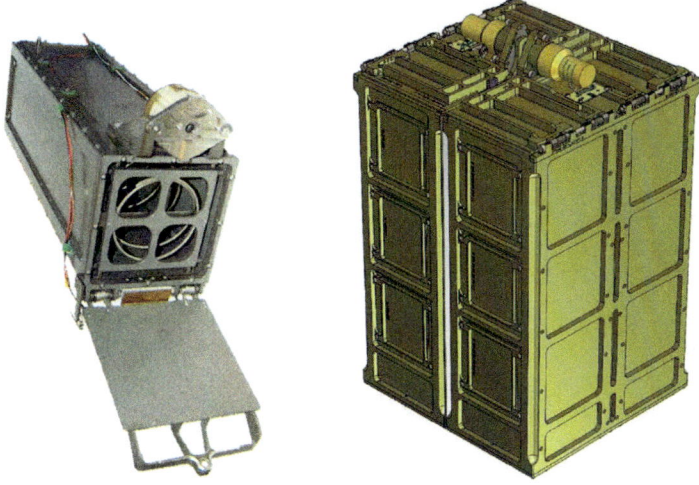

**Fig. 15.8** The original cal poly launcher and a 6 U CubeSat launcher

## 15.3 Secondary Payload Environment

The payload spacecraft must meet, and be able to survive, the launch vehicle vibration, acoustic, pressure, RF, and other requirements.

### 15.3.1 Vibration Levels

Vibration requirements have already been described. In case no information is yet available for a specific adapter, design to the GEVS environment, Fig. 15.9b. Note that what is given as the axial acceleration may become the lateral acceleration for horizontally mounted payloads, such as those mounted on an ESPA ring. The GEVS and the Dnepr vibration environments are shown in Fig. 15.9a, b.

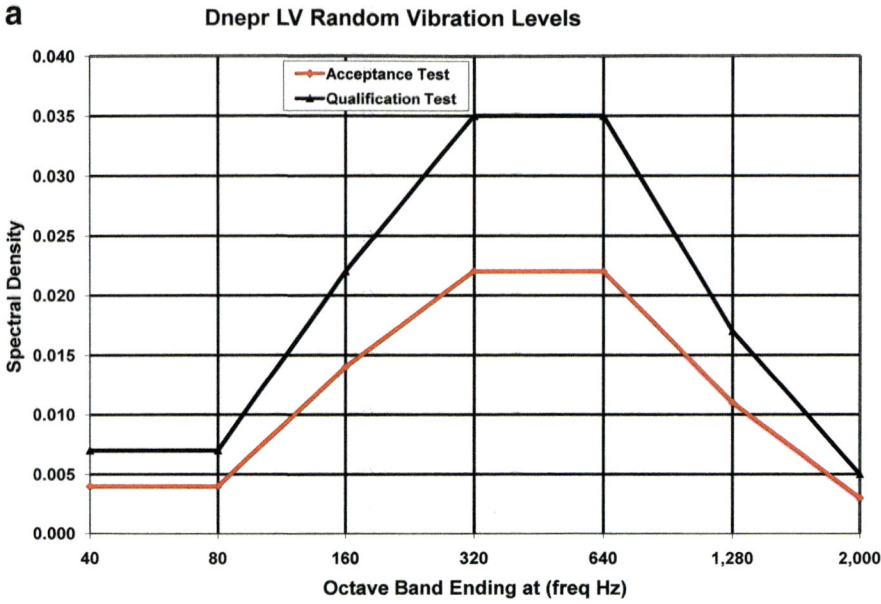
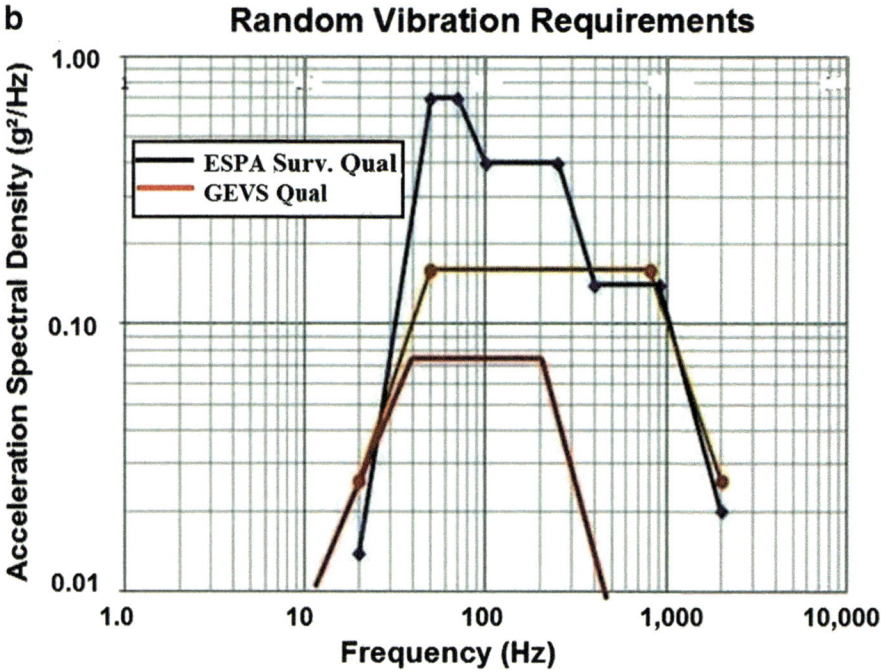

Fig. 15.9 (a) Dnepr vibration spectra. (b) GEVS and ESPA vibration spectra

15.3  Secondary Payload Environment

If no other information is available, design to 9 G axial and 3 G lateral accelerations. These can be relaxed when the launch vehicle organization is able to provide more specific information. The above spectral levels are tabulated in Fig. 15.10.

GEVS

| Frequency (Hz) | | | Qual Level | Acceptance |
|---|---|---|---|---|
| 20 | | g2/Hz | 0.026 | 0.013 |
| 20 | 50 | dB/oct | 6 | 6 |
| 50 | 800 | g2/Hz | 0.16 | 0.08 |
| 800 | 2000 | dB/oct | -6 | -6 |
| 2000 | | g2/Hz | 0.026 | 0.013 |
| Overall | | g RMS | 14.1 | 10 |
| Duration | | sec | 60 | 60 |

ESPA

| Frequency (Hz) | | | Qual Level | Acceptance |
|---|---|---|---|---|
| 20 | | g2/Hz | 0.014 | 0.007 |
| 20 | 50 | dB/oct | 12.85 | 12.85 |
| 50 | 70 | g2/Hz | 0.7 | 0.35 |
| 70 | 100 | dB/oct | -4.72 | -4.72 |
| 100 | 250 | g2/Hz | 0.4 | 0.2 |
| 250 | 400 | dB/oct | -6.72 | -6.72 |
| 400 | 900 | | 0.14 | 0.07 |
| 2000 | | g2/Hz | 0.02 | 0.01 |
| Overall | | g RMS | 16.18 | 11.44 |
| Duration | | sec | 60 | 60 |

**Fig. 15.10** GEVS and ESPA qual and acceptance vibe levels

### 15.3.2  Mass Properties

Center of gravity must be within typically ±0.25″ of the vertical axis of the spacecraft (to minimize tipoff at launch vehicle-spacecraft separation).

The CG position must also be known to within ±0.25″.

The center of gravity must not be above a certain launch vehicle-dependent distance from the launch vehicle-payload interface level.

Moments of inertia must be known to an accuracy of typically ±5% or ±0.5 slug-ft$^2$.

The lowest resonant frequency must be greater than 50 Hz for some launch vehicles and greater than 20 Hz for others. The first spacecraft bending mode frequency should be above 8 Hz.

### 15.3.3  Insertion, Separation and Recontact

Typical launch vehicle insertion accuracies are ±18 km in altitude and <0.2° in inclination. Launch vehicle spacecraft separation velocity is usually 2–3 ft./sec. Tipoff is usually less than 1° per second.

The possibility that the spacecraft will recontact the launch vehicle must be analyzed. Often the launch vehicle makes a small maneuver after releasing the spacecraft to avoid recontact.

### 15.3.4  RF Environment

Usually, the spacecraft is not permitted to radiate RF energy at the launch site. For this reason, provision should be made to radiate into a dummy load on the spacecraft for all ground tests at the launch site. Some spacecraft are launched without testing the RF link at the launch site.

Since there are intentional and unintentional RF emissions from various sources at the launch site, the spacecraft must be able to withstand these RF emissions. Typically, unintentional RF emissions may be as high as 114 dBµV/m in the 14 kHz to 18 GHz frequency range. Intentional emissions maybe as high as 160 dBµV/m from launch vehicle transmitters at specific (typically S-Band) frequencies.

Each launch site publishes its RF environment, and these publications are readily available.

Spacecraft RF emissions must be less than 114 dBµV/m in the 14 KHz to 350 MHz range, less than 140 dBµ/m between 350 MHz and 1 GHz and less than 169 dBµV/m above 1 GHz.

### 15.3.5 Acoustic Environment

Often, acoustically-induced spacecraft vibrations can drive the vibration environment. The launch vehicle will provide the acoustic spectrum seen by the spacecraft. An example (for Atlas V) is shown in Fig. 15.11.

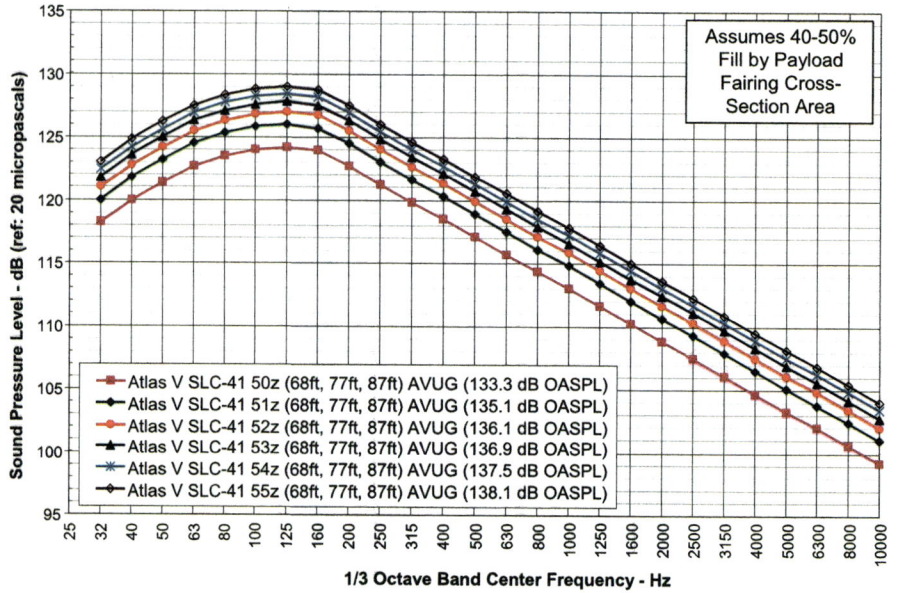

**Fig. 15.11** A typical payload acoustic environment
(Permission by United Launch Alliance)

### 15.3.6 Shock Environment

Typical shock the spacecraft may be subjected to is shown in Fig. 15.12.

## 15.4 Analyses, Documentation and Other Factors

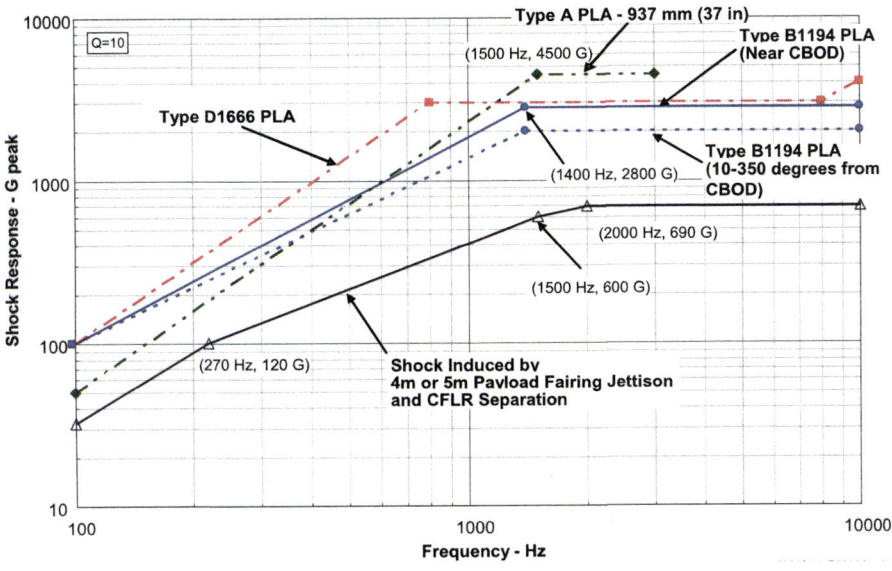

**Fig. 15.12** Typical shock environment (Permission by United Launch Alliance)

### 15.3.7 Additional Spacecraft Environmental and Other Factors

In addition to the above environmental matters, there are additional factors that the spacecraft must take into consideration. These include:

- EMI/EMC
- Contamination
- Spacecraft Fueling
- Liquid Fuel Slosh Frequency (accurate to 2 Hz)
- Payload Access Door Location on the fairing and umbilical cable routing

In addition, the spacecraft should be designed to have removable lifting fixtures, and, if horizontally mated to the launch vehicle, there should be a mating fixture that can turn the spacecraft from a vertical to a horizontal position.

## 15.4 Analyses, Documentation and Other Factors

Many analyses and documents are required to satisfy the launch vehicle organization. Examples of some are listed below:

- Range Safety
- Coupled Loads Analysis (performed by the launch vehicle from FEM provided by the spacecraft)

- Finite Element Model of the spacecraft
- Spacecraft-Launch Vehicle Interface Control Drawing (ICD), containing:
  - Mechanical ICD
  - Electrical ICD
- Mission Analysis and Payload Integration
- Safety Reviews
- Launch Readiness Review

# Chapter 16
# Ground Stations and Ground Support Equipment

Ground stations (1) command the spacecraft(s), (2) collect, display and analyze spacecraft telemetry about its state of health and (3) command payloads and retrieve data from them.

Some agencies and customers of spacecraft require the use of, and compatibility with, existing ground stations and two-way communication systems. There are several good ground station software packages available that require little tailoring to adapt to a specific spacecraft. Most of them use numerical rather than graphic displays. Regardless of whether one of those or a design-from-scratch ground station is used, most of the requirements are the same. These will be described next.

## 16.1 Ground Stations

*Ground Station Requirements* Experience has shown that ground station (GS) requirements can easily be given in terms of questions a visitor to the GS may ask. The first question he or she may ask: "Where is the spacecraft now?" The GS should be able to show a map of the world and the instantaneous position of the spacecraft on it. The communications range footprint of the spacecraft should also be shown.

Some numeric data should describe the altitude of the spacecraft and, when in range of a GS, the elevation and azimuth angles to the spacecraft. These quantities are also used to drive GS antennas. Figure 16.1 shows such a display. Note that numerics concerning spacecraft elevation and azimuth are not shown because the spacecraft is not in range of the GS. The range entry indicates how far it is from the GS. Converted to minutes, the spacecraft is 56.05 minutes away.

The next question the visitor may ask is: "How is the spacecraft doing?" That is, what is the state of health of the spacecraft? Therefore, telemetry should be collected from the spacecraft, and the subset of TTM data relevant to spacecraft (and payload) health and status should be displayed (numerically and graphically). Relevant data includes: (1) temperatures at key points, (2) ON/OFF states of

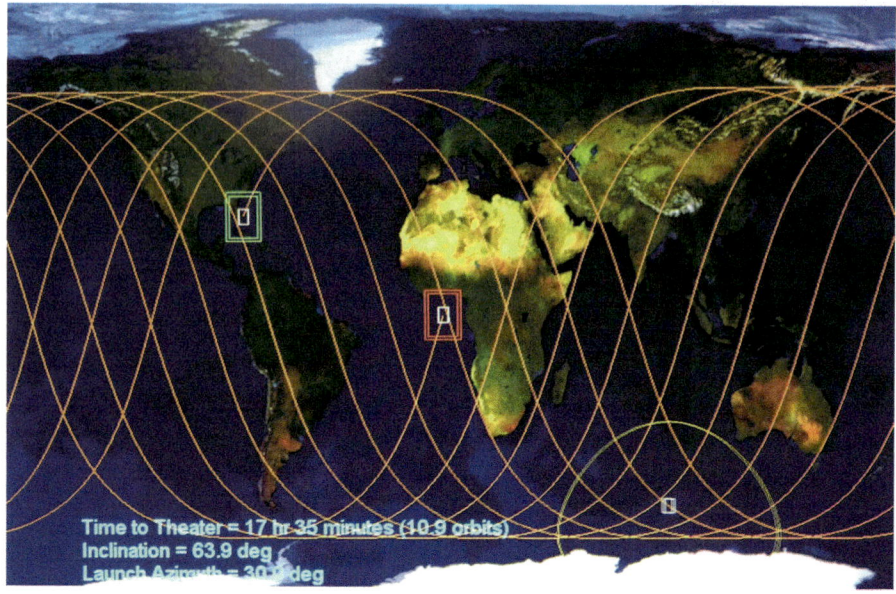

| Satellite ID | Sat 1 | GS | Bos | SC Posit | SC Visibility | No |
|---|---|---|---|---|---|---|
| Orbit No | 6598 | Lat | +38.92° | -16.57° | Elevation° | |
| Date | 07/21/2016 | Lon | -77.23° | +69.42° | Azimuth° | |
| time | 14:37:28 | Alt | 26 m | 568 km | Range km | 23,542.7 |

Fig. 16.1 Spacecraft position and range, elevation and azimuth to GS

hardware components, (3) voltages and currents of selected components, (4) spacecraft attitude and ADACS telemetry, (5) battery charge state for each string, (6) parameters that are close to anomalous condition and (7) payload-related quantities. To enable the GS operator to quickly review the large amount of information relevant to state of health, graphic displays of these quantities should be used, and the display should span several of the most recent orbits, rather than just the present values of the quantities.

The visitor's next question may be: "When is the spacecraft going to be in communication range with this and other ground stations?" The table of contact access times shown in Fig. 3.3b, repeated here as Fig. 16.2, answer this question. It should be displayed at the GS.

The next question may be: "What is the spacecraft going to do next?" There should be available from spacecraft telemetry a read-back of commands sent to the spacecraft earlier that answers this question.

In addition to these TTM displays, the GS must also command the spacecraft. For this reason, there should be command generation dialog boxes from which the set of commands to the spacecraft could be assembled. A great many commands are needed to operate a spacecraft and its payload, however, most are commanded

| Access | Start Time (UTCG) | Stop Time (UTCG) | Sec |
|---|---|---|---|
| 1 | 31 Aug 2016 04:26:44.649 | 31 Aug 2016 04:40:06.258 | 801.610 |
| 2 | 31 Aug 2016 06:07:21.438 | 31 Aug 2016 06:18:33.846 | 672.408 |
| 3 | 31 Aug 2016 18:34:18.880 | 31 Aug 2016 18:43:05.994 | 527.113 |
| 4 | 31 Aug 2016 20:11:13.687 | 31 Aug 2016 20:24:34.033 | 800.346 |
| 5 | 31 Aug 2016 21:52:18.323 | 31 Aug 2016 22:04:21.323 | 723.000 |
| 6 | 31 Aug 2016 23:35:46.300 | 31 Aug 2016 23:44:24.396 | 518.095 |
| 7 | 01 Sep 2016 01:18:11.672 | 01 Sep 2016 01:26:49.920 | 518.248 |
| 8 | 01 Sep 2016 02:58:14.680 | 01 Sep 2016 03:10:17.845 | 723.165 |

**Fig. 16.2** Spacecraft access times at a ground station

repetitiously. For this reason, the repetitious commands should be scripted to ease the workload of the GS operator.

Commands are generally of the form: What? When? and How? It is best if the commands were assembled at the GS at a high level (rather than micro-managing the spacecraft). Let the spacecraft convert the high-level command to the detailed actions that must occur to execute the command.

For example, the high-level command of: *Transmit Message A at 11:05 at 2.065 GHz* is converted by the spacecraft as follows:

1. Power UP the transmitter in the standby mode at 11:043
2. Tune it to 2.065 GHz
3. Retrieve Message A from memory
4. Set the Power Level and Bit Rate
5. At 11:05 Key the transmitter and send the message

Not so many years ago, spacecraft were not commanded in this way. Spacecraft used to be commanded, step by step, to implement each action that had to be undertaken. With the advances in computer technology, it is no longer necessary to do so. The spacecraft should be able to interpret high-level commands and detail these out automatically on the spacecraft.

*Anomaly Resolution and the Safehold Mode* When the spacecraft experiences an anomaly, the spacecraft or the GS should have scripted sets of commands to send up, while the spacecraft is still within line of sight, to resolve the anomalies, if possible. Often the first step in the process of resolving anomalies is to put the spacecraft in the Safehold Mode. This is a mode that minimizes power consumption and only operates the most needed spacecraft components so as to gain time to figure out what to do next. The spacecraft and the GS design should anticipate all kinds of possible failures and should have scripts ready to execute when one occurs. In many cases, the spacecraft can monitor the relevant points to determine if an anomaly has occurred and can invoke the scripted set of steps to correct, or work around, the anomaly without GS interaction. Then, the GS merely needs to be informed by the spacecraft what anomaly occurred and what the spacecraft did about it.

## 16.2 Ground Support Equipment

Ground Support Equipment, or GSE, consists of Mechanical Ground Support Equipment (MGSE) and Electrical Ground Support Equipment (EGSE). Together, these support the spacecraft on the ground and at the launch site.

MGSE typically include mechanical fixtures to lift, turn and move the spacecraft in the laboratory, in transport or at the launch site.

EGSE is in effect a paired down version of the ground station. With the EGSE connected by an umbilical cable, the spacecraft can be operated as if it were in space, with the exception that the GS antenna would not be used. In an ideal world, the EGSE can operate with the spacecraft as hardware-in-the-loop, just as the Dynamic Simulator in Chap. 14 could operate with the spacecraft ADACS as hardware-in-the-loop.

Testing all functions of the spacecraft repeatedly in the laboratory while the spacecraft is in development or in test is the main function of the EGSE. Since this is a time-consuming and repetitious function, the EGSE should be semi-automated so that it could act as Automatic Test Equipment (ATE), speeding up test and verification of satellite functions.

The cost and time required to develop good EGSE are usually underestimated. This is not the place where one should try to save money. Once built, the EGSE can be adapted to other spacecraft missions, enabling one to spread the cost of EGSE over several programs.

## 16.3 Ground Station Manual and Operator Training

A Ground Station Manual for operating the spacecraft should be prepared. It should contain:

- Description of the spacecraft and how it operates
- A list of all commands possible
- A listing of all the telemetry that it sends to the ground
- A keystroke by keystroke description of how to command the spacecraft
- A description of the ground station hardware and software
- A listing of anomalous conditions, how to recognize them, and what to do about them
- A description of how to archive TTM and CMD data

An Operator Training Manual/Course should be prepared. It should instruct the future ground station operators on all of the above capabilities contained in the Ground Station Manual. The Operator Training Course should enable the future operators to gain hands-on experience. For this reason, the operator training course may be given during the last stages of spacecraft checkout, so that the actual spacecraft and the EGSE simulation of the ground station are available.

Preparation of a good Ground Station Manual and an Operator Training Course are not small matters. The time and effort required to prepare these should not be underestimated.

## 16.4 Other Ground Station Matters

*Spacecraft RF Acquisition and Tracking* It is assumed that the spacecraft receiver is ON, but the transmitter is not when the spacecraft comes up over the horizon. Based on the satellite ephemeris, the ground station antenna can be slewed to the required direction. If the spacecraft and the ground station antennas are relatively low-gain antennas, the spacecraft can be tracked by the ground station *open loop*, that is, by only using the azimuth and elevation angles obtained from the spacecraft orbital elements. If the ground station antenna gain is high, the antenna system must first search the uncertainty area in azimuth and elevation until the spacecraft is acquired and tracked (*closed loop*). For this to happen, the spacecraft transmitter must be ON and transmitting something, typically telemetry.

In communicating with the spacecraft, the instantaneous doppler frequency should be used to pre-doppler shift the ground station transmitter frequency in the opposite direction. Similarly, in receiving spacecraft communications, the ground station receiver should be pre-doppler shifted to eliminate the time it would otherwise take for a doppler search.

*Use of Multiple Ground Stations* A single ground station will have access to the spacecraft only 4–5 times a day. All the data it collects must be downloaded during the few minutes of these access intervals. However, if the spacecraft is in a Polar or near-polar orbit, use of high latitude automated ground stations can increase satellite access and the amount of data that can be downloaded per day. There are several satellite systems that employ such ground stations. For US satellites, positioning such stations in Alaska is a way of increasing satellite throughput.

# Chapter 17
# Spacecraft Operations

Here, the methods of operating a spacecraft and its payload to (1) schedule events, (2) display health and status telemetry, (3) command changes in the setup of spacecraft or payload parameters, (4) operate the propulsion system, if any and (5) resolve and correct spacecraft operating anomalies will be discussed.

Once on orbit, manpower required to operate a satellite is the highest cost of ownership. In the following, we will describe methods of designing the ground station(s) and of operating the satellite(s) to maximize automation and minimize manpower requirements.

## 17.1 Ground Station Functions for Spacecraft/Payload Operation

The functions that must be performed in real-time operation of a spacecraft and payload from a ground station are listed below:

- Mission Planning - plan out what the spacecraft should do for the next period of time
- (Optionally) run a simulation of the mission plan on a satellite simulator
- Downloading and updating the spacecraft orbital parameters
- Scheduling - set up communications contact time instances (for the next few days)
- Review the spacecraft-ground station timeline of events
- Prepare the Command Uploads
- When in communication contact with the spacecraft, capture the telemetry (TTM) and data downloads
- Review the TTM to ensure the spacecraft (and payload) health are OK
- If there are anomalies, (1) resolve them, or (2) command the spacecraft into the Safe-Hold Mode

- Pass on the payload data to the payload customer
- Review and archive TTM data.
- Decide if anything different has to be done

The ground station hardware and software with which these functions are implemented is shown in the generic ground station block diagram, shown in Fig. 17.1.

The GS has many different displays. The most important displays are: (1) a map showing the spacecraft location, (2) TTM data alphanumeric and graphic displays, and (3) spacecraft command generation display.

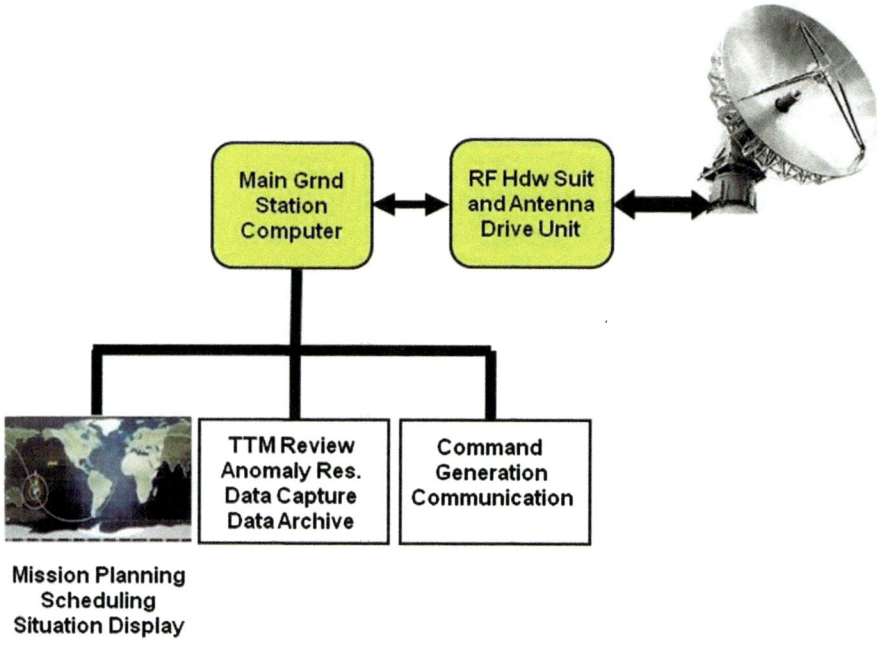

**Fig. 17.1** A generic ground station block diagram

## 17.1.1 *The Map and Access Time Interval Display*

To gain an understanding of what is going on, a ground trace of the satellite orbit over the next few hours (or days) is required. The display shown in Fig. 17.2 illustrates part of an orbit, the ground station location (Washington, DC), the present location of the satellite and elevation contours of 0°, 15° and 30°. This display was generated by the *Satellite Tool Kit* (STK) of Analytical Graphics. Several other programs (such as *NOVA* of the Amateur Satellite Society) are also able to generate such displays.

The passes to use, on which to downlink telemetry and on which to send commands to the spacecraft, can be selected from this display. Typically, nighttime and

## 17.1 Ground Station Functions for Spacecraft/Payload Operation

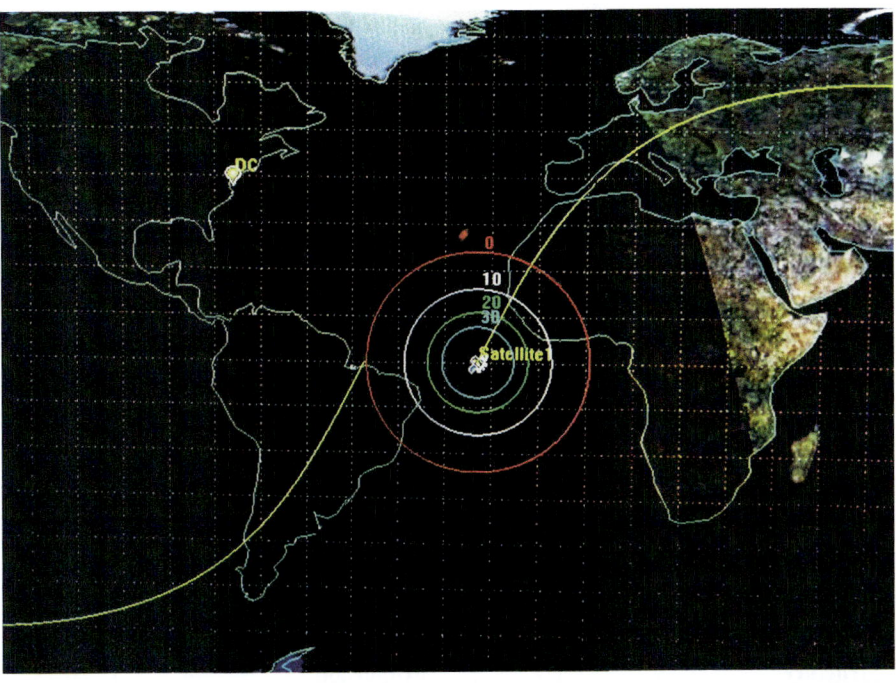

| Access | Start Time (UTCG) | Stop Time (UTCG) | Seconds |
|---|---|---|---|
| 1 | 31 Aug 2016 04:26:44.649 | 31 Aug 2016 04:40:06.258 | 801.610 |
| 2 | 31 Aug 2016 06:07:21.438 | 31 Aug 2016 06:18:33.846 | 672.408 |
| 3 | 31 Aug 2016 18:34:18.880 | 31 Aug 2016 18:43:05.994 | 527.113 |
| 4 | 31 Aug 2016 20:11:13.687 | 31 Aug 2016 20:24:34.033 | 800.346 |
| 5 | 31 Aug 2016 21:52:18.323 | 31 Aug 2016 22:04:21.323 | 723.000 |
| 6 | 31 Aug 2016 23:35:46.300 | 31 Aug 2016 23:44:24.396 | 518.095 |
| 7 | 01 Sep 2016 01:18:11.672 | 01 Sep 2016 01:26:49.920 | 518.248 |
| 8 | 01 Sep 2016 02:58:14.680 | 01 Sep 2016 03:10:17.845 | 723.165 |

**Fig. 17.2** Satellite location, communication footprints and communication access times

low elevation angle passes should be avoided. However, if the ground station(s) are automated or are at a high latitude so that they can see the satellite more often, then each pass can be used for telemetry downlink and command uplink.

### 17.1.2 Telemetry Monitoring

There is a tendency to want too much telemetry. Telemetry (except for that required to diagnose and correct anomalies) should give an overview of the spacecraft condition over the last few orbits, and should alert the operator to anomalies or to approaching anomalies. Sampling telemetry points every (say) 1 minute provides a plot of the behavior of each point with 90–96 samples per orbit. For a typical 100 telemetry points in a LEO satellite, 18,000 samples are enough to plot two orbits of behavior. If each sample is one Byte, the total telemetry downlink consists of 144 Kbits. At a data rate of (say) 19.2 kbps, it takes about 7.5 seconds to downlink all of the telemetry needed. For a specific 3-axis stabilized imaging satellite, described in Sect. 2.2, the telemetry points are Fig. 17.3:

| **Telemetry Points** | **Bits** | **Telemetry Points** | **Bits** |
|---|---|---|---|
| **EPS** | | **Telescope/Camera** | |
| 2 Battery Temperatures | 16 | 3 Telescope Temperatures | 24 |
| 2 Battery Volts and Amps | 32 | 2 Camera Temperatures | 16 |
| 4 Solar Panel Currents | 32 | 1 Camera Current, Voltage | 16 |
| 3 DC/DC Converter I and V | 48 | 1 Image Processor Temp, I , V | 24 |
| 2 Battery Heater ON/OFF State | 16 | **RF Subsystem** | |
| **ADACS** | | 1 Transmitter Power | 16 |
| 3 Reaction Wheel Amps, Volts | 48 | 1 Enable Transmitter | 8 |
| 3 Reaction Wheel Speeds | 48 | 1 Enable Receiver | 8 |
| 3 Reaction Wheel and Temps | 24 | 1 Receiver Signal Strength | 16 |
| ADACS computer Temp, Amp | 16 | 1 Transmitter, Rcvr Temps | 16 |
| Attitude, orbit elem frm GPS | 144 | Xmitter, Rcver ON/OFF state | 8 |
| **Structure** | | **Propulsion** | |
| 8 Structure Temperatures | 64 | 2 Tank Pressures, Temps | 48 |
| Separation Switch State | 8 | 4 Thruster Temperatures | 24 |
| **Digital** | | 4 Thruster ON/OFF states | 24 |
| 1 C&DH Temp, I and V | 24 | Miscellaneous | 96 |
| | | **Total bits/sample** | **856** |
| | | No of TTM bits. 180 samples | 154,080 |

Fig. 17.3 Typical telemetry points and telemetry data in a LEO spacecraft

Telemetry data can be displayed in tabular form or graphically as two orbit plots for each telemetry point, organized in several graphs (EPS, ADACS, Propulsion, etc.). The graphical display allows the operator to see at a glance if everything is okay without having to read the volumes of data.

Each telemetry point is assigned limits within which the telemetry reading is normal. Readings out of this range are potential anomalies that are brought to the operator's attention.

In addition to the *Routine Telemetry* described above, the spacecraft should be able to downlink more detailed *Engineering Telemetry* on request. For example, the spacecraft attitude vs. time during a maneuver is of interest during the spacecraft on-orbit checkout phase. Propulsion thruster activity during a station-keeping maneuver is another situation where more detailed telemetry is needed. The spacecraft should be able to provide such detailed telemetry. Commands to the spacecraft to collect and downlink such telemetry would specify (1) the point or points that should be sampled, (2) the time interval over which samples should be taken and (3) the sample rate that should be used.

17.1 Ground Station Functions for Spacecraft/Payload Operation

## *17.1.3   Spacecraft Command Generation*

Every piece of hardware and each spacecraft (and payload) function should be commandable from the ground station. Command Generation should be implemented with menus with default parameters corresponding to the usual values of the command.

With increasing automation, higher level commands or command sequences should be prepared to make it easier to command the spacecraft. Higher level commands are generally of the form: What, When and How? For example, to command the spacecraft to send TTM at a specific time should only require the ground station to command the spacecraft to: Send TTM at 09:35:35. The spacecraft should then interpret this command and convert it to the sequence of actions needed to implement the command. These might be: Power up the transmitter into the Standby Mode at 09:35:15. Transfer the TTM data from the C&DH computer to an output buffer at 09:35:30. Key the transmitter and send the TTM data at 09:35:35. The more scripted high level commands exist, the more human operator errors can be avoided.

When special telemetry is required to examine something in more detail, it should be a grounds requested data set. The normal TTM data should not be cluttered with matter normally not looked at.

## *17.1.4   Anomaly Discovery and Resolution*

Anomalous behavior can be discovered by looking at the TTM data. For example, if the transmitter current is zero or low after the transmitter was commanded to power up, there is something wrong. Either the transmitter is not functioning correctly, or the voltage and current monitors are not working. In either case, an anomaly occurred.

Most anomalies can be discovered by observing the TTM. For this reason, the discovery of an anomaly can be automated at the ground station. Since this can be done from TTM alone, the spacecraft software could contain the same anomaly detection logic so that it could Self-Discover an anomaly.

A few years ago, one could not talk about automating anomaly discovery at the ground station, no less in the spacecraft. Now, with the advances in the capabilities and speed of computers and microprocessors, the time is ripe for automating as much as possible. This can also help to reduce manloading at the ground station.

To resolve and correct anomalies, a list of possible anomalies and corrective actions should be prepared, and the ground station software should have a library of corrective actions as a set of scripted commands ready to be uploaded.

Just as in the case of automating anomaly discovery on the spacecraft, some anomaly resolution corrective actions can also be automated on the spacecraft.

The spacecraft should put into its downlinked telemetry what the anomaly was and how the spacecraft corrected it.

### 17.1.5 Archiving TTM and Data

Not much needs to be said about this. It is pretty obvious that this should be done.

## 17.2 Data and Data Rate Limitations

The amount of data that can be collected from the spacecraft at the ground station is limited, and the mission must be planned with these limitations in mind. While downlinking telemetry only takes a few seconds, downlinking payload data may be another matter.

For example, an imaging spacecraft taking 16 MP images with 24 bits/pixel collects 384Mbits per image. In a 6.4 minute pass at 1 Mbps, only 1 image can be downlinked. Even if the transmitter power or antenna gain were increased to permit transmitting at a higher bit rate, the number of images per pass that can be downlinked is quite small. If JPG compression is used, it could reduce the number of bits per image by a factor of (say) 10. This would enable to downlink 10–40 images per pass. This is much more acceptable. Since there are only about 4–5 passes per day at a ground station, 40–160 images per day would be the limit of what a single ground station could collect.

If there were additional ground stations (such as a high latitude automatic ground station), there would be 15 passes a day on which image data could be captured. This would increase the capability of the spacecraft system in that it could capture 150–600 images per day.

An alternative method of increasing the spacecraft system throughput is to use a geostationary satellite relay. Theoretically, the number of pictures collected per day could become very large. System throughput limitation would probably be transferred to the electric power system that may limit the amount of time the transmitter can be ON, or how long other components participating in the image taking and transmission process can be powered.

## 17.3 Other Ground Station Operations

### 17.3.1 Post Launch and Checkout

Immediately after launch of the spacecraft, the ground station must establish first contact, receive and interpret first telemetry and upload an initial schedule. After that, the spacecraft checkout process can start. This process consists of giving each possible command to the spacecraft and verifying that it correctly executed the command. This can be a very time-consuming process, as the ground station has access to the spacecraft only a few times per day. For more complicated spacecraft, this process can take several weeks.

While it is not yet done much today, automation could be introduced into the on-orbit checkout process. A very long string of spacecraft commands can be uploaded, and from telemetry, the performance of the satellite in response to these commands can be assessed almost automatically, significantly reducing checkout time and reducing manning requirements.

### 17.3.2  Test Plans and Reports

There are a great many (on the order of 20) plans and documents required for the development and operation of a ground station. It is not enough that a spacecraft should work well, the ground station must plan and execute spacecraft and payload operations according to preplanned scenarios and operations to achieve success. Compliance with data preparation and reporting requirements should not be taken lightly, as success can be jeopardized by lack of preparedness.

### 17.3.3  Manning the Ground Station

Years ago, it took a whole army to operate a spacecraft. That is no longer the case. Even multiple spacecraft can be operated by a single person if everything goes well. However, if it does not, one needs to have rapid access to the engineers who designed the spacecraft and wrote its software. The one full-time equivalent person can rapidly grow to five or more.

### 17.3.4  Cost of Spacecraft Operations

The spacecraft has to be developed only once, and it is launched only once, but spacecraft operations last for many years. Therefore, the cost of manning ground station operations is a key element of the total (or life cycle) cost of a spacecraft system.

The more automation is built into the spacecraft and the ground station, the smaller the staffing and ground operations cost can become. The concept of automating spacecraft operations is emerging as the trend for new ground station (and spacecraft) design.

## 17.3.5 Operator Training and the Spacecraft Simulator

The Operator Manual and the Training Aids ensure that ground operations run smoothly and that the operators could do their job well.

Often, particularly in developing a set of corrective actions in response to an anomaly, it is necessary that the corrective actions be tried out on a satellite simulator, rather than on the actual spacecraft. For this reason, a satellite simulator is usually constructed and placed at the ground station. The satellite simulator is a "Flat Sat," constructed from the engineering model hardware that may have been developed during the design and construction of the spacecraft.

## 17.3.6 Mission Life Termination

There is an increasing amount of space debris. Thus, it is important, if possible, to terminate the spacecraft mission by disposing of the spacecraft. There are several ways this can be done:

- Controlled Reentry, keeping some fuel on board with which to lower the orbit altitude, even if only putting the spacecraft into an elliptical orbit that dips low into the atmosphere. The word "controlled" usually means that the position on Earth (preferably at some remote place over the ocean) where the spacecraft reaches a low altitude of (say) 50 km is controlled.

- Kick the spacecraft into a parking orbit where it is not doing any harm. This too takes propulsion, but for a LEO spacecraft usually less.

- Let it burn up. The melting or ablating temperature of materials in a spacecraft make this a low cost solution; but one must be careful, for not everything in a spacecraft will melt during deorbit.

## 17.3.7 Ground Station Development Schedule

The spacecraft design engineer likes to design the spacecraft. He or she usually delays the start of the design of the ground station. Management must ensure that this does not happen. A hastily developed ground station usually is also hastily conceived. And it may not be as efficient as it could be, or as it needs to be.

# Chapter 18
# Low Cost Design and Development

In 1995, I wrote a chapter on the RADCAL spacecraft for *Reducing Space Mission Cost*, a book in the *Space Technology Series*. The book was edited by James Wertz and Wiley Larson. In the chapter, I wrote a section on the practices used to come up with a very capable and complex spacecraft at a very low cost. The schedule (from contract award to launch) took one day less than one year. In reviewing that chapter now, after having developed 34 spacecraft, I must say that each procedure for keeping costs down described in that chapter is true today. For this reason, this chapter is largely taken from that 1995 book. It is just as true today as it was then.

## 18.1 Approach to Low Cost

There is no magic formula or trick to achieving low cost. The secret, if any, is to do what every good program manager has known how to do since the beginning of the space age:

- Assemble a capable, small team of people
- Set a short schedule
- Make major trade-offs and technical decisions rapidly and decisively
- Practice judicious concurrency between fabrication and design
- Do not procrastinate or analyze unnecessarily
- Do not let anyone slow you down
- The Program Manager should run the program in a project vs. a matrix organization
- Maintain a harmonious relationship with the customer

The secret, if any, to producing a high-technology, quality product is to take bold and innovative approaches, but implement them ultra-conservatively, and test the finished product exhaustively under realistic operational conditions.

## 18.2 The Contract Should Focuses on Functional Rather than Technical Specifications

The contract document should be short and should focus on functional rather than technical specifications. This enables the design team to make major trade-offs to minimize cost while adhering to the contract. One can trade off characteristics of one subsystem against another. This facilitates selecting an implementation that meets cost goals. It permits "design-to-price." Unduly detailed technical specifications tie the hands of designers. This is one of the most leveraged factors responsible for low cost.

The single disadvantage of loosely defined contractual requirements is that it is easy for the customer to say during the inevitable "requirement creep" that occurs throughout the development process that the additional requirements are within scope. If it were not for the continuing ability to make system level tradeoffs, this could lead to a cost overrun.

The key to having harmonious relations with the customer is to keep the customer informed of everything that is going on, anticipate any concerns or questions, make sure that satisfactory answers exist to those possible questions, and run faster than questions can be posed, so that the customer cannot slow you down.

## 18.3 Experienced, Small Project Team

Organize as a project team. Do not use a matrix organization where the people working on the program may not be under full control of the Program Manager. A program starts by assigning an experienced core project team. Each member of the team is responsible for his or her subsystem from initial conception to the launch pad. Do not have separate systems engineers, preliminary designers, engineers to develop the hardware, integration and test engineers and another team for the launch campaign. Each of the team members should not only be a subsystem specialist but also a systems engineer. Because of his or her experience of having taken other satellites "from cradle to grave," less experienced staff members can be assigned to the experienced team leaders to learn. On the next program, the now-experienced team member is assigned subsystem responsibility.

The small team is augmented by technicians, assembly, machine shop and other needed functions on an "as required" basis. At any given time, there are typically three times the number of people working on a program than the average number of Full-Time Equivalent people. This practice keeps costs down. Instead of assigning a large full-time staff to the program, the average spending level is much lower. People are removed from the program and assigned elsewhere when no longer needed.

## 18.4 Vertical Integration

Build in-house as much as possible. This has two advantages. One is to lower the cost of components, while the other is improved availability. Since we do not have to make a living by building, for example, attitude control system components for sale, we do not have to support a whole company making these, and we do not have to maintain a full engineering, marketing, program management, and manufacturing organization for each subsystem (as would a subsystem vendor). Thus, price can be substantially lower for the same subsystem. The other advantage of vertical integration is the ready availability of the component. If a component is needed in a hurry, expedite it in your own shops.

## 18.5 Short Schedules and Concurrency of Development and Manufacturing

A short schedule is a blessing in disguise to keep costs low. One simply does not have enough time to spend much money. It also forces practicing concurrency between engineering development and manufacturing.

In the customary development process, very senior engineers start the program doing systems engineering and front end design. The same team then prepares for the Preliminary Design Review, eliminates deficiencies identified during the review and then continues to detailed design and the Critical Design Review. Only then does development and fabrication begin in earnest. This is an expensive way of doing business. By delaying start of fabrication until the design is completed, the heavy front-end design effort comes to an end, and the people who did it have little to do after Critical Design Review. So, they invent additional things to do, refine, study, or redesign. In any case, this effort costs money and stretches the schedule.

Operate differently. First assess which subsystems or components will not change during detailed design and release those to production right away. In this way, while the engineers complete the system design or the design of new components, already mature components are in fabrication and ready for checkout and subsystem integration when an engineer becomes available from system design. Knowing that work is waiting whenever his or her front-end engineering is completed disincentives procrastination and eliminates prolonging unnecessary analyses. Thus, much of the hardware may be completed by Critical Design Review while other parts of the system have not yet been designed.

This method of operation would not be possible if it were not for very experienced engineers who have, over several programs, developed a "feel" for knowing when the risk of proceeding with fabrication is low, and when more analysis is needed before it is prudent to proceed. Concurrency is not a four letter word if practiced by seasoned engineers.

The expenditure profile versus time of the program run in this way tends to be linear; it does not exhibit the usual S-shape, resulting from a slow buildup, an accelerating middle, and a phase-down toward the end. There are two reasons for this. One is that more of the components are built in-house. So, instead of ordering components that have to be paid for when they are delivered (giving rise to a deferral of expenditures and the S-shape), components are built in-house, expending labor as soon as a component is released to production. The second is the short schedule. There is no time for buildup. We have to begin with a running start.

## 18.6  Make Major Technical and Cost Trade-Offs Rapidly and Decisively

The Program Manager (or company management) should do a "cost at completion" estimate every 2 to 4 weeks and monitor cost performance personally. At the earliest sign of trouble (technical, cost or schedule), the Program Manager must immediately take the long view. Nearly all problems of cost and schedule are rooted in technical problems. In case of technical trouble, the manager must apply his or her best judgment to assess how long it is going to take to solve the problem, how much it is going to cost, and then add 30% to his estimate. If the grand total is not within the budget, alternative technical approaches must be rapidly brainstormed. These technical trade-offs must be made quickly, they must be sweeping in scope, and the decision to continue or to change must be made decisively. Small changes do not result in large savings.

## 18.7  Production Coordinator to Expedite Manufacturing

The key to low-cost manufacturing is the assignment of a production coordinator to each program. The production coordinator assures that the schedule is kept, that nothing falls through the cracks and that problems are solved to meet production schedule requirements. The production coordinator participates in engineering meetings. He or she is thoroughly familiar with the product before it is built. He or she expedites releasing engineering drawings, reminds people of their schedule obligations, makes sure that all the parts needed have been ordered, tracks the status of each subassembly and moves the work to the next process step. The production coordinator participates in weekly program status reviews where he or she accounts for each subsystem to the team leaders and identifies shortages and problems. This disseminates information among project team members, solves problems on the spot and permits each member of the team to speak up if a proposed change would adversely affect him. The production coordinator is the key to meeting a tight manufacturing schedule.

## 18.8 Do Not Try to Save Money in Testing

It is possible to hurry design and fabrication. It is foolish to hurry testing. Spend a great deal of time testing and exercising the system under simulated scenario conditions. It is only through use that problems become apparent that did not show up in normal checkout. These tend to be mostly conceptual problems. It is only through running mission scenarios that it is discovered that "as designed" is not necessarily "as intended." It often takes five months from the time the spacecraft is fully assembled with all subsystems checked out to the time it is ready to be shipped to the launch site.

## 18.9 Holding Program Budget Responsibility Tightly

Budget management practices and the views expressed here are nearly diametrically opposite to those of customary management teachings and practices. It is customary to divide the program into work packages and to assign to each major work breakdown structure (WBS) element a cognizant manager, give each a schedule and a budget, and then hold each responsible for both during the performance of the program. This traditional approach almost guarantees a cost overrun for two reasons:

- Some of the work packages may have been underbid by mistake and their budgets will have to be increased. Assigning budgets and cognizance for each work package ties the hands of the Program Manager; it is difficult for him or her to reduce the budget of a work package manager who has been doing a good job to augment another who is underfunded.

- Typically, only a small percentage of work package managers are good money managers. Those who are not will overrun their budgets, some very badly. Those who are good will bring in their tasks within budgets, but not much under budget. The net result is that the overall program will overrun.

By contrast, the Program Manager should keep the entire budget to him or herself. The program manager allocates this budget and a reserve between the work packages, but does not tell subsystem managers what the budgets are. If, while discussing the technical approach, schedule and cost of a work package with its team leader, the cost estimate is higher than anticipated, the team leader is asked to think of other ways of doing that task. This is done iteratively until the budgetary goal is reached or the Program Manager is convinced that the budget must be increased. If the estimate comes in lower than planned, then after assuring that it is valid, the estimate is accepted, adding to the management reserve.

In this way, the work package managers can perform their individual tasks according to their own approach, while the Program Manger can build up a management reserve that will invariably be needed later. The management reserve must be large. Norm Augustine has said (based on some actual data) that "a program manager always underestimates his 'cost-to-complete' by 30%." He was correct.

## 18.10 Conclusion

There is no magic or technology advance needed to make dramatic reductions in the cost of any spacecraft. Tight management, small project core teams, short schedules, opportunity to make large cost-technical trade-offs and concurrency between engineering and production are the keys to low cost. The keys to a quality high technology product are making bold technical decisions, implementing them ultra-conservatively and subjecting the finished product to extensive testing.

# Chapter 19
# Systems Engineering and Program Management

## 19.1 Introduction

The Program Manager has the responsibility to deliver a spacecraft that meets customer functional requirements and specifications, and is constructed within available funds and within a specified schedule. He or she also ensures that the spacecraft meet launch vehicle and ground station interface requirements, and that the spacecraft, when on orbit, can perform the specified mission.

Systems Engineering starts with the customer functional specifications of the spacecraft and ground systems, flows these down to spacecraft subsystem and component specifications to develop a spacecraft design that meets customer specifications, mission requirements and cost targets, and can be developed and built within the specified schedule. Systems Engineering also performs the tradeoff analyses among alternative implementations to determine the technical approach that best fits program and customer needs. It also performs sustaining and supporting analyses during the development program. Systems Engineering maintains documentation of the hardware, software, test plans and test results.

In the following, the above functions are described in more detail in the context of designing and building an imaging spacecraft system, like the one described in Sect. 2.3.

## 19.2 Top Level Requirements

Top level systems requirements are usually specified by the customer. They should state functional rather than technical requirements so as to permit the design team the widest possible tradeoff space. For example, for an imaging spacecraft system consisting of, perhaps, multiple satellites, the top level requirements should state the geographical areas the system should cover, the resolution (GSD) requirements, the revisit time, the spacecraft system agility, the number of pictures per unit time or angular separation that the system should provide, the size or coverage of each picture, the distance between satellites and the time of day the spacecraft should be over specified areas.

Top level specifications should not get into any more technical detail than necessary. However, which top level requirements are firm and which can be relaxed and "negotiated" with the customer should be identified. This is particularly important if the spacecraft will be a secondary payload of some launch vehicle, because the precise orbit inclination, altitude and eccentricity will be controlled by the Primary Payload.

## 19.3 Requirements Flowdown

From the top level requirements, Systems Engineering should develop orbit and subsystem technical requirements. This is the flowdown process from which a set of spacecraft technical specifications is obtained.

For example, from the top level requirement of the size and resolution of the images the satellite should transmit to the ground, the number of pixels per image can be computed. From the number of images per pass over areas of interest, the downlink bit rate can be obtained. Then, computing the RF link equation for the specified ground station antenna size, the satellite transmitted EIRP (sum of transmitter power and spacecraft antenna gain) can be obtained for any specified frequency band. If a reasonable transmitter power and low gain antenna can close the link, the transmitter power and downlink bit rate are readily specified. If, on the other hand, the required EIRP is very large, a high gain spacecraft antenna and/or high transmitter power may be required. This introduces additional complexities of (a) developing mechanisms for spacecraft antenna pointing and slewing, (b) using more sophisticated modulation and forward error correction (FEC) methods that require lower received signal-to-noise ratios for the same bit error rate, BER, (c) on-board storage of digitized images for later transmission to the ground or (d) use of a geostationary relay satellite to lengthen the time available to downlink images.

Another example of flowdown of top level requirements to subsystem specifications would be the propulsion subsystem technical requirements derived from the desired distance between satellites, the rapidity with which satellites have to be moved and the station keeping operations the propulsion system must perform. It is here that some "negotiation" with the customer on top level requirements may be needed. As shown in Chap. 11, moving the spacecraft rapidly to put it on station within a few hours of launch requires a great deal of fuel, while reducing the requirement to a few days, substantially reduces the amount of fuel needed, reduces the size of the required fuel tanks, reduces weight, and it simplifies the method of mounting the fuel tanks in the spacecraft. Since the size, weight and cost of the spacecraft can be affected in a major way by the required speed of satellite

19.5 Trade Studies

deployment, the above arguments would deal the spacecraft Program Manager a strong hand in negotiating with the customer.

## 19.4 Multiple Approaches

There are often multiple approaches in flowing down top level requirements to subsystem specifications. The above example of how to obtain the required EIRP is an instant in point. It can be achieved by (a) a high power transmitter and a low gain antenna or by (b) using a lower power transmitter and a higher gain antenna. The former simplifies the antenna system, and reduces pointing accuracy requirements. On the other hand, it increases electric power requirements (leading to larger solar arrays and larger DC/DC converter power handling capabilities).

Within the electric power subsystem, the electric power could be distributed from a single DC/DC converter for each required component voltage (with separate switches at the inputs of the components for turning it ON or OFF), or separate DC/DC converters could be used to power each component independently, turning ON/OFF each separate converter as its use is required. The latter distributed system reduces the power handling requirement of each converter, and it provides an additional measure of overall system reliability. On the other hand, it may increase system complexity and weight.

To achieve a given GSD, the spacecraft may be flown at a higher altitude to prolong satellite lifetime, driving up the required telescope aperture size. Alternatively, a lower orbit altitude may permit use of a less expensive telescope, but may require a small propulsion system for drag makeup. Since a propulsion system is required for putting the satellite on station and for station keeping, the additional fuel required for drag makeup may be acceptable. Flying higher with a larger aperture telescope may be much more expensive than the cost of additional fuel for drag makeup.

Multiple approaches to satisfying the mission and functional requirements of the spacecraft may exist in many subsystems. For this reason, the requirement flow-down process may require doing a number of tradeoff studies.

## 19.5 Trade Studies

The trade studies should identify multiple approaches to the design of the spacecraft, and should compare the technical risk, cost, schedule and performance risk or reliability of these approaches. These are the objectives of the trade studies. The trade studies should be performed as rapidly as possible, for completing these studies holds up the design of the spacecraft, and impacts cost and schedule.

## 19.6 Selection of a Point Design

Since the design of the spacecraft cannot be fixed or frozen until the trade studies are completed, it is important to only perform the most important tradeoffs first, the outcome or conclusions of which have the largest impact on the satellite design and mission performance capabilities, as well as the cost and schedule.

Other less important trade studies can be performed concurrently with the spacecraft detailed design process.

Once the selection process is completed from the results of the tradeoff studies, the satellite baseline design can be finalized, and the technical specifications of each subsystem can be developed.

In the selection of the point design, the experience of the Program Manager and the design team is critical in choosing a robust and simple design that can be implemented within budget and schedule.

## 19.7 Concept of Operations

The Concept of Operations is a document that describes how the spacecraft-ground station system will operate from launch to the end of the mission. It describes the way the system will be commanded to perform each of the system functions, and how the mission will be performed. It describes in detail a "typical day" in the life of the system. It is perhaps one of the most important initial outputs of Systems Engineering, because from it, potential conceptual system problems can be identified and corrected.

As the system design matures, the Concept of Operations should be updated periodically.

## 19.8 Preliminary Design Review (PDR)

When the spacecraft design reaches a point where the preliminary design is completed and most of the important tradeoffs have been made, a Preliminary Design Review (PDR) is scheduled with the customer. In this design review, the design team describes how subsystem specifications were derived from top level requirements, and how each subsystem will be designed to meet subsystem specifications. Short of the actual design of the hardware and software, the PDR fully describes the spacecraft. It also shows how the Concept of Operations is implemented by the spacecraft.

## 19.9  Interface Control Documents (ICDs)

Interface Control Documents define the mechanical, electrical, signal and communication protocol interfaces between subsystems. They are essential documents when designing a spacecraft with a large number of people or with many subcontractors responsible for different subsystems.

Each ICD is like a contract between subsystems or subcontractors, where each can work on its own as long as they satisfy the pertinent ICDs.

Satisfying the requirements of an ICD also exhibits technical, cost or other difficulties that designers of a subsystem encounter. This may require renegotiating the ICD between subsystems to permit realizing a working and cost-effective design within budget. The function of the Program Manager is to officiate over these ICD negotiations, if they are required, since they usually involve changes to the budget or schedule.

## 19.10  Detail Design

After the PDR identified the technical approaches and the basic design of the spacecraft, its detail design can start. In some instances, where the subsystem design is mature, construction may also commence.

As the design progresses, the cost-to-complete and the schedule are monitored and actions are taken and decisions are made to assure that total project costs remain within budget. This is a difficult task for cost estimating is an imprecise process, and it depends to a large extent on the proficiency of the design team.

## 19.11  Critical Design Review (CDR)

When the detail design of the spacecraft is completed, a Critical Design Review (CDR) is scheduled. The CDR, like the PDR, describes the design of the satellite and all of its subsystems, and provides the detailed backup of how or why specific technical approaches were selected.

The CDR generates the final design baseline of the satellite, describing what will be built. It is the most important document that defines the design.

On some large projects, it is also the point in time when the build process begins. However, as pointed out earlier, a prudent measure of concurrency between design and fabrication can significantly reduce schedule and cost.

## 19.12 System and Mission Simulations

While not mandatory, a simulation of system performance during the detail design process can identify problems or increase confidence in the design. For this reason, a software simulation environment of the system, converted later into a "hardware-in-the-loop" simulation tool, is often created. This simulation is used to exercise the system in different scenarios, and to observe system performance during the mission. System simulation is an important tool for ensuring that the final system can perform to the mission requirements.

## 19.13 Test Bed and "Flatsat"

Development of the spacecraft is facilitated by the use of hardware test beds in which individual subsystems can be tested and subjected to realistic flight conditions.

For example, the electric power test bed, with programmable power supplies, generates all solar panel output currents as a function of time. These are then supplied to the spacecraft electric power subsystem hardware where the batteries, charge regulators, DC/DC converters, power switches and the EPS telemetry collection subsystem are exercised and are used to power the spacecraft components. The components are simulated by electric loads that match the power consumption of the actual spacecraft components.

The test bed for the entire spacecraft, the *Flatsat*, is ultimately populated with actual (or duplicate) spacecraft hardware to convert the *Flatsat* to a hardware-in-the-loop simulation system.

## 19.14 Statement of Work

The Statement Of Work (SOW) describes in detail how the spacecraft system is going to be designed, built and tested, and it describes each specific activity the design and build team will undertake. To make it easier to describe and perform the Statement Of Work, the work is divided into work packages in a Work Breakdown Structure (WBS). The WBS is the collection of each of these specific tasks. Below, a WBS for the imaging spacecraft is described.

## 19.15 The Work Breakdown Structure

The WBS contains work packages related to the Launch vehicle, Integration with the Launch Vehicle, Launch Site Activities, etc. Here the part of the WBS related to the building of the spacecraft is detailed. In this example, the WBS is carried to three levels.

1.0 **The Spacecraft**

1.0.1 **Program Management** – Overall cognizance of the program, its technical, cost and schedule performance over the life of the program is controlled and monitored by the Program Manager. The Program Manager is also the main interface between the customer and the program.

1.0.2 **Systems Engineering** – In this task, customer requirements are flowed down to subsystem technical specifications. Technical, cost and schedule tradeoffs are carried out, and from these, the point design that meets the technical, cost and schedule requirements of the program is selected. Continuing analyses are performed during the design and build phase to support hardware and software development. Systems Engineering also maintains a record of the spacecraft design, and updates this data base frequently.

1.1.0 **Structure**

1.1.1 **Structure Design** – Design the spacecraft structure to accommodate spacecraft electronic components, deployable solar arrays, star tracker(s) (with clear Fields of View to the star field), a propulsion system with multiple thrusters, all required antennas (with assurance that the structure should not interfere with the antenna patters), and the telescope. Coordinate with the thermal design to ensure that the telescope temperature range should be small enough to keep the telescope in focus. Provide autofocus capabilities if the telescope temperature range is predicted to exceed the focus limits. Plan to use a Lightband separation system. Lay out the electronic components of the satellite to keep the CG within 0.25" of the axial center of the satellite (a Launch Vehicle requirement), and the CG-CP distance to be minimum (to keep atmospheric drag from tipping the spacecraft in flight). Maintain a weight statement of the spacecraft. Develop CAD drawings of the structure.

1.1.2 **Structural Analysis** – Develop a Finite Element Model of the structure and analyze it to determine resonant frequencies, margins, mass properties, CP and CG offset of the deployed spacecraft as propulsion fuel is used up. Special emphasis should be placed on mounting the telescope to minimize vibrational loads. Iterate the design until a satisfactory solution of low weight is obtained. Provide a NASTRAN file to the launch vehicle contractor for coupled loads analysis. Continue to support the program during the development phase with analyses, as required.

1.1.3 **Structure Fabrication** – Prepare the drawings from which the structure is to be built, send these to the machine shop and monitor fabrication. Once completed, assemble the structure and verify its weight. In the assembly, use Mass Mockups in place of electronic components not yet available.

1.1.4 **Mass Mockup Fabrication** – Design and fabricate the Mass Mockups, and ensure that their mechanical properties agree with the mechanical parts of the component ICDs (weight, CG location, resonant frequencies).

1.1.5 **Structure Testing** – Develop Static Loads and Vibration Test Plans and support the tests. Perform the tests and analyze the test results by comparing them with predicted the performance. Adjust the structure model to ensure that the model and test results agree.

1.2.0 **Telescope & Camera**

1.2.1 **Determine the Optical System Technical Requirements** – Determine the size of the required telescope aperture for the specified GSD and orbit altitude. Design the telescope as a folded optical system with a Modulation Transfer Function sufficiently large to achieve the required GSD everywhere in the Field Of View of the optical system. Perform the telescope mechanical design so that it should survive launch vehicle vibrations as represented by the Launch Vehicle supplied spectrum. Also, by proper use of zero temperature coefficient materials, design the telescope mechanical structure to ensure that its focal length should remain fixed over the expected temperature range. The mechanical design of the telescope should also block stray light and out of FOV light interference. If required, design an aperture door to block solar heat from the telescope, and a door closing mechanism to shut the aperture door when the telescope is in direct sun incidence. Perform a radiometric analysis to determine the minimum exposure times for satisfactory images and to determine the interval in a day when the telescope will receive sufficient light intensity to generate a useful image. Select the camera to provide sufficient sensitivity and small enough pixel size to support GSD requirements and minimum exposure times. The mechanism for mating the camera to the telescope should also be designed. Provide in the design space for a motorized focusing mechanism to permit increasing the temperature dynamic range over which the telescope-camera combination will remain in focus. Work with the telescope subcontractor to accomplish the above objectives.

1.2.2 **Procure the Telescope** – Place the telescope on contract, monitor the contract, support a PDR, CDR, and monitor the telescope and camera manufacturing and testing process.

1.2.3 **Camera** – Review available cameras and select one with short exposure time, small pixel size and high sensitivity. Ensure that the camera could be easily mated to the telescope. Procure two copies of the camera, one for in-house use, the other for integration with the telescope. Thus the image capture and processing digital hardware can be tested and exercised with the in-house camera while the telescope is being built.

1.2.4 **Image Capture & Processing Electronics** – Design, build and test the image processing subsystem. It should capture the images from the camera, should JPG encode images to reduce their file size, encrypt each and append each with the time instant when the image was taken and with the spacecraft position and attitude at the instant of image capture. This permits subsequent calculation of image locations on the Earth. The image capture and processing electronics might be a separate digital processor, or its functions might be included in the C&DH computer. The design of the digital image processor should include Forward Error Correction and the communication protocol. In this way the output data stream is directly suitable for transmission by the spacecraft downlink transmitter. The image processor should accept imaging commands from the C&DH.

1.2.5 **Telescope/Camera EGSE** – Design, construct and test hardware and software to test the spacecraft imaging payload (Telescope-Camera). This EGSE may include a collimator to produce calibrated target images for the telescope to image in a laboratory environment, and hardware to vary the light intensity and contrast of the targets to simulate actual conditions.

1.3.0 **C&DH**

1.3.1 **C&DH Hardware** – Select the C&DH hardware to provide the required speed, word size, number of input and output ports, memory size, radiation hardness and software operating system to meet mission requirements. The hardware should be within the cost budget allocated for this subsystem. Determine if the candidate C&DH has sufficient reliability, SEU sensitivity and redundancy to meet the multi-year mission requirements at the operating altitude. Ensure that its power consumption is within the power allocation for the C&DH in the PDR. The mechanical design or enclosure of the hardware should provide some radiation protection. If not built in-house, procure the hardware and monitor progress of its development and fabrication at the subcontractor.

1.3.2 **C&DH Software Development** – Develop the C&DH software specifications, and generate from these software module descriptions and equations. Code the software modules, test the code, and assemble the modules into the C&DH software to test it with simulated inputs at the C&DH and other component interfaces. As part of testing, introduce anomalies to test the ability to recover from these. Document the software and maintain and update the software configuration. Since the C&DH software development task is very large, it is usually broken down into many smaller subtasks.

1.4 **ADACS**

1.4.1 **ADACS Design and Analyses** – Develop an initial ADACS subsystem specification, a flowdown from the top level system requirements. Conduct tradeoff analyses to compare the performance of several ADACS alternatives. Compare three-axis stabilization systems in which the actuators are three reaction wheels with others where the actuators are control moment gyros. For attitude sensing the tradeoff studies should compare system performance using a single or two star trackers. The use of an inertial measuring unit (IMU) to track spacecraft attitude during maneuvers should be studied. Based on the results of the tradeoff analyses, select a baseline ADACS configuration that meets the technical cost, weigh, power consumption and schedule requirements of the program. A key output of the tradeoffs is the size of the reaction wheels, driven usually by spacecraft agility requirements. Since agility is a major cost driver, it is important to minimize reaction wheel torque and momentum requirements. Since the spacecraft will have a propulsion system for station keeping and for getting on station rapidly, The requirements of propulsion activities on the ADACS must be examined. The tradeoff studies have to compare the performance of the spacecraft Determine attitude knowledge and control accuracies of KE Block II during imaging, propulsion activity and during other operations including rapidly slewing between different off-nadir targets. Determine the sensor and actuator suites required to achieve these accuracies. Simulate the SC under various maneuvers to determine what the actual performance of the sensor and actuator suit is going to be during typical SC missions. Upgrade the present Proprietary MAI Dynamic Simulator to enable performing the required simulations. Continue to support the development program with such other ADACS analyses as required.

1.4.2 **Procure or Build the ADACS Hardware.**

1.4.2.1 **Procure a 3-Axis Magnetometer** – Specify and purchase a 3-axis magnetometer and perform experiments to determine how far the magnetometer has to be from the components of the spacecraft that generate magnetic fields (ADACS torque coils, motors, etc.). Modify the spacecraft mechanical design, if required, to ensure that magnetometer is positioned so that magnetic readings should not be affected by spacecraft activities. In addition, devise ways of operating the magnetometer when spacecraft magnetic activities are low or zero (like in between torque coil activities).

1.4.2.2 **Procure Torque Rods or Coils** – Conduct a tradeoff analysis to determine if torque coils or rods should be used, and to determine the magnetic moment that they should produce. If the spacecraft is large enough, often torque coils provide lighter actuators, and a magnetic field without hysteresis. Torque coils or rods should be operated intermittently in a bang-bang manner. This simplifies the torque coil driver, and it provides for periods when the torque coils or rods produce no magnetic field, making these periods available for magnetometer readings. After a decision is made and torquer specifications are generated, the torque coils or rods should be procured.

1.4.2.3 **Procure the Reaction Wheels or CMGs** – Having specified the reaction wheels or CMGs, procure them, and monitor performance of the subcontractor. Since the Reaction Wheels or CMGs are usually long lead items, it is important to commit to a purchase early in the program. Usually CMGs are selected to increase the control authority of the ADACS system for rapid slewing. Reaction wheels are selected if lower power consumption and larger momentum storage is required. Since there are relatively few companies making CMGs compared to the number who make reaction wheels, lack of timely availability of the selected actuators may require changing, and staying with the more readily available reaction wheels. A decision based on these considerations is made, and the reaction wheels or control moment gyros are built or procured.

1.4.2.4 **Sun Senor System** – The sun sensor system (its sensors, electronics and software) provide a coarse measure of the sun vector direction in SC coordinates during the sunlight portion of the orbit. The type of sun sensor elements and the placement of its components on the spacecraft must be determined. A mockup should be constructed, the sensor system and its software should be fabricated and evaluated in the laboratory. The flight hardware should also be built and procured.

1.4.2.5 **Inertial Measurement Unit** – An IMU will be used to measure in real time spacecraft attitude variations. This estimate will be used in the ADACS during maneuvering, but will also be needed in the propulsion system which must accomplish long duration thrusts to accelerate the spacecraft in a specified direction. During thrusting, the misalignment of the thrust vector from the spacecraft CG will cause tumbling or drifting in the wrong direction. The IMU will provide a continuous measure of spacecraft attitude to sense the attitude error during thrusting. This information will provide the basis for an algorithm to adjust the burn durations of the different thrusters to cause the net thrust to be through the CG and to thus eliminate tumbling or drifting in the wrong direction. The IMU will be procured and electronics and software to interface it with the ADACS and/or the propulsion system will be constructed.

1.4.2.6 **Star Tracker(s)** – One or two star trackers are required to achieve the attitude knowledge accuracy required day and night. An analysis has to be performed to determine the number and orientation of the star trackers on the spacecraft, the directions to which they must point (in spacecraft coordinates) and the dependence of star tracker pointing, depending on the orbit in which the spacecraft will fly. Baffles may be needed to shield the star trackers from direct sun or moon incidence. Star Tracker locations and directions will be adjusted in accordance with baffle dimensions. The star trackers will be procured.

1.4.3 **ADACS EGSE** – A critical tool for evaluating and testing overall ADACS performance is the ADACS Electronic Ground Support Equipment. The Dynamic Simulator, described in Chap. 9, is a key component of this EGSE. It simulates the space environment, provides signals to the spacecraft attitude sensors, flies the spacecraft in position and attitude in accordance with the orbit propagator and the activities of the spacecraft actuators. An ADACS EGSE should be built. It could be combined with the propulsion system EGSE.

1.4.4 **Assemble, Integrate and Test ADACS** – Assemble the ADACS subsystem from its components. After assembly, test the system in accordance with the ADACS Test Plan using the Dynamic Simulator to fly the spacecraft. Separately test the star tracker in outdoors on clear nights.

1.4.5 **ADACS Software, Validation and Verification** – Develop the ADACS software and test it in a hardware-in-the-loop configuration, including the spacecraft and the ADACS EGSE. After each of the software and hardware modules perform their respective functions correctly, from time-to-time software verification will be run. This is a shortened scenario in which most of the spacecraft capabilities are exercised. Telemetry data is collected and compared with the data that should exist if everything worked as designed. The comparison is automated, and the Verification test results in a Pass-No Pass score.

1.5.0 **RF Communications**
1.5.1 **RF Analyses** – In addition to the LOS (light-of-sight) communication of telemetry, TTM and Imagery and LOS communication of spacecraft commands, CMDs in the assigned frequency bands, consider the alternative (or additional) use of a geostationary satellite relay for spacecraft communications to enable continuous communications between spacecraft and Ground to increase satellite throughput. This communication, most likely through Inmarsat, is at L-Band. Determine the required spacecraft antenna gain, transmitter power and spacecraft antenna elevation and azimuth steering requirements. Consider alternative spacecraft antennas (phased arrays or mechanically steerable antennas). Determine the impact of the different antennas on topside real-estate, power consumption, and on spacecraft attitude control, the need to deploy and steer the antenna, the cost and schedule to select the most cost-effective antenna configuration. Perform the preliminary design of the antenna configuration. Analyze the link margins for the most likely size of the Ground Station antenna, with a view toward operations with mobile antennas. Choose the modulation type (FSK, BPSK, QPSK, O-QPSK, etc.) best suited to the achievement of a robust link that is compatible with existing ground station capabilities. The legacy FSK modulation has many operational and hardware advantages. However, it is not the most optimum modulation type.
1.5.2 **Image Transmitter** – The RF transmitter on the spacecraft used to transmit the high data rate imagery data may or may not be the same as the one used to transmit the low data rate telemetry data. For image transmission usually a high power transmitter at a very high frequency is selected, as compared with the telemetry transmission requirement that can be satisfied with low power lower frequency transmitters with Earth coverage antennas. The choice of spacecraft transmitter and antenna combination must adhere to the FCC requirement of maximum ground illumination density. This usually means that the maximum spacecraft transmitter power will be in the 10–20 W range. The communication link calculations in Chap. 5 indicate that a 10–20 W transmitter will suffice. The transmitter frequency, depending on compatibility requirements with the ground station, is likely to be between S-Band and X-Band, although the trend is for even higher microwave frequencies to capitalize on the larger bandwidth available. However, if we use a commercial Geostationary Relay satellite, the uplink frequency most likely will have to be changed to L-Band. Therefore a new Image Transmitter must be selected. Since having redundant transmitters is an important aspect of reliability, consideration should be given to using a second transmitter for telemetry. Select the most cost-effective solution and procure the transmitter. Monitor its development frequently.
1.5.3 **Telemetry Transmitter** – The above discussion covers this subject also. Purchase the TTM transmitter.
1.5.4 **Command Receiver** – The command receiver, usually operating at a lower frequency and low data rate, can be a SDR (Software Defined Radio) or a

legacy radio receiver. It should include the decryption capability required to receive and decode encrypted commands; however, it should be possible to bypass encryption to provide a measure of reliability. The antenna through which uplink CMDs are received by the satellite should be omni-directional to permit communicating with the satellite even if it is in a momentarily arbitrary attitude. Purchase the CMD receiver and decryption hardware or software.

1.5.5 **Antennas and Power Dividers** – The satellite antennas should be selected to provide the required gain in all directions. In the case of Earth coverage downlink antennas their peak gain is likely to 3–4.5 dB. The antenna should be mounted on the spacecraft so as not to distort the antenna pattern. This is usually ensured by building a satellite mockup, mounting the antennas on the mockup, and performing antenna tests, on the combination of antenna and spacecraft mockup, either in an anechoic chamber or at an open-range test site. If use of a satellite relay is planned, the antenna must be pointing up, and must have considerable beam steering capability in both azimuth and elevation. A selection of the antenna combination should be made, and the antennas should be procured, monitoring development during the procurement cycle. Power dividers may be used if it is decided that the same transmitter should transmit image and telemetry data. In this case the transmitter is fed to the high gain antenna for image data transmission, and it is also fed (through a power divider) to an omni-directional antenna for transmitting low data rate telemetry.

1.5.6 **Communications Processing** – The transmitted signals/data must be controlled tunable in frequency, the data must be encrypted and randomized (to make its DC component zero) and made compatible with the HDLC protocol. This is accomplished by a communications processor. The required digital processing may be included in the C&DH, although it may be best to separate the C&DH from the communications functions. The processor and the software must be developed, and the communication function must be tested with a simulated ground station.

1.5.7 **SDR for the Ground Station** – There may be several ground stations. New, transportable ground stations will probably use the modern Software-Defined-Radio, while existing ground station with which compatibility is required probably use legacy radios. If a geostationary relay is used to increase the allowable data rate, the receiver for the ground station probably already exists, and the spacecraft design only needs to concern itself with being compatible with the ground station demodulator.

1.6.0 **Electric Power System (EPS)**

1.6.1 **EPS Analyses** – Compute the OAP (Orbit Average Power) consumption of the spacecraft. The power consumption of each subsystem and its duty cycle, aggregated over all subsystems, determine the OAP and Peak Power requirements. EPS requirements will be analyzed and battery WH requirements will be determined. From these the size and deployment of the solar panel will be determined. The EPS architecture will be chosen to select the various DC/DC converters and their switching. In addition, the number of degrees for deploying the solar arrays for optimum electric power generation for the most likely selected orbit will be determined. The result will be a total analysis of the electric power system and the specification of its parts. Coordination with the mechanical design is required to ensure that the planned solar arrays and their deployment mechanisms are compatible with the spacecraft structure design. From the EPS architecture the configuration of charge regulators and DC/DC converters will be selected. There will be a continuing effort to support the program during the development phase with EPS analyses and with resolution of emerging issues.

1.6.2 **Solar Array** – The solar array resulting from the above analyses will be designed by designing panels and methods of populating them with solar cells to provide the required panel voltages and diode protection from reverse currents. The solar panels will also contain additional wiring through which the panel current flows to minimize the magnetic field produces by the panels The deployment mechanisms, other mechanical parts and hinges will be designed and fabricated.

1.6.3 **Battery System** – The battery system required Watt-Hours (WH) will be determined. This is the WH required to operate the spacecraft during the eclipse, plus the charge-discharge inefficiency of the batteries, plus the margin to allow for requirement creep, plus degradation of battery performance with temperature and age. The resulting required WH is usually much larger than that calculated from operations during eclipse only. Then the type of batteries to be used will be determined, and, considering redundancy also, the battery system can be defined and the batteries can be procured.

1.6.4 **Charge regulator(s)** – Multiple charge regulators (equal to the number of "strings" in the battery system) will be designed, built or procured and tested.

1.6.5 **Electric Power Distribution** – The electric bus voltage needs to be switched to the different DC/DC converters to provide switched power for each SC component. The converters will be designed or specified, the method of switching them will be determined, and the power distribution system will be fabricated or procured. To dissipate the heat generated by the power distribution function, the chassis of PCB will be heat sinked to the spacecraft structure, and the heat will be conducted away and radiated out of the spacecraft.

1.6.6 **EPS Computer and Software** – Control over the electric power system and collection of EPS telemetry is accomplished by a small computer and associated software. The computer accepts commands from the C&DH, implements these commands by turning EPS components ON or OFF, monitors telemetry points, and collects telemetry data for subsequent downlink to the ground. This computer will be selected, the required EPS software will be written and the hardware and software will be incorporated into the EPS.

## 19.15 The Work Breakdown Structure

1.6.7 **EPS AIT** – When all the components of the entire electric power system are completed, the EPS will be assembled and the components and the software will be integrated into an EPS system; and the system will be tested with the EGSE. The test needs to include use of a solar array simulator and a means for simulating the electrical loads on the EPS DC/DC converters.

1.7.0 **Spacecraft Software** – All software functions will be described and broken into software modules. The inputs, equations and outputs of each software module will be generated to become the total software description. Then coding each module will begin, and the code will be archived and updated whenever changes are made. The software will be tested in the target hardware (C&DH, EPS Processor, Communications Processor, ADACS computer or Image Processor). The six different computers each have their independent software, and communications between them is by file transfer. The software will be tested first on a module by module level, later on a subsystem level, and finally on an all-up spacecraft level.

1.8.0 **Propulsion**

1.8.1 **Propulsion Analyses** – Compute the delta V requirements of the propulsion system. Take into account the top level requirements of how much time is permitted to deploy the satellite to its station, requirement for station keeping, the requirement regarding how often (or seldom) station keeping maneuvers are permitted, any orbit altitude changes that may have to be implemented, and drag makeup requirements. Then calculate how much fuel is required to obtain this delta V. For small satellites, a simple cold gas propulsion system is sufficient. However, if the required delta V is large, a hydrazine propulsion system may be needed. Next the size and pressure in the fuel tanks are computed so that tank size should be manageable and should fit into the spacecraft. Finally the pressure reducer and the thrusters are selected, and the thrusters are positioned on the spacecraft. The thrust vector should pass through the CG of the spacecraft. If this cannot be guaranteed, or if depletion of fuel changes the position of the CG, multiple thrusters should be used, so that intermittent unequal duration thrusts from multiple thrusters should cause the net thrust to pass through the CG and should continue to do so as fuel is exhausted. An alternative way of configuring the thrusters should also be examined. In this approach the spacecraft flies horizontally when a delta V maneuver is executed. In fact, the spacecraft may be flown horizontally at all times except when the imaging mission is to be executed. Then the horizontal spacecraft must be turned so it becomes nadir-pointing. The potential advantage of this system is that spacecraft drag is reduced, the CG and CP are nearly co-located, minimizing drag-induced tip. The thrusters must be placed on the bottom of the spacecraft, and multiple thrusters, purposefully slightly misaligned, can be operated so that the average thrust vector passes through the CG. This spacecraft flight configuration may also increase mission life, for the drag area of the spacecraft is reduced in horizontal flight. On the other hand, this system is more complicated. The alternative propulsion systems will be analyzed and a baseline propulsion system will be defined.

1.8.2 **Propulsion System Design** – The propulsion system specified by the above analyses will be designed, and the components will be selected. The design of each propulsion subsystem component is described in subsequent sections.

1.8.2 **Fuel Tanks** – There are different fuel tanks from which a choice will be made. For cold gas Nitrogen systems, high pressure tanks are available. The highest tank pressure is typically 6000 psi, but most often 4500 psi tanks are used because they are more readily available. The lightest one of these are made of thin aluminum foil, and they are overwrapped by graphite epoxy webbing to enable them to withstand the high internal pressure. Other high pressure tanks are made of steel, and they are heavier for the same internal pressure. For hydrazine propulsion the tanks contain internal bladders to separate the hydrazine from the pressurant. The tanks will be selected and procured.

1.8.3 **Thrusters, Valves and Pressure Reducers** – The thrusters, shut off valve, thruster control valves, pressure reducers and the plumbing to assemble the entire cold gas propulsion system will be selected and purchased. The choice of thrusters influences the propulsion system concept of operation. Since cold gas thrusters open or close in about 5 ms, thrusting granularity is limited to about 5 ms. For this reason the thrusters should be relatively small, so that thrust durations much in excess of 5 ms should be required. Also, while the operating pressure of the thrusters may be high, care must be taken to select thrusters that operate at low pressure also, because this is the pressure in the fuel tank when the propulsion system stops working. The fuel left in the tank at this pressure goes to waste, requiring that a somewhat larger tank be selected.

1.8.4 **Propulsion Electronics** – The electronics required to operate the valves and the thrusters will be designed and built. Consideration will be given to incorporate a small computer into the propulsion system to give it more autonomy and to relieve the computational load from the C&DH or ADACS computers. The thruster electronics must drive the thrusters and shutoff valve. The thrusters usually require relatively high current at 28 V. For this reason separate DC/DC converters may have to be used in the propulsion system electronics.

1.8.5 **Algorithms Relating to Propulsion System** – The algorithms for performing the maneuvers for turning the SC around whenever thrusting in the opposite direction is needed, algorithms to keep the net thrust through the SC CG and in the correct direction are required. To sense the instantaneous direction of flight data from the IMU in ADACS will be used. The algorithm senses changes in direction (or attitude), and determines which of multiple thrusters must be activated, and for how long, to bring the attitude back to the desired value.

1.8.6 **Propulsion Software** – Software will be written to accomplish the algorithms outlined above. This software will be installed into the target hardware and will be tested.

1.8.7 **Propulsion System AIT** – The propulsion system will be assembled by a qualified contractor who will perform welding the stainless steel parts. When assembled and integrated with the propulsion electronics and software, the entire propulsion system will be tested under simulated conditions where the dynamic simulator will introduce thrust aiming errors. The propulsion system will be a stand-alone subassembly that can be tested without the rest of the SC.

1.9.0 **Miscellaneous**

1.9.1 **Harnessing** – A harness will be fabricated to interconnect the SC hardware components. The wire list for this hardware will be prepared by electronics engineers familiar with all the SC electronics. The harness will then be fabricated and installed into the SC, and the harness will be laced and tied down to the chassis and to selected components. The harness will be checked out for wiring errors before turning power on.

1.9.2 **Lab and Flight Spares** – An analysis of the spare requirements will be made, and, in accordance with constraints of the budget, an appropriate number of spares will be procured or built.

1.10.0 **Testing**

1.10.1 **Functional Tests** – Component level testing is done mostly at the vendor who supplies the component, or by the spacecraft team, for those components that are built in-house. Once the SC is assembled, functional testing will commence. This consists of operating the SC in every mode possible, and testing its performance under all scenario conditions. Test scenarios and telemetry or engineering telemetry responses will be collected so that expected and actual performance could be compared. These tests will be performed iteratively at room temperature.

1.10.2 **Test Plans** – Test Plans will be written for functional, spacecraft level thermal, vibration, magnetic grooming, propulsion and thermal vacuum testing. In addition, scenarios for mission simulation testing will also be generated. The Test Plans will specify the objectives of the test, the test procedures and limit values of the test results which, when exceeded, are test failures.

1.10.3 **Component Level Thermal Tests** – While overall thermal testing will be performed on the all-up spacecraft level, certain components, for which the manufacturer of that component has not performed thermal testing prior to delivery, will be thermal tested prior to incorporation into the SC. These include the ADACS, which is an in-house built subsystem, the digital subsystem and the telescope. Thermal testing over a $-40$ °C to $+55$ °C temperature range will be performed.

1.10.4 **SC Level Thermal Testing** – The entire spacecraft will be placed in a temperature chamber and tested over the temperature range mentioned above. Three failure free test sequences are required for the spacecraft to pass thermal testing. During the test the spacecraft will execute a scenario that will exercise all functions of the spacecraft. Thermal testing is one of the most important tests for it uncovers latent hardware problems. Seldom is there a thermal vacuum test failure if the spacecraft passed thermal testing.

1.10.5 **SC Level Vibration Testing** – The spacecraft will undergo sine sweep and random vibration tests in all three axes. The sine sweep test will determine the spacecraft resonant frequencies that will be compared with those predicted by the structural analysis. Random vibration testing will be performed to the GEVS vibration input levels, unless the (launch vehicle) LV permits a lower level of vibrational excitation. The spacecraft will be inspected for deformation or cracks after the vibration test, and the optical performance of the telescope-camera combination will be retested to assure that there has been no degradation.

1.10.6 **Magnetic Grooming** – The spacecraft will be put into a large Helmholz coil facility where the residual magnetic moment of the spacecraft will be determined, and the bias will be removed by compensating permanent magnets. Also, the magnetometer will be operated to ensure that its readings are correct and not biased due to any residual spacecraft magnetic flux.

1.10.7 **Propulsion Testing** – The propulsion system and its control electronics will be tested to verify that the system is able to produce long duration (60 s) of thrust in a direction that passes through the CG of the spacecraft, as determined by the spacecraft net angular acceleration. In addition, the total thrust level will be measured and compared with the nominal thrust required of the propulsion system.

1.10.8 **Thermal Vacuum Testing** – The spacecraft will undergo 3 cycles of failure free thermal vacuum testing while the spacecraft will operate and implement a realistic mission-like scenario.

1.10.9 **Mission Simulation Testing** – At room temperature the spacecraft will undergo testing under a variety of mission scenarios that exercise all functions of the spacecraft. The commands from the EGSE will be uplinked (through RF) to the spacecraft, which will execute the commands, and will communicate to the EGSE by RF (images and telemetry). The Dynamic Simulator will be connected to the spacecraft through the EGSE to "fly" the spacecraft in accordance with the torque and reaction wheel activity of the spacecraft ADACS, and will generate corresponding simulated spacecraft attitude sensor information, which it will present to the spacecraft via the EGSE.

1.11.0 **GSE and Ground Station**

1.11.1 **Mechanical GSE** – Fixtures to support the spacecraft in various attitudes in the laboratory, to rotate it automatically for sun sensor testing, and fixtures for horizontal or vertical integration with the LV or the LV Adaptor will be constructed. Removable lifting fixtures and a transportation container will also be built to permit handling and transporting the spacecraft.

1.11.1 **Electronic GSE** – EGSE will power, command and communicate with the spacecraft in the laboratory and at the launch facility. The EGSE will be designed and constructed. The EGSE will also include the Dynamic Simulator to enable testing with hardware-in-the-loop. The EGSE will also enable performance of a complete automatic verification test, complete from scenario upload to the automatic collection of test data and its comparison with the expected data. This is an important piece of hardware, for it accelerates the repetitious test of the spacecraft, and thus saves a lot of money.

1.12.0 **Stand-Alone Ground Station** – Often the customer wants to operate the spacecraft independently of other organizations. In this case a stand-alone ground station is required. A ground station from which the constellation of spacecraft can be controlled and operated will be designed, built, tested and integrated with the spacecraft. Based on the link studies, the size of the ground station antenna has been determined. Let us say it was 3.5 m. The Stand-Alone Ground Station will be able to generate spacecraft commands, receive and recover their telemetry and image data transmissions, program the spacecraft of the constellation to operate in any area of the world based on the commands received from the stand-alone ground station. In response to these imaging requirements, the Ground Station will be able to command the various spacecraft to perform maneuvers to image the required targets.

1.11.3 **Ground Station Transmitter** – A transmitter capable of integration with the Ground Station will be procured and integrated with the 3.5 m antenna and receive subsystem to complete the RF hardware of the Ground Station. The heart of this ground stations is the Software Defined Radio (SDR) which generates the transmitted modulated RF signal for transmission to the spacecraft and which receives, demodulates and decrypts the received RF signals to input to the Ground Station computers and displays. A SDR will be constructed for the Ground Station.

1.11.4 **Ground Station Antenna** – The 3.5 m ground station antenna, the antenna feed, the diplexer, and the Low Noise Amplifier (LNA) will be specified and procured. If communications with the spacecraft via a GEO relay is used, the ground station antenna system will have to operate at L-Band, or a separate L-Band antenna system has to be procured.

## 19.16 Cost

Project cost is built up by estimating the cost in man-hours and materials or subcontractor cost of each WBS element. The total cost of the project is then the sum of the burdened labor plus material cost of all WBS elements. To this the Program Manager should add his management reserve to obtain the total program cost, which must be within his budget.

When first generated at the beginning of the program, the program cost arrived at in this way, usually exceeds available funds. It is then the task of the Program Manager to alter technical approaches and manpower estimates to bring program total cost within budget. This is a difficult process, but one that must be done rigorously and repeated frequently.

## 19.17 Scheduling

In addition to costing each WBS element, the duration of performing each WBS must also be estimated and the resulting bar representing the performance period must be placed on the schedule chart at the time the task is to be initiated. WBS elements that depend on completion of other WBS elements are placed at consecutive intervals on the schedule chart.

At this time the names of the people who will perform the work defined by the WBS are also identified. The WBS performance periods must be so scheduled that the same people should not (unintentionally) be assigned to other WBS elements as well, for this would create a schedule conflict, resulting in a schedule slip.

Having identified the people who will work on each given WBS, the cost of each WBS can be recalculated to improve overall program cost accuracy.

The cost as a function of time can be constructed by adding up the sum of WBS cost elements versus time. In this way the cost expenditure profile of the program can be exhibited and compared to the costs incurred to date.

Use of the Gantt chart (available by many software manufacturers) is a simple way to perform scheduling and costing the manpower portions of each WBS, and of the entire program. The Gantt chart lays out each WBS on successive lines of the chart, while calendar time is on the horizontal axis. It allows for specifying the start and completion of each WBS and assuring that interdependence of WBS schedules are observed and maintained. It also provides an easily understood graphic display of the time line and of significant events and milestones in the development process. It also displays the cost profile of the program.

## 19.18 Critical Path

The total program performance period is determined by the Critical Path. This is the path of consecutively performed WBS tasks that take the longest, and determine the minimum duration of the program.

The Critical Path, when the schedule is first constructed, may indicate that the program cannot be performed within the allocated schedule, or that the start times of certain WBS must be moved to an earlier date to permit completing the program on schedule.

The Program Manger must ensure that the schedule be realistic and that performing the tasks according to the schedule will bring in the total program within the required time period.

To facilitate constructing such a schedule there must be slack periods built into the realistic schedule. These are periods that can be used up by the designers without impacting the overall program schedule. A schedule without slack periods is probably an unrealistic schedule.

## 19.19  Schedule Slack

There are sequences of WBS elements (such as the design, fabrication and testing of the electric power subsystem) where the performance period of the sequence of tasks is short enough that by starting the sequence at a given time will result in the completion of the subsystem before it is needed in the overall spacecraft build, integration and test process.

Therefore the development of this subsystem contains a schedule slack, a period of time that can be used up before it impacts the overall schedule.

The Program Manager should assure that, if possible, each sequence of interdependent tasks should contain schedule slack. This, of course, is an unrealizable dream. However, some slack should be built into the schedule of tasks to minimize the risk of overall schedule slip.

## 19.20  Earned Cost

It was mentioned above, that the predicted program cost profile can be obtained from the cumulative sum of the costs of each WBS versus time. The earned cost of the program can be calculated from the cumulative cost of the completed WBS elements and from the estimated cost earned by the partially completed WBS costs.

In other words, a completed WBS earned the cost of the cost assigned to that WBS. If the actual cost was larger than programmed, the additional cost is an overrun. If the actual cost is less than programmed, the cost difference adds to the management reserve of the Program Manager.

Continuous computation and monitoring of the earned cost and its comparison with the predicted cost profile must be performed very frequently. Without this comparison the Program Manager has no way of knowing where he stands on bringing the program to completion within budget.

If the earned cost is significantly smaller than the predicted cost profile, the Program Manger must develop means for reducing the cost forward of the remaining WBS elements. This is a challenging task.

## 19.21  Cost to Complete Calculation

Another computation that is important to perform every once in awhile is the computation of the Cost-To-Complete.

The cost-to-complete reexamines each not yet completed WBS element and re-estimates its cost from its present status or knowledge. The cost-to-complete is then the sum of the new WBS costing versus time from now to the end of the project.

This must be done carefully and accurately. Normal Augustine, then Deputy Secretary of the Army, used to say, based on Army data, that a Program Manager usually underestimates the cost-to-complete at any time by some 30%. The spacecraft Program Manger should make sure that he is not this type of a Program Manger.

## 19.22 Requirements Creep and Engineering Change Proposal.

During the performance of the program, changes are often proposed. Some of these are within the scope of customer requirements, but others exceed the requirements scope. The program should not perform the out of scope changes without receiving compensation for their costs.

The Engineering Change Proposal is the vehicle by which the cost of out of scope proposed changes are brought to the customer for adoption and funding, or for cancellation of the requested change.

The customer is often enthusiastic about "nice to have" additional capabilities until the cost of these additional capabilities is brought to his attention. He then must often back down and give up on the nice-to-have performance enhancements because of limitations of his own budget.

## 19.23 Reallocating Budgets, Cost Management

During program performance it is often realized that specific subsystems and their associated WBS costs and schedules were underestimated, while the costs of others were overestimated. The Program Manager must then reallocate budgets to provide more money and time to some and take away money from others.

Taking away money is a difficult task that often alienates those from whom money is taken away. Yet it must be done.

The budgets of tasks must be changed dynamically by the Program Manger to keep from overrunning the program. This is contrary to usually accepted program management practices. Yet without this program management flexibility it is difficult to run a program successfully.

## 19.24 Documentation

Documentation requirements of a satellite development program are extensive. They start with the Technical Proposal and continue with tradeoff documentation, Preliminary Design Review and Critical Design Review documentation. The set of

Test Plans, Test Reports and the reports for the specific requested Technical Interchange Meetings (TIM) add to the volume of documentation. Interface Control Documents between subsystems and the launch vehicle, Preship Review and Launch Site Procedures complete the list of documents.

## 19.25 Test Plans and Test Reports

Preceding tests of the spacecraft subsystems, and the spacecraft as a system, test plans should be prepared. The test plans include:

(a) Mechanical static load test. This test subjects the spacecraft structure (often with mass mockups of the electronic components) to loads that correspond to the maximum forces elements of the structure will see. The test exhibits the strains and deformations of the structure. Test results may require modifying some structural elements to increase the static loads margins.

(b) Electric Subsystem test plan. This test plan is constructed to expose the electric power subsystem to a realistic mission environment where solar panel outputs are simulated as a function of time. The electric power stored in the batteries and the power generated at various voltages in the system are monitored over a temperature range corresponding to what the EPS subsystem will experience.

Many other test plans are prepared. These include, but are not necessarily limited to:

(c) Digital Subsystem Test Plan
(d) Attitude Determination and Control Subsystem Test Plan
(e) Antenna Test Plan
(f) RF Subsystem Test Plan
(g) Propulsion Subsystem Test Plan
(h) Vibration Test Plan
(i) Thermal Test Plan
(j) Acoustic Test Plan
(k) Thermal Vacuum Test Plan
(l) Magnetic Grooming Test Plan

The test plans often include the expected results and the test values that, when exceeded, indicate that the subsystem or test article failed the test.

The results of individual tests should be included in reports in detail to permit assessing the adequacy of the subsystem or the full-up spacecraft.

# Chapter 20
# A Spacecraft Design Example

This example of a spacecraft design problem is given so that the reader can put to work what he or she has learned by reading this book. While the "Requirements" are clear, they are not very specific. There is not one good answer or design to meet the requirements.

## 20.1 The Spacecraft Mission Requirements

In this hypothetical case, we assume that several organizations of the news media got together and decided that they would jointly fund a spacecraft that would be on orbit all the time, and that could be commanded to image news events taking place anywhere on Earth. They heard that one can buy such a spacecraft for under $10 million (plus launch costs). It turns out that the $8 million spacecraft the media heard about had a 1 m GSD. The customer stated that the resolution should be sufficient to generate good colored pictures that could be published in newspapers, and on-orbit life should be at least 5 years. To assist the spacecraft designer, the customers agreed that they could send humans to places readily accessible by plane and automobile. They wanted the satellite primarily to image events happening in remote or politically inaccessible areas.

## 20.2 Derived Technical Requirements

The first thing to decide is how the customer-stated requirements can be translated to technical requirements. What should be the spacecraft resolution? It is known from photogrammetry that the relationship between target dimensions and the number of pixels needed follow approximate rules given in Fig. 20.1. Applying these to some of the likely targets, we obtain the table shown in Fig. 20.2.

| Detection | Orientation | Recognition | Identification | Technical Description |
|---|---|---|---|---|
| 5 pixels | 18 pixels | 40 pixels | 200 pixels | 500 pixels |

**Fig. 20.1** Number of pixels on target for various degrees of identification

| Item | Length m | Width m | Detect m/pixel | Recognize m/pixel | Identify cm/pixel | Describe cm/pixel |
|---|---|---|---|---|---|---|
| 5-Ton Truck | 7.9 | 2.5 | 1.5 | 0.20 | 4.0 | 1.5 |
| Car | 4.5 | 1.8 | 0.9 | 0.11 | 2.3 | 0.9 |
| Small Bldg | 10.0 | 10 | 2.0 | 0.25 | 5.0 | 2.0 |
| Tank | 8.5 | 3.7 | 1.7 | 0.21 | 4.3 | 1.7 |
| Small A/C | 40 | 30 | 8.0 | 1.00 | 20.0 | 8.0 |
| Crater | 138 | 138 | 27.6 | 3.45 | 69.0 | 27.6 |
| Destroyer | 122 | 10 | 24.4 | 3.05 | 61.0 | 24.4 |

**Fig. 20.2** Resolution needed for ground targets and various identification criteria

It is seen that, ideally, resolution should be on the order of 20–50 cm to recognize the various targets. This is a very tall order and one that is extremely expensive to fill. It has been stated that the cost of an imaging system varies approximately as the 2.5th power of the resolution. So, if the customer can afford a 1 meter resolution system, going to 50 cm would increase the cost by a factor of 5.65, and reducing resolution to 25 cm would increase cost by a factor of 56. So, obviously, the customer's budget would be exceeded; he needs to decide what resolution is acceptable and affordable.

The customer is shown a set of targets imaged with different resolutions to give him a feel for what the pictures from the spacecraft might look like. For example, the image of the ship shown in Fig. 20.3 was taken with 1 meter resolution. It might be quite acceptable for a newspaper photograph. One can clearly see from that image that there are no people on deck, that there is no damage to the ship, and one can make out the name of the ship. Clearly, 25 cm resolution would be much better, but if the 1 meter resolution spacecraft cost $8 million, would the customer pay $45 million for one with 25 cm resolution?

Let us say that the customer decided that he can afford to pay 20.5 million and get a 75 cm resolution imaging spacecraft. He states with seriousness that this is the most he can afford to pay. Of course, the customer realizes that this spacecraft is going to be quite a bit heavier than the 1 m resolution spacecraft, so the launch costs would also increase significantly.

Next, the orbit should be selected. Since the spacecraft must be able to image any place on Earth, a Polar orbit or a Sun-Synchronous orbit come to mind. Imaging will be a daytime operation, so a Sun-Synchronous orbit is selected.

The telescope aperture must satisfy the equation in Fig. 2.4. For 0.75 m GSD = $1.22*\lambda*H/D$. Therefore, for $\lambda = 5.5*10^{-6}$, H/D must be 111,773. For a 35 cm aperture, the orbit altitude should be 391 km. There is too much atmosphere at 391 km; and a 5-year mission life cannot be achieved at that altitude without a drag makeup propulsion system. So the altitude should be raised to at least 550 km to achieve 5-year life without propulsion (see Fig. 3.15). At that altitude the telescope

## 20.2 Derived Technical Requirements

**Fig. 20.3** A ship image taken with 1 m GSD

aperture should be 49.2 cm. This might increase the cost of the system by a factor of 2.3 and would break the budget.

If the spacecraft is flown at 391 km to keep the aperture at 35 cm (the baseline cost), a propulsion system is needed for drag makeup with an estimated $\Delta V$ of 50 m/s. While having a propulsion system complicates the spacecraft, its cost impact is relatively small; certainly the cost growth does not compare with the cost increase due to a larger aperture.

Launch vehicle selection, while never easy, is not too difficult to a Polar or Sun-Synchronous orbit. With a 2-3 year leadtime, it should be possible to get manifested to a suitable mission as a secondary payload.

To flow down the customer requirements to the ADACS subsystem, it is recognized that there was nothing said by the customer that would suggest that the spacecraft should be very agile. That is good, as agility also means a larger set of reaction wheels, greater power consumption and more cost.

Since the spacecraft would have to slew to the target to image it, a 3-axis zero-momentum system with 3 reaction wheels is indicated. If the camera CCD has 5.5 μm pixels, and it takes 16 MPixel images with an aspect ratio of 3/2, then each image will be 3265 × 4899 pixels. At 75 cm per pixel, the image will be 1633 meter × 3674 meter (at nadir). This is probably much larger than any area the customer wants to image. Therefore, the pointing accuracy can be relaxed, as long as the target remains in the FOV of the image. Assume that pointing accuracy of 400 meters is sufficient. The target will be within about 14% of the center of the image. Pointing accuracy of 400 m at 391 km altitude is ±0.058° accuracy. Thus attitude sensing to an accuracy of about 0.028° is required and suggests the use of 2 star trackers.

The resulting set of preliminary Technical Requirements is shown in Fig. 20.4.

| Orbit Altitude | 391 km |
|---|---|
| Inclination | Polar or Sun Synch |
| Imager | |
| Telescope Aperture | 35 cm |
| Focal Length | TBD (#f/8 to #f/10) |
| Camera | 16 MPixel, 5.5 μm/pixel |
| ADACS | |
| 3-axis zero momentum stabilized | 3 Reaction Wheels |
| Pointing Accuracy | ±0.058° |
| Attitude Sensing | ±0.028° with 2 Star Trackers |
| Propulsion | ≈50 m ΔV Cold Gas or Hydrazine |
| Dimension s | TBD |
| Weight | TBD |

**Fig. 20.4** Preliminary specifications of the imaging spacecraft

## 20.3 Preliminary Design

Before the detailed design of the spacecraft can begin, the solar array configuration and the method of thrusting should be considered.

In a sun synchronous orbit, the sun is always in the same direction relative to the spacecraft. Therefore a set of fixed deployable panels, like those shown in Fig. 20.5 can be used. The set would supply electric power to the spacecraft very efficiently.

It is not simple to figure out where the propulsion thrusters should be so that the thrust vector would go through the CG of the spacecraft. The 4- thruster configuration shown in Fig. 11.6, and a thrusting regimen like that shown in Fig. 11.15, however, is one arrangement that works for this case.

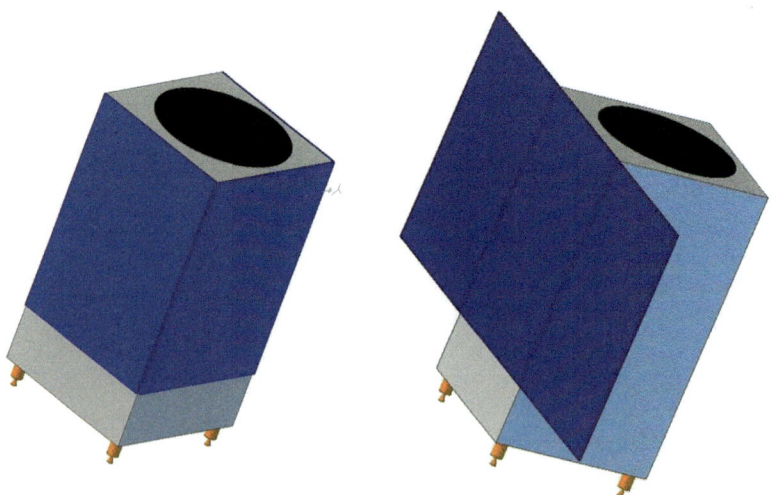

**Fig. 20.5** The spacecraft configuration with fixed deployable solar panels shown stowed and with panels deployed

## 20.4 Design Steps

Now that the customer requirements have led to a more specific set of technical requirements, the reader should undertake the design of the spacecraft as an exercise. The design steps are outlined below:

1. Draw a preliminary block diagram of the spacecraft
2. Determine the preliminary requirements of each of the components
3. Choose the digital architecture (one computer does it all, or federated computer architecture?)
4. Estimate the power consumption of each component in each mode and determine OAP needed
5. Study and choose the solar panel configuration
6. Select the structure
7. Determine the spacecraft agility required (sizes the reaction wheels)
8. Develop a preliminary *Weight Statement* with the appropriate margins
9. Perform a preliminary thermal design and structural analysis
10. Put together the entire design in a *Preliminary Design Document*

With this last step, a coherent design package is completed.

# Chapter 21
# Downloadable Spreadsheets

There are many mathematical relationships of interest in spacecraft design. Some of the more important ones are between the spacecraft and its orbit, pass duration at various altitudes above given minimum elevation angles, between the RF Signal-to-Noise ratio and ground elevation angle, and others. These relationships have been expressed as spreadsheets that could aid the spacecraft engineer. A list of these spreadsheets and their brief description is given below.

| 1  | Earth Magnetic Field Properties | Earth Magnetic Field vs. Latitude and Altitude, Dip Angle vs. Magnetic Latitude |
|----|----------------------------------|--------------------------------------------------------------------------------|
| 2  | Acceleration of Gravity, Gravity Gradient, Period of Oscillation | Gravity vs. Altitude, Gravity Gradient vs. Altitude, Gravity Gradient Stabilization, Period of Oscillations vs. Tip Mass, Boom Length and Spacecraft Weight |
| 3  | Diffraction and Geometrical Resolution | Diffraction Limit, Geometrical Resolution vs. Altitude, Focal Length, Aperture |
| 4  | Exposure Time vs. Pitch Rate and Pointing Error | Required Pitch Rate vs. Exposure Time for Different Altitudes, Along Track Error vs. Pitch |
| 5  | Longitude Range between Orbits | Longitude Range between orbits vs. Altitude at Various Latitudes |
| 6  | Pitch and Yaw Looking at a Target | Pitch and Yaw, Looking at a Target vs. Range, for Different Altitudes and CPA |
| 7  | Pass Properties | Elevation vs. Range to CPA, Pass Duration vs. CPA Range, |
| 8  | Pass Durations | Pass Duration vs. Minimum Elevation Angle for Various Altitudes |
| 9  | Various Quantities vs. Elevation | RF Link Equation vs. Elevation, Slant and Ground Range vs. Elevation for Various Altitudes |
| 10 | Lithium Battery Life | Battery Life vs. Depth Of Discharge |
| 11 | Spacecraft OAP Requirements | Spreadsheet of SC components, operating modes, power consumption resulting in SC OAP Requirements |

**Electronic supplementary material**: The online version of this chapter (https://doi.org/10.1007/978-3-319-68315-7_21) contains supplementary material, which is available to authorized users.

| | | |
|---|---|---|
| 12 | Two Turn Quadrifilar Antenna Patteren | Antenna Pattern |
| 13 | Dish Antenna Gain | Dish Antenna Gain vs. Frequency, Antenna Diameter |
| 14 | Turnstile Antenna | Turnstile Antenna Gain |
| 15 | 2,3 GHz Link Equations | Eb/No vs. Elevation Given Antenna and Xmitter Power |
| 16 | BER vs. Eb/No | Eb/No for Various Modulations, vs. Elevation |
| 17 | Link Margin with Quadrifilar | Link Margin vs. Elevation with Quadrifilar SC Antenna |
| 18 | Link Margin with Full Wave Quadrifilar | Link Margin vs. Elevation with Full Wave Quadrifilar SC Antenna |
| 19 | Pith Bias Momentum Stabilization | Sizing the Reaction Wheel for Pitch Bias Momentum Stabilization |
| 20 | GEVS Random Vibe | GEVS Random Vibration Requirements |
| 21 | Weight Statement | Weight Statement, CG Position for positions and weights of SC Components |
| 22 | Propulsion Delta V | Delta V required to Get on Station and for Station Keeping |
| 23 | Propulsion Relationships | Rocket Equation, Station Keeping, Hohmann Transfer Orbit, Constellation Spacing |
| 24 | Radiation Shielding | KRAD Total Dose vs. Shield Thickness, KRAD dose for Given Number of Years for Various Shield Thicknesses and Orbit Altitudes |
| 25 | Reliability | Reliability vs. MTBF and Duty Cycle Use |
| 26 | Azimuth vs. Range | Azimuth vs. Range for Various CPA |
| 27 | Beta vs. Time | Beta vs. Time for a Year |
| 28 | SC Pitch and Roll | SC Pitch and Roll vs. Time when Looking at a Target on the Ground |
| 29 | OAP vs. Beta | OAP as percentage of installed power for various solar panel configurations |

# Appendix 1: Tensile Strengths of SS Small Screws

| Screw size | Tensile strength lbs | Torqued to in-lbs | Clearance hole (in) | Tap drill size |
|---|---|---|---|---|
| 4–40 | 360 | 5.2 | 0.116 | 43 |
| 6–32 | 550 | 9.6 | 0.144 | 36 |
| 8–32 | 850 | 19.8 | 0.169 | 29 |
| 10–24 | 1050 | 22.8 | 0.196 | 16 |

# Appendix 2: NASA Structural Design Documents Accessible at http://standards.nasa.gov

| | |
|---|---|
| ANSI/AIAA S-080-1998 | Space Systems – Metallic Pressure Vessels, Pressurized Structures, and Pressure Components, September 13, 1999 |
| ANSI/AIAA S-081A-2006 | Space Systems – Composite Overwrapped Pressure Vessels (COPVs), July 24, 2006 |
| JSC 65829 | Loads and Structural Dynamics Requirements for Spaceflight Hardware |
| JSC 65830, rev. 2 | Interim Requirements and Standard Practices for Mechanical Joints with Threaded Fasteners in Spaceflight Hardware |
| NASA-STD-5019 | Fracture Control Requirements for Spaceflight Hardware |
| NASA-STD-5012 | Strength and Life Assessment Requirements for Liquid Fueled Space Propulsion System Engines, Baseline, June 13, 2006 |

# Appendix 3: Temperature Coefficients of Materials

| Material | $10^{-6}$ m/(m°C) | $10^{-6}$ in/(in°F) |
|---|---|---|
| Aluminum | 22.2 | 12.3 |
| Beryllium | 11.5 | 6.4 |
| Brass | 18.7 | 10.4 |
| Bronze | 18 | 10 |
| Cadmium | 30 | 16.8 |
| Cast iron gray | 10.8 | 6 |
| Chromium | 6.2 | 3.4 |
| Copper | 16.6 | 9.3 |
| Copper, beryllium 25 | 17.8 | 9.9 |
| Epoxy, cast resins & compounds, unfilled | 45–65 | 25–36 |
| Glass, hard | 5.9 | 3.3 |
| Glass, Pyrex | 4 | 2.2 |
| Glass, plate | 9 | 5 |
| Gold | 14.2 | 8.2 |
| Inconel | 12.6 | 7 |
| Indium | 33 | 18.3 |
| Invar | 1.5 | 0.8 |
| Iridium | 6.4 | 3.6 |
| Iron, pure | 12 | 6.7 |
| Iron, cast | 10.4 | 5.9 |
| Iron, forged | 11.3 | 6.3 |
| Kapton | 20 | 11.1 |
| Lead | 28 | 15.1 |
| Lithium | 46 | 25.6 |
| Magnesium | 25 | 14 |
| Nickel | 13 | 7.2 |

(continued)

| Material | $10^{-6}$ m/(m°C) | $10^{-6}$ in/(in°F) |
|---|---|---|
| Nylon, type 11, molding and extruding compound | 100 | 55.6 |
| Paraffin | 106–480 | 58.7–265.8 |
| Phosphor bronze | 16.7 | 9.3 |
| Plastics | 40–120 | 22–67 |
| Polycarbonate (PC) | 70.2 | 39 |
| Polyester | 123.5 | 69 |
| Polyvinyl chloride (PVC) | 50.4 | 28 |
| Porcelain, industrial | 6.5 | 3.6 |
| Solder lead – tin, 50 – 50% | 24 | 13.4 |
| Steel | 12 | 6.7 |
| Steel stainless austenitic (304) | 17.3 | 9.6 |
| Tungsten | 4.3 | 2.4 |
| Zinc | 29.7 | 16.5 |

# Appendix 4: Hohmann Transfer Orbit

The Hohmann Transfer Orbit is the minimum energy maneuver to raise orbit altitude. The equations are derived below, and an example is given. The example starts with a 600 km orbit, and raises it to a 700 km. A Delta V of 53.768 m/s is required. Raising the orbit in 20 km steps are also given.

# Appendix 4: Hohmann Transfer Orbit

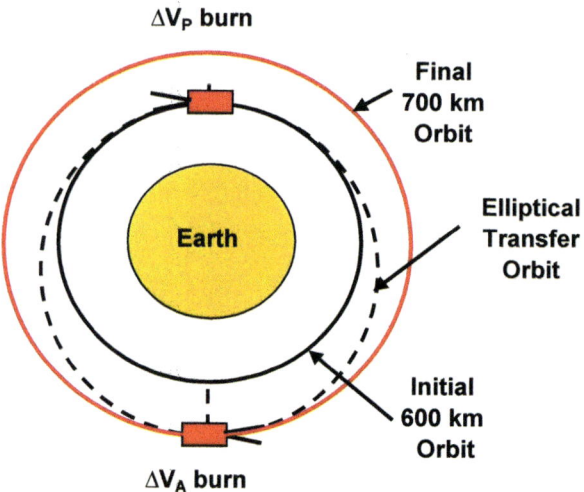

| Property | Equation | Value | Units |
|---|---|---|---|
| Earth Radius | R | 6,378.137 | km |
| Initial Orbit Altitude | $H_P$ | 600 | km |
| Final Altitude | $H_A$ | 700 | km |
| Initial Orbit Period | $P_I = 0.00016587*(R+H_P)^{\wedge}1.5$ | 96.689003 | min |
| Initial Orbit Velocity | $V_I = 2*\pi*(R+H_P)/P_I/60$ | 7.55772399 | km/sec |
| SemiMajor Axis | $SM = R+(H_P+H_A)/2$ | 7,028.137 | km |
| $\Delta a$ | $\Delta a = (H_A - H_P)/2$ | 50 | km |
| $\Delta V_P$ | $\Delta a = 2a\ \Delta V_P/V_1;\ \Delta V_P = \Delta a*V/2a$ | 26.884 | m/sec |
| Time to Apogee | $P_E/2$ | 48.865 | min |
| Time to Apogee | $P_E/2/60$ | 0.814 | hours |
| Period of Higher Orbit | $P_2 = 0.00016587*(R+H_A)^{\wedge}1.5$ | 98.77483 | min |
| Final Velocity Required | $V_F = 2*\pi*(R+H_A)/P_2/60$ | 7.5041463 | km/sec |
| Period of Higher Orbit | $P_2 = 0.00016587*(R+H_A)^{\wedge}1.5$ | 98.77483 | min |
| Velocity of Higher Orbit | $V_F = 2*\pi*(R+H_A)/P_2/60$ | 7.5041463 | km/sec |
| $\Delta V_A$ | $\Delta V_A \approx \Delta V_P$ | 26.884 | m/sec |
| Total $\Delta V$ | $\Delta V \approx 2*V_P$ | 53.768 | m/sec |
| Raise Orbit from 600 to | | | |
| | 600 | 0 | m/sec |
| | 620 | 10.815 | m/sec |
| | 640 | 21.815 | m/sec |
| | 660 | 32.353 | m/sec |
| | 680 | 43.075 | m/sec |
| | 700 | 53.768 | m/sec |

# Appendix 5: Elevation and Azimuth from Spacecraft to Ground Target for Various CPA Distances

Ground Range to CPA is the independent variable. Find the spacecraft Pitch and Azimuth look angle from the spacecraft to the target. Also find the elevation angle from the target to the spacecraft.

First solve for the CPA subtended angle, **c**, then for the ground range to CPA subtended angle, **φ**. From these the Off-Nadir angle, α, can be obtained. The pitch and azimuth to target is computed, and the elevation angle is also obtained. The equations are shown below, as is an example for 600 km orbit altitude, ground range of 2000 km and CPA distance of 500 km.

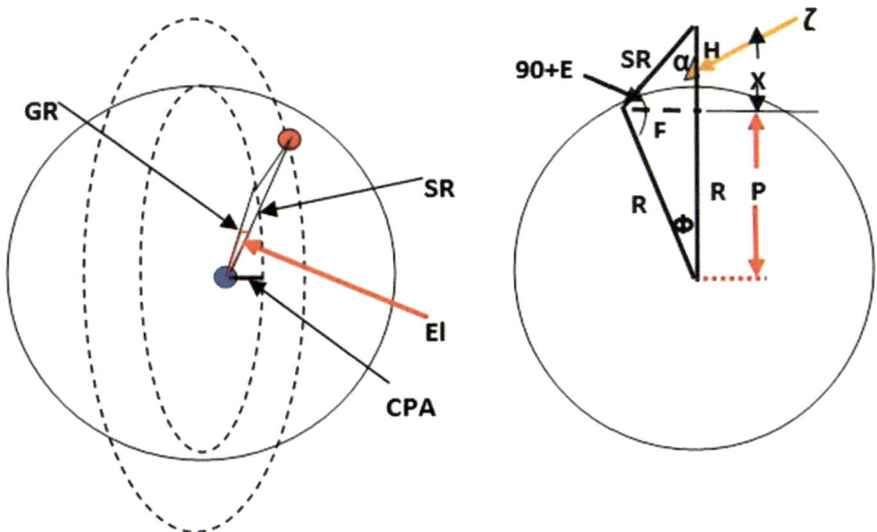

| Item | Symbol | Units | Quantity | Equation |
|---|---|---|---|---|
| Earth Radius | R | km | 6,378.3 | |
| Altitude | H | km | 600.0 | Input |
| CPA | CPA | km | 500.0 | Input |
| Grnd Range to CPA | $GR_{CPA}$ | km | 2,000.0 | Input |
| CPA Subtended Angle | c | deg | 4.491 | $c = 180/\pi \ast (CPA/R)$ |
| $GR_{CPA}$ Subtended Ang | φ | deg | 17.966 | $\varphi = 180/\pi \ast (GR_{CPA}/R)$ |
| | P | km | 6,067.3 | $P = R \ast \cos(\varphi)$ |
| | F | km | 2,068.2 | $F = R \ast \tan(\varphi)$ |
| | X | km | 911.0 | $X = H + R - P$ |
| Off-Nadir Angle | α | deg | 66.228 | $\alpha = \operatorname{atan}(F/X)$ |
| | α | deg | | $\alpha = 180/PI() \ast \operatorname{atan}(R \ast \tan(GRcpa/R)/(H+R-R \ast \cos(GRcpa/R)))$ |
| Pitch Angle | p | deg | 23.772 | $p = 90 - \alpha$ |
| Azimuth to Target | az | deg | 14.036 | $az = 180/\pi \ast \tan(CPA/GR_{CPA})$ |
| Off-Nadir Angle | α | deg | 66.228 | $180/PI() \ast \operatorname{atan}(R \ast \tan(GRcpa/R)/(H+R-R \ast \cos(GRcpa/R)))$ |
| Elevation Angle | ε | deg | 5.806 | $\varepsilon = 90 - \alpha - \varphi$ |

# Appendix 6: Beta as a Function of Time (Date) Inputs are Orange

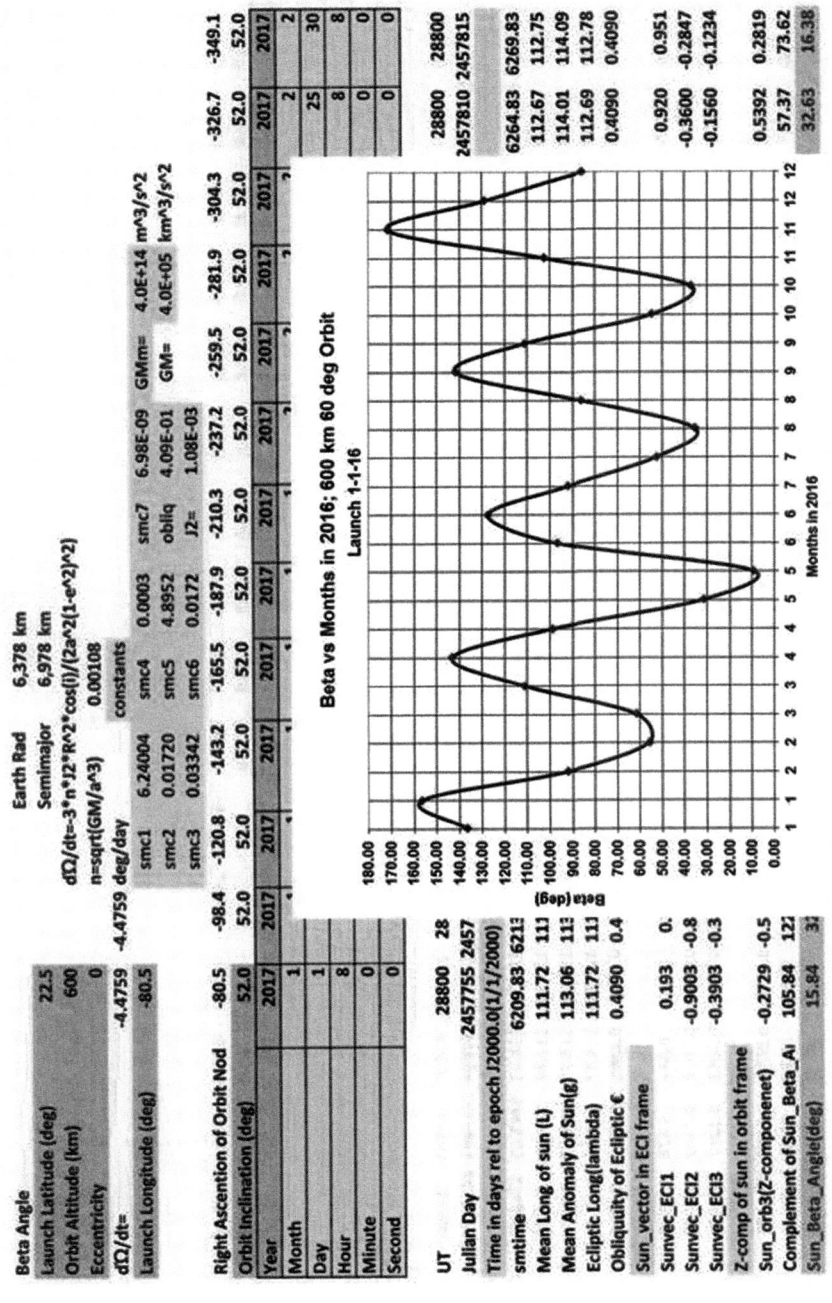

© Springer International Publishing AG 2018
G. Sebestyen et al., *Low Earth Orbit Satellite Design*, Space Technology Library 36, https://doi.org/10.1007/978-3-319-68315-7

# Appendix 7: Eclipse Duration

The duration of the eclipse is a function of Beta, and is given in Fig. A7.1 and in Fig. A7.2.

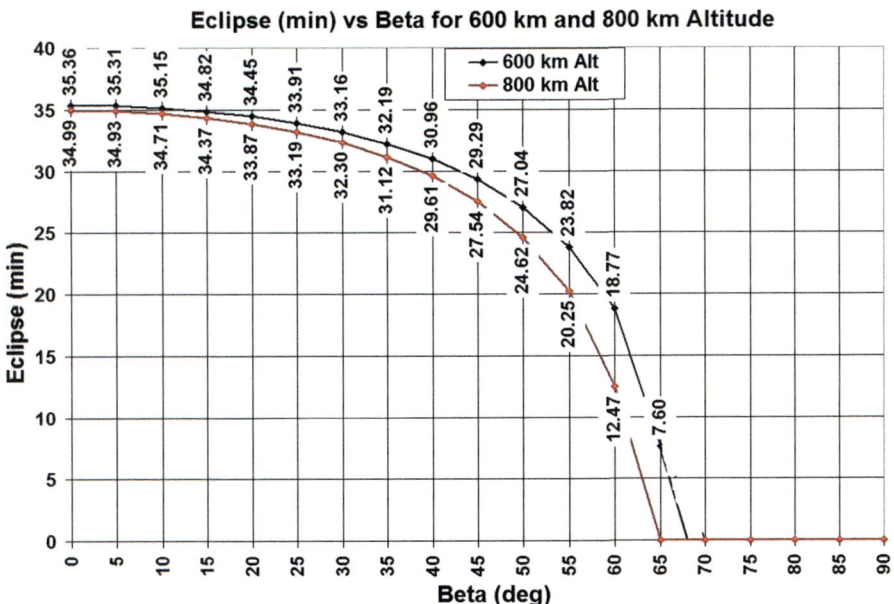

**Fig. A7.1** Duration of the Eclipse vs. Beta for 600 and 800 km orbits

Period (min)=0.00016557 x $(R_E+Alt)^{1.5}$

| Alt (km) | 600 | 600 | 800 | 800 |
|---|---|---|---|---|
| Period (min) | 96.5175 | Anomaly | 100.6964 | Anomaly |
| ß° | Ecl (min) | deg | Ecl (min) | deg |
| 0.0 | 35.36 | 131.89 | 34.99 | 125.09 |
| 5.0 | 35.31 | 131.70 | 34.93 | 124.88 |
| 10.0 | 35.15 | 131.11 | 34.71 | 124.09 |
| 15.0 | 34.82 | 129.87 | 34.37 | 122.88 |
| 20.0 | 34.45 | 128.49 | 33.87 | 121.09 |
| 25.0 | 33.91 | 126.48 | 33.19 | 118.66 |
| 30.0 | 33.16 | 123.68 | 32.30 | 115.48 |
| 35.0 | 32.19 | 120.07 | 31.12 | 111.26 |
| 40.0 | 30.96 | 115.48 | 29.61 | 105.86 |
| 45.0 | 29.29 | 109.25 | 27.54 | 98.46 |
| 50.0 | 27.04 | 100.86 | 24.62 | 88.02 |
| 55.0 | 23.82 | 88.85 | 20.25 | 72.40 |
| 60.0 | 18.77 | 70.01 | 12.47 | 44.58 |
| 65.0 | 7.60 | 28.35 | 0.00 | 0.00 |
| 70.0 | 0.00 | 0.00 | 0.00 | 0.00 |
| 75.0 | 0.00 | 0.00 | 0.00 | 0.00 |
| 80.0 | 0.00 | 0.00 | 0.00 | 0.00 |
| 85.0 | 0.00 | 0.00 | 0.00 | 0.00 |
| 90.0 | 0.00 | 0.00 | 0.00 | 0.00 |

**Fig. A7.2** Eclipse duration and range of anomaly Angles vs. Beta

# Glossary

**Acceleration of gravity** The acceleration of gravity times the mass of an object is the force with which that object is pulled toward the center of the Earth. Centrifugal acceleration times the mass of an object is the force with which an object spinning around the Earth is trying to escape the Earth. The orbit altitude is the altitude where the acceleration of gravity is equal to the centrifugal acceleration of the satellite at the orbital velocity. The acceleration of gravity varies inversely with the square of the altitude. At the surface of the Earth it is 9.8 m/s$^2$, or 32.174 ft/s$^2$.

**ADACS** Attitude Determination And Control Subsystem of a spacecraft. With its optical, magnetic and infrared sensors the ADACS can determine the spacecraft attitude. Using a GPS receiver on the spacecraft, it can also determine spacecraft positions. It contains a computer that calculates the best attitude from the multiple sensors, and it computes torque rod and reaction wheel commands to execute commanded attitude changes. It also interfaces with these actuators to implement the required attitude changes.

**Albedo** Reflected infrared radiation from the surface of Earth.

**Baffle** A light shield that shields an optical system from light entering it from outside the field of view (FOV).

**Ballistic coefficient** The ratio of the spacecraft mass to its drag. The drag is atmospheric density times the drag coefficient. For spacecraft the drag coefficient is usually about 2.2. The ballistic coefficient is a measure of how resistant the object is to deceleration due to atmospheric drag. Satellites typically have ballistic coefficients in the 60–200 kg/m$^2$ range.

**Beta (ß)** The angle between the sun vector and the plane of the orbit. For Beta = 90° the sun is normal to the orbit, and the sun incidence on the spacecraft is maximum.

**Camera pixels** The camera image sensor is composed of Y rows of X sensors, providing X*Y spots or pixels of which an image is composed. There are typically two methods of arranging pixels in a camera. In a "push-broom" camera the pixels are all in 1 or 2 rows, and the image is taken, like a push broom, by letting the satellite sweep out a rectangular area over time. The other method of arranging pixels is in a staring sensor, like in ordinary commercial cameras, where the pixels are arranged in a rectangular array.

**CMOS sensor** Complementary Metal-Oxide Semiconductor is a technology often used for constructing image sensors, instead of the more often used CCD (Charge Coupled Device) sensors.

**CMG** The control moment gyro is an attitude control device used in spacecraft. The rotor of the gyro spins continuously, and when a change, or slew, of spacecraft attitude is required, a torque is applied at right angles to the shaft of the rotor, causing the gyro to presses, which, in turn, causes the spacecraft to change attitude also. The CMG is often used instead of a reaction wheel, for it may improve the agility of the spacecraft by offering higher torque at the expense of higher power consumption and mass.

**COTS** Commercial Off-The-Shelf.

**CPA** Closest Point of Approach.

**CubeSat** Small satellites that conform to a set of specifications and are 10 cm cubes (1 U), 10 cm × 10 cm × 20 cm (2 U), 10 cm × 10 cm × 30 cm (3 U) or larger, up to 6 U (at this time).

**Diffraction limit** The minimum angular separation between images that can be resolved. When expressing the ground resolution capabilities of a space telescope, the diffraction limit is the minimum distance on the ground that can be resolved by a space telescope of given aperture. The diffraction limit is $DL = 1.22*\lambda*H/D$, where $\lambda$ is the wavelength of the light, H is the orbit altitude and D is the diameter of the optical aperture.

**Disturbance torques** Small environmental torques acting on a spacecraft. They include gravitational field variations, aerodynamic drag forces or torques caused by an offset between the Center of Gravity and Center of Pressure of a spacecraft, magnetic torques induced by residual magnetic moments, solar radiation torques and torques induced by leaks of fuel or other matter on the spacecraft.

**Earth horizon sensor** Used on spacecraft to determine the direction of the Earth horizon in spacecraft coordinates. There are at least two different kinds of Earth horizon sensors. The legacy sensors are mechanically rotating infrared detecting narrow beams. The axis of rotation is close to horizontal and normal to the orbit plane. The line of sight is skewed, causing a conical scan path. When the receiving beam intersects the Earth limb, the IR signature of the Earth causes the received signal to rise; when the receiving beam no longer intersects the Earth, the signal drops. The bisector of the resulting pulse is when the beam pointed straight down. This is used as the pitch attitude in pitch bias momentum stabilized satellites. The other type of Earth horizon sensor uses a set of individual IR sensors covering an arc on the satellite that straddles the Earth limb. The direction of the Earth limb is determined from the specific individual detectors, depending on whether they receive or do not receive Earth IR energy. The accuracy of Earth horizon sensors is typically in the range of 0.5–1.5°.

**Earth oblateness** Earth is not spherical. The radius to the North and South Poles is shorter than the radius to the Equator by a factor of 1/297. This causes orbit nodal precession and Earth sensor attitude estimation errors.

Glossary

**Eclipse** The fraction of the orbit when the sun is not visible to the satellite. At low orbit and Beta = 0° the maximum eclipse is about 35 min. At Beta = 90° there is no eclipse.

**EGSE** Electrical Ground Support Equipment consists of the electrical or electronic equipment used to support spacecraft operations while it is on the ground. It includes means for charging batteries, commanding the spacecraft, collecting telemetry from it, and otherwise testing and exercising the spacecraft.

**EPS** The Electric Power Subsystem consists of the hardware and software on a spacecraft that is pertinent to powering and switching all onboard electronics and charging the batteries. It usually consists of solar panels to generate power, charge regulators to charge batteries, DC/DC converters to convert the battery bus voltage to all the voltages required by the spacecraft components, and switches to turn power ON or OFF for each satellite component. The EPS also often contains a small computer to manage the EPS, to collect EPS telemetry and to accept commands from the C&DH.

**FEC** Forward Error Correction includes methods of introducing additional bits in a data stream to make the reception of the data stream impervious to one or more bit errors. There are many FEC algorithms.

**FEM** Finite Element Model calculates the stresses and strains at each point in the satellite from a CAD (Computer Aided Design) of the satellite structure. It also computes the resonant frequencies. The FEM is the main mathematical tool in structural analysis.

**FlatSat** The spacecraft electronics assembled on a table and used in the development and testing the spacecraft components and the spacecraft system prior to the assembly of the actual spacecraft. Its utility is to accelerate development and through its use gain confidence about the spacecraft design.

**Geometrical resolution** The size of a spot on the ground that can be resolved by a space telescope at altitude H, of Focal Length F, and of camera pixel size P. The GSD = P*H/F, as long as the GSD obtained this way is larger than the Diffraction Limit of the telescope.

**GEVS** NASA General Environmental Vibration Specifications (for spacecraft).

**Graphite epoxy** A material of great strength and light weight, from which spacecraft structures and telescope cylinders are made on a spacecraft. It is almost completely non-conducting.

**Gravity gradient** The gravity gradient boom makes use of the decrease of the acceleration of gravity with increasing altitude, the gravity gradient, to stabilize a spacecraft equipped with a long vertical boom and tip mass. The gravity gradient over the length of the boom generates a difference in force acting on the tip mass versus on the spacecraft body. This creates a pendulum from the spacecraft and tip mass which causes the spacecraft to assume an Earth pointing attitude. Vertical pitch and roll pointing accuracy of a few degrees can be achieved.

**GSD** Ground Sample Distance is the resolution of the system on the surface of the Earth.

**Heat pipe** A conductor of heat typically containing an aluminum pipe and a working fluid inside it. The pipe conducts heat from one point to another more efficiently than using metal conductors.

**Hydrazine** $N_2H_4$ is a liquid propellant used in spacecraft applications where cold gas propulsion does not provide enough thrust. Its $I_{SP}$ is 160, compared to that of $N_2$, which has an $I_{SP}$ of around 70.

**Hohmann** The Hohmann transfer orbit from one orbit altitude to another is the minimum energy maneuver for changing orbit altitudes.

**ISO9000** A quality control system that consists of a set of procedures for manufacturing that is regularly reviewed by the government.

**ISP** Specific Impulse is a measure the spacecraft propulsion system fuel efficiency. It is the total impulse (change in momentum) delivered per unit of propellant consumed. Typical fuel $I_{SP}$ values range from a low of about 77 (for compressed Nitrogen gas) to about 290 (for rocket fuel).

**IGRF** International Geomagnetic Reference Field. A mathematical model of the secular magnetic field. By comparing on-board magnetometer readings with the model values, the spacecraft can determine its attitude and navigate.

**IMU** Inertial Measurement Unit is a solid state gyro, used to aid the spacecraft ADACS by providing accurate attitude during agile maneuvers when spacecraft angular rates exceed the ability of the star trackers to image stars. It is also used in the propulsion system for the same reason.

**Invar** A metal alloy of iron and Nickel. When the percentage of Nickel is about 36%, the metal has nearly zero temperature coefficient. It is for this reason the Invar 36 is often used as structural elements in space telescope design.

**ITU** International Telecommunication Union.

**J2000** The Earth Centered Inertial (ECI) coordinate system origin is at the center of the Earth. In a specific ECI coordinate system, the J2000 coordinate system, the x axis points in the direction of the mean equinox. The z axis is in the direction of the Earth spin axis, and the y axis is 90° East from the celestial equator. This coordinate system is used in spacecraft ADACS.

**Kalman filter** A computational method that determines the best least squared estimate (of attitude) on orbit from the various attitude sensors on a spacecraft. Usually, the star tracker and the IMU estimates of attitude are used to obtain a more accurate attitude estimate.

**Apogee kick motor** Usually a solid propellant rocket motor (like the ATK STAR series of motors) used to circularize the orbit of a satellite after the booster took it to the correct altitude. Since the satellite is separated from the launch vehicle at the apogee of the orbit, the motor is called apogee kick motor.

**Link margin** The number of dB above the amount of RF signal that must exist at the RF receiver output to achieve signal reception with a given bit error rate. For example, for receiving incoherent FSK signals at a BER (Bit Error Rate) of $10^{-5}$, about 15 dB more signal than the per bit signal versus noise energy is required. If the signal is larger than that, the link margin is the excess signal, expressed in dB.

**LVLH** The Local Vertical, Local Horizontal reference frame is used in Earth pointing spacecraft. The +X direction is in the direction of orbit velocity, +Y is the negative orbit normal, and +Z is to the nadir.

**Marman band** Used to secure the spacecraft to the launch vehicle. The Marman Band consists of two semicircular machined bands, hinged together at one end and held together with a frangible bolt at the other end, with which the satellite and the launch vehicle are secured to each other. The satellite and launch vehicle mating surfaces are beveled, and so are the Marman Band semicircular rings, so that when the band is secured, it squeezed the spacecraft and the launch vehicle together. When on orbit, and the spacecraft needs to be separated from the launch vehicle, the frangible bolt is severed, usually with an explosive cutter, so as to separate the spacecraft from the launch vehicle.

**MLI** Multi Layer Insulation. These are blankets, consisting of multiple layers of insulating materials and reflective or absorptive finishes. They are used to cover the exterior surfaces of parts of a satellite, as part of its thermal control system.

**MGSE** Mechanical Ground Support Equipment. The mechanical fixtures used to lift, turn or otherwise handle the spacecraft are collectively called MGSE.

**MTF** Modulation Transfer Function is a measure of the spatial resolution of an optical system. Like in electronics, where the transfer function expresses the amplitude and phase changes from input to output of electronic circuits at each frequency, the MTF similarly expresses the input-output relationship of an optical system as a function of the granularity of the input image. MTF = 1 is perfect transmission, while MTF = 0.1 provides a marginally acceptable picture.

**NASTRAN** A Finite Element mathematical model, used to compute structural vibrational modes. A NASTRAN model of the spacecraft is usually given to the launch vehicle organization for them to compute coupled loads and resonant frequencies of the spacecraft when mated to the launch vehicle. NASTAN software is available from many companies.

**nT** Nano Tesla is a unit of the magnetic field. At the Magnetic Equator on the surface of the Earth the magnetic field is about 30,000 nT, while in the Polar region the magnetic field intensity is about 60,000 nT.

**OAP** Orbit Average Power. This is the average power generated or consumed in an orbit.

**Orbit Propagator** A computational tool for predicting the position (and velocity) of a spacecraft at any time from the six orbital elements of the spacecraft (or from a sequence of GPS data).

**Perigee** The altitude of the lowest point of an orbit.

**Polar orbit** An orbit of 90° inclination. Such an orbit passes over both the North and the South Poles. Its utility is that it permits surveillance of the entire Earth.

**QPSK** Quadrature Phase Shift Keying is a type of modulation where the a + 1 bit of information results in a 90° phase shift of the carrier, while a − 1 bit of information results in a − 90° phase shift.

**Paraffin actuator** Paraffin, when heated, expands up to about 25% of its initial volume. This is used to build an actuator. The expanding paraffin pushes out a pin, causing the release of a deployable device on the satellite.

**Radiometric analysis** Calculation of the amount of light incident on the camera pixels for specified aperture, integration time, time of day, sun elevation angle and other parameters.

**Ram pointing** Pointing in the direction of the orbit velocity vector (or forward).

**Reaction wheel** A rotating wheel of large moment of inertia, driven by an electric motor. At high RPM the reaction wheel can store a large angular momentum, causing the spacecraft to have a stable attitude. When accelerating, the resulting torque creates an angular momentum, which causes the spacecraft to change attitude (by conservation of the system's angular momentum).

**Recontact** When two objects in space are released or expelled from each other, then there are two bodies with very slightly different velocities in the same orbit. It is possible that some time in the future (usually an integer number of orbits later) the two bodies may collide (or recontact) each other. For this reason a recontact analysis is conducted to ensure that this does not happen.

**RAAN** Right Ascention of the Ascending Node. Indicates the geocentric right ascension of a satellite as it intersects the Equator traveling in a Northerly direction.

**Satellite pass** The part of the orbit visible to an observer on the ground. The satellite rises above the local horizon, progresses to the highest ground elevation point. It then sets by dropping below the horizon. Pass durations for LEO satellites are typically a maximum of about 12 min.

**Separation system** The mechanism used to secure the spacecraft to the launch vehicle, and to separate the two on command. The most frequently used separation system is the Marman band, described elsewhere in this Glossary. A motorized Lightband, built by Planetary Systems Corp. does not use explosive bolts to release the band, and therefore imparts less shock to both the spacecraft and the launch vehicle.

**SEU** Single Event Upset. A change in state of a digital word in the spacecraft computer memory, caused by an ionizing particle striking a sensitive point in the circuitry. The ionizing particle may cause a SEL (single event latchup), causing irreparable harm. The SEU may only cause an error that can be corrected by resetting the computer.

**Software Defined Radio** An RF receiver constructed by software operations on the received radio transmissions. It consists of digital mixers, filters, decryption and forward error decryption software.

**SADA** Solar Array Drive Assembly. The mechanism on the spacecraft that rotates the solar panels to keep them normal to the sun, thus maximizing the electric power generating capability of the solar panels.

**South Atlantic Anomaly** An area of the Earth, in the South Atlantic, where satellites are exposed to higher than normal levels of radiation. It is caused by the non-concentricity of the Earth and its magnetic dipole. It is the area where the Van Allen radiation belt is nearest to the Earth Surface, as low as 200 km. It is located approximately between $-70°$ and $+30°$ in Longitude and between $0°$ and $-45°$ in Latitude.

**Spin stabilization** A method of stabilizing spacecraft by making use of the gyroscopic stabilization property of a spinning body. The early Hughes communication satellites were spin stabilized.

**Star tracker** An optical and digital processing system where the optical system looks at the star field, projects it onto a CCD camera, from which the angular distances among stars are calculated and matched to an internally stored star catalog. The instrument can determine the attitude of the spacecraft on which the star tracker is mounted with great precision.

**Star catalog** A listing of brighter stars and their angular positions in the sky. It is used by star trackers to determine spacecraft attitude very precisely. A Star Catalog for use by star trackers is available from the Goddard Space Flight Center.

**Sub-satellite point** The point on the Earth surface directly under the spacecraft. That is, it is a point on the line between the satellite and the center of the Earth where the line intersects the Earth surface.

**Sun synchronous** An orbit where the satellite is over a given point of the Earth at the same time each day. Generally, a sun synchronous orbit has an inclination slightly greater than 90°.

**Thermal vacuum test** The test of a spacecraft component or the entire spacecraft in a TVAC (Thermal Vacuum Chamber), equipped with heaters and coolers is tested. The chamber can simulate the environmental conditions found in space. The spacecraft is operated while in the TVAC chamber. To pass the test several failure free cycles of temperature variations must be passed with the satellite operating successfully/.

**TRL** Technology Readiness Level is the state of development of a system, subsystem or component. There are 9 TRL levels. They are:

TRL 1 Transition from scientific research to applied research

TRL 2 Technology concept and/or application formulated

TRL 3 Analytical or experimental proof of concept is demonstrated

TRL 4 Component or subsystem validation in the laboratory environment

TRL 5 the System, subsystem or component is validated in a relevant environment thorough testing

TRL 6 System or subsystem model is demonstrated

TRL 7 prototype is demonstration in an operational environment

TRL 8 the system is completed and qualified through test in an operational environment

TRL 9 the system "mission proven" through successful mission operations

**TTL** Telemetry is a digital data stream, transmitted by the spacecraft via its RF downlink communications transmitter. It represents the numerical values of all relevant spacecraft parameters as a function of time.

**Three axis magnetometer** the 3-axis magnetometer (TAM) contains three orthogonal coils with which components of the Earth magnetic field are measured. Placed on a satellite, it is a coarse spacecraft attitude sensor that works at night as it works during the daylight.

**Watchdog Timer** A timer, usually implemented in software, which causes a computer to perform a set of operations that, if correctly executed, indicates that the computer is operating properly. If the computer comes up with incorrect results, then the computer is declared to be disabled, requiring that some action be taken (like resetting the computer) to correct the situation.

**Umbra** Same as eclipse. It is the portion of the orbit where the Earth blocks the sun.

**Universal Time** UTC is a modern version of Greenwich Mean Time (GMT), and it is the time standard used by spacecraft.

# References

1. Atlas V User Guide, downloadable from the Atlas V web site
2. R.R. Bate, *Fundamentals of Astrodynamics* (Dover, New York, 1971)
3. L.V. Blake, *Antennas* (John Wiley and Sons, New York, 1982)
4. G. Born, B. Tapley, B. Schutz, *Statistical Orbit Determination* (Elsevier Academic Press, Amsterdam, 2004)
5. A. Cambriles V. Monsalve, D. Taboada, Galileo stationkeeping strategy
6. M. Dean, et al., Configurable fault-tolerant processor for space based application, Naval Post Graduate School, Small Satellite Conference, 2015
7. F. de'Lima, Designing single event upset mitigation techniques for large SRAM-based FPGA devices, Porto Alegre, 2002
8. Digital Globe and Digital Globe Data Sheet – WorldView 2 Overview from the Digital Globe web site
9. A. Dissanayaka, *Rain Attenuation Modeling*, (COMSAT, Washington, DC, 1998)
10. EENS 6340/4340 The Earth Magnetic Field
11. R. Farley, *Spacecraft Deployables* (University of Maryland)
12. FEMAP FEC Software
13. R. Fischell, Gravity gradient stabilization of earth satellites, APL Technical Digest, 1964
14. T. Flatley, J. Forden, D. Henretty, G. Lightsey, L. Markley, On- board attitude determination and control algorithms for SAMPEX. Flight mechanics/estimation theory symposium 1990, NASA-GSFC, pp. 379–398
15. G.F. Fowles, *Introduction to Modern Optics* (Dover Publications Inc, New York, 1975)
16. J. Furumo, Cold gas propulsion for satellite attitude control, station keeping and deorbit, University of Hawaii
17. E.M. Gaposchkin, A.S. Coster, Analysis of satellite drag
18. GSFC-STD-7000A, General Environmental Verification Standard (GEVS) for GSFC Flight Programs and Projects, 2013
19. D. Gibbon, The review of butane as a low cost propellant, in *Space Propulsion*, (Surrey Satellite Technology, San Sebastian, 2010)
20. D. Gibbon, C. Underwood. Low cost butane propulsion system for small spacecraft, USU Small Satellite Conference, 2001
21. E. Gill, J. Guo, Enabling intersatellite communications and ranging for small satellites, Delft University of Technology, August 2016
22. D.G. Gilmore, *Satellite Thermal Control Handbook* (Aerospace Corp, El Segundo, 1994)
23. C. Hall, *Gravity Gradient Stabilization* (Virginia Tech)

24. B. Huettle, C. Willey, Design and development of miniature mechanisms for small spacecraft, Utah Small Satellite Conference, 2000
25. J. Hussein, G. Swift, Mitigating single event upsets, XILINX, 2015
26. International Geomagnetic Field Model. The Enhanced Magnetic Field Model (EMM 2015) is downloadable from the NOAA web site
27. Iridium website
28. ISO/WD/27852 Determining Orbit Life
29. ITU Rain Attentuation Model, Recommendation IRU-R P.838–3
30. T.R. Kane, *Spacecraft Dynamics* (McGraw Hill, New York, 1983)
31. M.H. Kaplan, *Modern Spacecraft Dynamics and Control* (Wiley, New York, 1976)
32. V. Khoroshilov, A. Zakrhvevskii, Deployment of long flexible element on spacecraft with magnetic damper. J. Model. Simul. Identif. Control (2014)
33. S. Koontz, Space radiation effects on spacecraft materials and avionics systems, NASA/MIT Workshop, 2012
34. I. Martinez,, Space environment, lecture on July 11, 2008 on the web
35. D. Mortari, M.A. Samaan, C. Bruccoleri, J.L. Junkins, The pyramid star identification technique. NAVIGATION, Journal of The Institute of Navigation **51**(3), 171–184 (Fall 2004)
36. T. Murphy, Overview of deployable structures, AFRL, 2008
37. G. Ning, B. Popov, Cycle life modeling Li-ion batteries, U. of South Carolina. J. Electrochem. Soc. **151**, A1584–A1591 (2004)
38. S. Ogden, et al., Review on miniaturized paraffin phase change actuators, valves and pumps, microfluidics and nanofluidics, 2013
39. J. Opiela, et al., Debris assessment software version 2.0 user's guide, NASA Johnson Space Flight Center, 2012
40. M. Ovchinnikov, et al., Spin stabilized satellite with three stage active magnetic attitude control system, Russian Academy of Science, 2011
41. V. Pisacane et al., *Stabilization System Analysis and Performance of the GEOS-A Gravity Gradient (Explorer XXIX)* (Johns Hopkins University APL, Maryland, 1967)
42. J. Prowald Santiago, Large deployable antennas, ESA ESTEC, Cal-Tech Large Space Aperture Workshop, 2008
43. Quadrifilar Helix Antennas, Antenna Development Corporation, 2008
44. RADC Reliability Engineer's Toolkit, 1988
45. Radiation Dose and Deose Rate, NASA Spacemath GSFC
46. Radio Frequencies for Space Communication, Australian Space Academy, 2016 web site
47. Reliability of Systems with Various Element Configurations Chapter 26
48. Satellites to be Built and Launched by 2023, Euroconsult Research Report, 2014
49. S-Band Horn, Surrey Satellite Technology Ltd. web site
50. A. Siahpush, et al., A study for semi-passive gravity gradient stabilization of small satellites, Utah State University, USU conference on small satellites, 1976
51. Singe Board Computer, OBC 750, Surrey Satellite Technology Ltd Spec, 2015
52. W. Skullney et al., Structural design of the MSX spacecraft. J. Hopkins APL Tech. Dig. **17**(1), 59 (1966)
53. Small Spacecraft Technology State Of the Art, Ames Research Center, 2014
54. Space Micro 200k DSP Processor Board Spec Sheet on the web
55. Spacecraft-Deployed Appendage Design Guidelines, NASA, GSFC
56. SpaceQuest spec sheet of Turnstyle Antenna on the web
57. Spectrolab Solar Cell spec sheet for 28.3% Efficient UTC GaAs Solar Cells
58. Structural Design Requirements and Factors of Safety for Spaceflight Hardware, NASA JSC, 2011
59. R. Struzak, *Basic Antenna Theory* (ICTP, Trieste, 2007)
60. Surrey Satellites Product Spec Sheets
61. B. Tankersly, B. McIntosh, Largest Mesh Deployable Antenna Technology, in *The Space Progress Proceedings*, (Harris Corp, 1982)

# References

62. Understanding Series and Parallel Systems Reliability, Selected Topics in Assurance Related Technologies, Vol 11, No 5
63. Understanding Space Radiation, NASA JSC, 2002
64. D.A. Valledo, *Fundamentals of Astrodynamics and Applications* (Microcosm Press, Hawthorn, 2013)
65. J. Vogt, Attitude determination of a 3-axis stabilized spacecraft using star sensors, Naval Post Graduate School, 1999
66. F. Wang, V. Agrawal, Single event upset, an embedded tutorial, 21st international conference on VLSI design, 2008
67. J. R. Wertz, W. J. Larson (eds.), *Reducing Space Mission Cost* (Space Technology Series, El Segundo, California, 1996)
68. J.R. Wertz, *Spacecraft Attitude Determination and Control* (Reidel, Dordrecht; Boston; London, 1985)
69. J. R. Wertz, D. F. Everett, J. J. Puschell (eds.), *Space Mission Engineering: The New SMAD* (Space Technology Library, El Segundo, 2011)
70. B. Wie, P.M. Barba, Quaternion feedback for spacecraft large angle maneuvers. J. Guid. Control Dynam. **8**, 360–365 (1985)
71. B. Wie, *Space Vehicle Dynamics and Control* (AIAA, Reston, 1998)
72. L. Wood, *Introduction to Satellite Constellations* (Cisco Systems, 2006)
73. World View Data Sheets on the web

# Index

**A**
Anomaly
   discovery, 245
   resolution, 91, 237, 245
Antennas
   deployable, 82, 164–165
   dish, 80, 84, 256
   horn, 79, 80
   N-Turn Helix, 76
   patch, 78–79, 86
   quadrifilar antennas, 77–78, 82, 286
   steerable, 81, 85, 266
   switching, 82–83
   turnstile, 78, 79, 286
Atmosphere
   density *vs.* altitude, 2–7, 9–12
   drag, 4, 106, 155, 184, 261
Attitude determination and control
   ADACS components, 99–108
   ADACS computer algorithm, 106–107
   ADACS integration and test, 112–114
   attitude control system design methodologies, 108–112
   attitude control *vs.* missions, 107–108
   attitude sensing, 92, 93, 263, 281
   axial launch acceleration, 128, 229
   dynamic simulator, 108–110, 112, 113, 212, 265, 272
   GPS receiver, 105–107
   gravity gradient stabilization, 93–95
   hysteresis rod, 92–95
   momentum unloading, 100–102
   on-orbit checkout, 91, 114
   performance flowdown, 91–93
   period of oscillation, 95
   pitch bias momentum stabilization, 92, 95–96
   pointing accuracy, 91–92, 112
   reaction wheels, 92, 99–100, 106, 107
   sensing, 92, 93
   spacecraft agility, 92, 263, 283
   spin stabilization, 98
   star tracker, 91–93, 102–108
   three axis zero momentum stabilization, 263

**B**
Ballistic coefficient, 49–51
Battery
   depth of discharge, 58, 285
   efficiency, 58
   lithium battery cycle life *vs.* DOD, 58, 285
Beta, 9, 51, 52, 60, 62–65, 188–191, 198, 199, 286

**C**
Camera pixel size, 15
Center of gravity (CG), 4, 14, 100, 124, 154, 155, 178, 184, 213, 216, 231, 261, 265, 269
Center of pressure (CP), 4–5, 14, 155, 184, 261, 269
Communications subsystem
   bit error rate (BER), 69–73
   data qantity, 29
   downlink tme, 29
   $E_b/N_o$, 72–75
   forward error correction, 72–73, 85, 122, 256, 262

Communications subsystem (*cont.*)
  frequency allocation, 69–71
  FSK, 72, 73, 75, 266
  globalstar constellation, 32
  GMSK, 72
  GPS satellites, 10
  intersatellite links, 37, 81
  link equations, 73–75, 256
  modulation types, 72, 73, 122, 266
  O-QPSK, 72, 266
  overlap or gaps of coverage, 33, 35
  PSK, 72
  QPSK, 72, 73, 75, 85
  RF block diagram, 9–24
  S-Band, 56, 71, 73–74
  station keeping, 37, 173, 174, 181–185
  Swath Width *vs.* SC in Constellation, 26–28
  throughput, 266
  UHF, 56, 69, 70
  VHF, 69, 70
  X-Band, 71, 79, 266
Constellation, 31–37, 106, 173, 181–185, 273, 286
Coverage gaps, 32–36

# D
Data rate, 29, 69, 71–73, 75, 82, 83, 86, 122, 244, 246, 266, 267
Data rate limitation, 246
Deployables
  antennas, 82, 164–165
  bolt cutters, 165–166
  constant speed governors, 170
  deployable booms, 162–164
  electric burn wires, 165–167
  fluid dampers, 169–171
  gravity gradient boom, 92, 93, 162
  hinges, 162
  motorized cams, 168
  paraffin pin pushers, 168
  separation system, 168–169
  solenoid pin pullers, 167, 171
  testing deployables, 171
Diffraction limit (DL), 15, 16, 285
Digital hardware
  architecture, 87–89
  spacecraft computer examples, 89–90

# E
Earth environment
  Albedo, 3
  atmosphere, 3–5
  coordinate system, 5–6
  earth axis, 7
  earth IR, 187
  earth solar flux, 189
  earth-Sun distance, 7
  sun spot activity, 5
Electric power subsystem design
  battery capacity, 57–59
  battery types, 59
  block diagram, 66–68
  design steps, 56
  OAP *vs.* Beta, 62, 63
  projection of Sun on plane on SC, 51–54
  required OAP, 52

# G
Geostationary satellites, 9, 29, 246, 266
GPS, 9, 10, 31, 32, 88, 105–107, 111, 116, 117, 123, 244
Ground station
  access time, 42, 236, 237
  acquisition and tracking, 239
  block diagram, 242
  elevation contours, 242
  functions, 241–246
  ground support equipment, 219, 235, 268
  manual, 238–239
  map display of spacecraft position, 236
  operator training, 238–239, 248
  requirements, 235–237
  spacecraft command generation, 242, 245
  staffing, 247
  telemetry monitoring, 243–244
  telemetry points, 244

# H
Horizon-to-horizon range, 182

# I
Imaging
  CCD pixel noise, 19
  integration time, 21–22
  modulation transfer function (MTF), 18, 262
  pitch rate, 21
  pitch *vs.* time when pointing to target, 107
  pointing to the target, 22, 24, 25
  quality, 18–19
  radiometric analysis, 19, 20
  resolution/GSD, 255
  roll rate when pointing to target, 30
  satellites, 37, 244

Index 315

scenario, 28, 30
spacecraft yaw and pitch, 23
swath width, 26–28
telescope, 14, 15, 18
yaw *vs.* time pointing at target, 22
Insertion, separation recontact, 231

**L**
Launch vehicles
acoustic environment, 232
Antares launch profile, 224
Atlas V, 224–226, 228
axial acceleration during ascent, 229
Dnepr, 224–226, 229, 230
ESPA, 225, 227, 229
payload interfaces, 223–234
present launch vehicles, 223–225
RF environment, 231–232
separation, 168, 225
separation systems, 227
shock environment, 232–233
vibration levels, 227
vibration spectrum, 129
LEO satellites
CubeSat, 10, 229
Lightband, 168
Low cost design and development, 249–254

**M**
Magnetic
dipole model, 1
International Geomagnetic Field Model, 1
magnetic field *vs.* altitude, latitude, 1

**O**
Orbital relations
acceleration of gravity *vs.* altitude, 39
azimuth *vs.* time, 45
CPA, 42–47
elevation angle, 42, 43
geometry of orbit and spacecraft, 39–54
ground range, 42–47
ground track, 41, 43, 47
horizon-to-horizon range, 182
inclined orbit, 9, 32
Molnya orbit, 9
off nadir angle, 15, 24, 28
Orbcomm, 31, 33
orbital elements, 5, 6, 40
orbit drift, 28
orbit life, 50, 51, 203

orbit period, 9, 27, 41
orbit velocity, 39
pass duration, 45, 46, 84
period, 41
polar orbit, 9, 11, 13, 35
satellite access time, 119
semi-major axis, 40–41
slant range, 42–47
spacecraft coordinate system, 5–6
sun synchronous orbit, 9
swath width, 26–28, 33, 34

**P**
Propulsion
burn, 124
changing altitude, 269
cold gas propulsion system, 176–178, 180, 269, 270
cold gas tanks and thrusters, 179
delta V, 124, 269, 286
drag force, 175–176
fuel ISP, 173
fuel mass, 173, 175, 176
getting on station, 183–185, 263
Hohmann, 185, 186
hydrazine, tanks, thrusters, 12, 124, 178–181, 184, 185, 187, 244, 261, 269, 270, 282
maneuvers, 181–185
Rocket equation, 173, 286
rocket motors, 179, 185

**R**
Radiation
single event upset, 203, 204
total radiation dose and hardening, 203–206
Reliability
MTBF, MTTF, 206, 207, 286
redundancy, 203, 208
reliability diagram, 206, 207
Resolution, 15, 16, 18, 19, 91, 237, 245, 256, 268, 279, 280, 285
Revisit time, 31, 256

**S**
Satellite imaging, 13–31
Satellite orbits
elliptical orbits, 9, 10, 40
orbital elements, 5
Polar orbit, 9

Satellites, 10–37, 41, 55, 58, 62, 69–71, 74, 79, 81, 84, 85, 90, 98, 103, 109, 119, 120, 122, 128, 133, 135, 170, 182, 183, 198, 223, 236, 238, 239, 241–244, 246–250, 255–259, 261, 266, 267, 269, 276, 279
  agility, 14–15, 28
Secondary payload interfaces
  cal-poly CubeSat launcher, 229
  ESPA, 225, 227–230
  secondary payload vibration environment, 229
  separation system, 227
  shock environment, 232–233
Separation systems
  lightband separation system, 169
  RUAG separation system, 228
Software
  architecture, 115–117
  executing
    communication functions, 116, 122–123
    housekeeping functions, 120
    managing on-board electric power system, 116, 120–121
    managing thermal control system, 121
    propulsion control system, 124
    scheduled events, 116, 118–119
    telemetry data collection, 120, 121
  and functions, 115–124
  development, 124–125
Solar array configurations, 60–64
  solar cell, 60
Solar panel tilt, 61, 63
Spacecraft design example
  derived requirements, 28–29
  design steps, 283
  spacecraft mission requirements, 279
Spacecraft operations
  anomalies, 241, 243, 244, 248, 263
  command generation, 242, 245
  cost of, 247, 248
  ground station displays, 235–237
  ground station functions, 241–246
  telemetry monitoring, 243–244
  test plans and reports, 247
  training, 248
Space environment
  Albedo, 3
  atmospheric pressure *vs.* altitude, 224–225
  gravity gradient, 5
  magnetic field, altitude, magnetic latitude, 1–3
  solar cell, 60, 65–66, 191, 192, 195, 198
  solar energy, 3
  sun spot activities, 4

Spreadsheets list, 52
Star catalog, 7, 102, 103
Structures
  acoustic environment, 232
  analyses, 139–154, 261, 272, 283
  axial acceleration, 128
  eigenvalue analysis, 146–148
  factors of safety, 129, 130, 138–139
  fastener analysis, 153–154
  FEM mesh, 141
  finite element model (FEM), 139–142, 144, 146, 148, 150, 153
  GEVS vibration spectrum, 129
  LCROSS, 133, 135
  mass properties, 231
  material properties, 136, 137, 143
  options, 130–135
  pressure loads, 130, 132
  requirements flowdown, 256–257
  vibrational modes, 146, 147
  vibration requirements flowdown, 256–257
  weight estimate and distribution, 154–159
Swath width, 25–28, 33–35, 42

**T**
Telemetry
  engineering telemetry, 121
  state of health telemetry, 121
Telescope, 14–19, 21, 168, 244, 257, 261–263, 271, 272, 280–282
Testing
  component level, 209–211, 271
  EGSE, MGSE, 211–212, 218
  environmental testing, 212–221
  Flat-Sat, 211
  spacecraft level tests, 211–212
  spacecraft simulator, 212, 213, 218
  thermal and thermal vacuum test, 199–200, 209, 220–221, 272, 277
  vibration tests, 211–218, 261, 272, 277
Thermal design
  albedo heat flux, 187, 189–192, 198, 200
  conduction across bolted joints, 194–195
  constructing a thermal model, 197
  heat absorption, 188, 191
  heat balance, 194
  heat by internal electronics, 187
  heaters, 188, 193, 196, 197, 200
  heat pipes, 193, 195–196, 198
  heat rejection, 192